THE SOCIAL DYNAMICS OF CARBON CAPTURE AND STORAGE

Carbon capture and storage (CCS) has emerged rapidly as a crucial technological option for decarbonising electricity supply and mitigating climate change. Great hopes are being pinned on this new technology but it is also facing growing scepticism and criticism. This book is the first to bring together the range of social and policy issues around CCS, shedding new light on this potentially vital technology and its future.

The book covers many crucial topics including the roles and positions that the public, and different NGOs, industries, political parties and media are taking up; the way CCS is organised, supported and regulated; how CCS is being debated and judged; how innovation, demonstration and learning are occurring and being conceptualised and promoted; and the role of CCS in the transition to a low carbon energy future.

The authors draw on a variety of approaches, concepts, methods and themes and provide a new understanding of innovation in the energy and climate change fields. It tackles the many issues in a way that speaks not only to those concerned to understand these developments, but to those who are involved in the scientific and technological work itself, as well as those charged with evaluating and making decisions relevant to the future of the technology.

Nils Markusson is a Research Associate at the School of Geosciences, University of Edinburgh, UK.

Simon Shackley is a Lecturer at the Carbon Capture and Storage Group at the University of Edinburgh, UK.

Benjamin Evar is a PhD student at the Scottish Carbon Capture and Storage research centre at the University of Edinburgh, UK.

Science in Society Series

Series Editor: Steve Rayner

Institute for Science, Innovation and Society, University of Oxford

The Earthscan Science in Society Series aims to publish new high quality research, teaching, practical and policy-related books on topics that address the complex and vitally important interface between science and society.

Animals as Biotechnology:
Ethics, Sustainability and Critical Animal Studies
Richard Twine

Business Planning for Turbulent Times:
New Methods for Applying Scenarios
Edited by Rafael Ramírez, John W. Selsky and Kees van der Heijden

Debating Climate Change:
Understanding Debate and Agreement
Elizabeth L. Malone

Democratizing Technology:
Risk, Responsibility and the Regulation of Chemicals
Anne Chapman

Dynamics of Disaster:
Lessons on Risk Response and Recovery
Rachel A. Dowty Beech, Barbara Allen

Genomics and Society:
Legal, Ethical and Social Dimensions
Edited by George Gaskell and Martin W. Bauer

Influenza and Public Health:
Learning from Past Pandemics
Tamara Giles-Vernick, Susan Craddock and Jennifer Gunn

Integrating Science and Policy:
Vulnerability and Resilience in Global Environmental Change
Edited by Roger E. Kasperson and Mimi Berberian

Marginalized Reproduction:
Ethnicity, Infertility and Reproductive Technologies
Edited by Lorraine Culley, Nicky Hudson and Floor van Rooij

Nanotechnology:
Risk, Ethics and Law
Edited by Geoffrey Hunt and Michael Mehta

Rationality and Ritual:
Participation and Exclusion in Nuclear Decision Making, 2nd Ed.
Brian Wynne

Resolving Messy Policy Problems:
Handling Conflict in Environmental, Transport, Health and Ageing Policy
Steven Ney

The Limits to Scarcity:
Contesting the Politics of Allocation
Lyla Mehta

Uncertainty in Policy Making:
Values and Evidence in Complex Decisions
Michael Heazle

Unnatural Selection:
The Challenges of Engineering Tomorrow's People
Edited by Peter Healey and Steve Rayner

Vaccine Anxieties:
Global Science, Child Health and Society
Melissa Leach and James Fairhead

A Web of Prevention:
Biological Weapons, Life Sciences and the Governance of Research
Edited by Brian Rappert and Caitrìona McLeish

THE SOCIAL DYNAMICS OF CARBON CAPTURE AND STORAGE

Understanding CCS Representations, Governance and Innovation

Edited by Nils Markusson, Simon Shackley and Benjamin Evar

Routledge
Taylor & Francis Group
LONDON AND NEW YORK

First published 2012
by Routledge
4 Park Square, Milton Park, Abingdon, Oxon OX14 4RN

Simultaneously published in the USA and Canada
by Routledge
605 Third Avenue, New York, NY 10017

Routledge is an imprint of the Taylor & Francis Group, an informa business

British Library Cataloguing in Publication Data
A catalogue record for this book is available from the British Library

Library of Congress Cataloging in Publication Data
The social dynamics of carbon capture and storage : understanding CCS representations, governance and innovation / [edited by] Nils Markusson, Simon Shackley, Benjamin Evar.
p. cm.
Includes bibliographical references and index.
1. Geological carbon sequestration--Social aspects. 2. Carbon dioxide--Environmental aspects. 3. Carbon dioxide mitigation--Social aspects. 4. Air quality management--Social aspects. 5. Climatic changes--Social aspects. 6. Environmental policy--Social aspects. I. Markusson, Nils. II. Shackley, Simon. III. Evar, Benjamin.
QE516.C37S63 2012
363.738'746--dc23
2011049430

ISBN: 978-1-84971-314-6 (hbk)
ISBN: 978-1-84971-315-3 (pbk)
ISBN: 978-0-203-11872-6 (ebk)

Typeset in Bembo
by Saxon Graphics Ltd, Derby

DEDICATION

This book was conceived in 2010 as a result of discussions between ourselves and Dr Stewart Russell. Stewart, Deputy Director of Science, Technology and Innovation Studies at the University of Edinburgh, was to have been one of the editors. As the project progressed, Stewart fell seriously ill and died in September 2011. We miss you greatly as a friendly soul and a stimulating colleague and collaborator! We dedicate this book to you – it could have been a better one with your insights.

CONTENTS

ILLUSTRATIONS

Figures

TABLES

BOXES

NOTES ON CONTRIBUTORS

Nils Markusson, PhD

Nils Markusson has 15 years' experience in the area of innovation and technology studies. After working for the Swedish government, he undertook a PhD at the University of Edinburgh on cleaner technology and firm organisation. He currently works as a Research Associate at the Scottish Carbon Capture and Storage research centre at the University of Edinburgh on several projects on aspects of CCS innovation. Throughout he has sought to understand how we use technology in society with as little environmental impact as possible, and what policy is needed to that end. He has published in journals such as *Global Environmental Change*, *Energy Policy* and *Journal of Cleaner Production*.

Simon Shackley, PhD

Simon Shackley has a long-standing interest in low-carbon technologies with a special interest in carbon dioxide capture, removal and storage – both via CCS and biochar. He has undertaken a number of public and stakeholder CCS perception and engagement studies in the UK and is interested in how to design projects to more effectively incorporate community preferences. He is also interested in the relationship between energy and climate policy, particularly the appropriate balance between market-based, community and government-led mechanisms.

Benjamin Evar

Benjamin Evar is a PhD student at the Scottish Carbon Capture and Storage research centre at the University of Edinburgh. His research focuses on emerging trends in carbon dioxide capture and storage governance, and the role of expert advice in technology policy. During his PhD he has worked as a research consultant for the Scottish Government on behalf of the Global CCS Institute and contributed to research on public perceptions of CCS for the International Energy Agency. He has previous experience as a management and research consultant and holds an MSc with distinction in Carbon Management from the University of Edinburgh.

Chiara Armeni

Chiara Armeni is an environmental lawyer and has been a Research Associate with the University College London Carbon Capture Legal Programme (CCLP) since 2009. She holds degrees in Comparative Law (LLB), Public International Law (LLM) and Environmental Law and Policy (LLM). She is currently Deputy Director of the CCLP. Since 2010 she has been leading the CCLP project on the implementation of the CCS Directive in selected European countries. Chiara's main research interests lie in international and European environmental law, with special focus on Carbon Capture and Storage and the law and policy of climate change.

Judith A. Bradbury, PhD

Judith Bradbury was born and initially educated in the UK, with a degree in sociology from the London School of Economics followed by a PhD in Public and International Affairs from the University of Pittsburgh. In her work, she has emphasised the relevance of social science insights and tools to the analysis and resolution of science policy issues. She has extensive experience in the research and practice of public involvement in hazardous technologies, including nuclear waste storage and transportation and chemical demilitarisation. Her most recent experience has been in managing public outreach activities for the Midwest Regional Carbon Sequestration Partnership Program, sponsored by the US Department of Energy.

Olaf Corry, PhD

Olaf Corry is Lecturer in International Studies at the Open University specialising in global environmental politics. He also teaches International Relations at the University of Cambridge and has recently completed a research project at Cambridge University Judge Business School on CCS communications and CCS and the European environmental movement. He has a PhD in International Relations from the University of Copenhagen and an MPhil in politics and sociology from the University of Cambridge. He previously worked as energy and climate policy advisor to the Danish Social Democratic Party.

Anders Hansson, PhD

Anders Hansson is Assistant Professor at the Unit of Technology and Social Change at Linköping University, Sweden. His research has mainly focused on critical analyses of Carbon Dioxide Capture and Storage. The aim of his doctoral thesis was to scrutinise the scientific and political efforts behind claims that CCS is a rational and viable solution to the global climate change problem.

Atsushi Ishii

Atsushi Ishii is an Associate Professor at the Center for Northeast Asian Studies, Tohoku University. His research interests include global environmental and resource politics, science and technology studies, and Japanese environmental politics. His recent publications include 'Toward policy integration: assessing carbon dioxide capture and storage policies in Japan and Norway', in *Global Environmental Change* and 'Scientists learn not only science but also diplomacy: learning processes in the European transboundary air pollution regime' in *Governing the Air: Science-Policy-Citizens Dynamics in International Environmental Governance* (MIT Press). Other articles have appeared in, among other journals, *Climate Policy*, *RECIEL*, and numerous Japanese books and journals.

Yue Liu

Yue Liu currently is a research assistant at the Centre for Research in Energy and Minerals Economics (CREME) at Curtin University, Australia. She has been awarded the Curtin Strategic International Research Scholarship and will start her PhD study from the beginning of 2012 at Curtin University. Her research interests focus on the economic and policy study of carbon dioxide capture and storage and the renewable energy technologies. She received her bachelor degree from China Agricultural University, Beijing, China, and obtained her MA degree from Clark University, Massachusetts, USA.

Hauke Riesch, PhD

Hauke Riesch received a PhD in Science and Technology Studies from University College London in 2008 and has since worked on the public understanding of risk and public and stakeholder perceptions of CCS at the University of Cambridge. Since August 2011 he has been working on public engagement with environmental sciences at the Centre for Environmental Policy at Imperial College, London.

Vivian Scott, PhD

Trained as a physicist, Vivian Scott has a PhD in climate modelling where he studied models of the ocean carbon cycle used in global circulation models to predict the future climate. He works on CCS as a climate mitigation technology, researching policy and commercial progress towards the large-scale testing and deployment of the technology. He is currently engaged in policy development relating to the delivery of the EU CCS demonstration programme as part of Europe's wider climate mitigation and energy strategies, collaborating with representatives of the EU Commission, national and regional governments, industry, environmental NGOs and academic colleagues.

Jennie C. Stephens, PhD

Jennie C. Stephens is an Assistant Professor of Environmental Science and Policy at Clark University's Department of International Development, Community and Environment. She is also a research affiliate of the Energy Technology Innovation

Policy Group at Harvard's Kennedy School. Jennie's teaching, research and community engagement focus on socio-political aspects of energy technology innovation, deployment of renewable energy, public perception of energy technologies and climate change education/awareness. She received her BA (1997) from Harvard University in Environmental Science and Public Policy and then earned her MS (1998) and PhD (2002) at the California Institute of Technology in Environmental Science and Engineering.

Mark Winskel, PhD

Mark Winskel is a Senior Research Fellow in the Institute for Energy Systems, University of Edinburgh, and the Research Co-ordinator of the UK Energy Research Centre. An applied social scientist, Mark's research addresses the dynamics of innovation in energy systems, especially for supply technologies, and he has held a number of research grants in this area, often working in cross-disciplinary research teams. His own education spans the physical, environmental and social sciences, and his doctoral thesis analysed technological change in the UK electricity supply sector. Outside academia, he has advised organisations such as the Carbon Trust and the International Energy Agency.

ACKNOWLEDGEMENTS

Many people have helped make this book possible – not least colleagues at the University of Edinburgh. We would like to thank the head of the SCCS research group, Stuart Haszeldine, for agreeing to us spending time on this. And thank you Leslie Mabon, for helpfully commenting on the draft conclusions chapter.

We are also grateful to the professionals in industry, civil society and beyond who helped us review the chapters and make them more relevant to CCS practitioners:

- Nicolas Molho, WWF
- Mike Farley, Doosan Power Systems
- Ross Loveridge, Scottish Government
- John Kessels, International Energy Agency Clean Coal Centre
- Aidan Whitfield, Environment Agency
- Duncan McLaren, former Chief Executive of Friends of the Earth Scotland

Finally, we want to thank Lindsay Terrell for great help with the formatting, and Moyra Forrest for superb indexing.

ABBREVIATIONS AND ACRONYMS

ACC	anthropogenic climate change
ACCSEPT	Acceptance of CO_2 Capture, Storage, Economics, Policy and Technology project
ACF	advocacy coalition framework
AEP	American Electric Power
AIDS	auto immune deficiency syndrome
ANT	actor network theory
BECCS	biomass energy with carbon dioxide capture and storage
BUND	Bund Für Umwelt und Naturschutz Deutschland (FoE Germany)
CATF	Clean Air TaskForce
CCGT	Combined Cycle Gas Turbine
CCJ	*Carbon Capture Journal*
CCS	carbon dioxide capture and storage
CCSA	Carbon Capture and Storage Association
CDM	clean development mechanism
CFC	chlorofluorocarbons
CH_4	methane
CHP	combined heat and power
CO_2	carbon dioxide
COP	Conferences of the Parties
CSIRO	Commonwealth Scientific and Industrial Research Organisation
CSLF	Carbon Sequestration Leadership Forum
CSP	concentrated solar power
DECC	Department of Energy and Climate Change
DNV	Det Norske Veritas
DOE	US Department of Energy
DTI	Department of Trade and Industry

EC	European Commission
ECBM	enhanced coal bed methane recovery
ECN	Energy Research Centre for the Netherlands
EEPR	European Energy Programme for Recovery
EGR	enhanced gas recovery
ELSI	ethics, legal and social implications
EMR	Energy Market Reform
ENDS	Environmental Data Service company
ENGO	environmental nongovernmental organisations
EOR	enhanced oil recovery
EPA	US Environmental Protection Agency
EPA/IRIS	US Environmental Protection Agency Integrated Risk Information System
EPS	emissions performance standard
EPSRC	Engineering and Physical Science Research Council
ERRG	Energy Research Review Group
ERU	Emission Reduction Unit
ETF	Environmental Transformation Fund
ETI	Energy Technologies Institute
ETS	emissions trading scheme
EUR	euro
FGD	flue gas desulphurisation
FoE	Friends of the Earth
G8	Group of Eight [major economies]
GCCSI	Global CCS Institute
gCO_2	gram CO_2
GET	The General Environmental Technos Ltd
GHG	greenhouse gases
GHGT	Greenhouse Gas Control Technology
Gt	gigatonne
GW	gigawatt
HCSTC	House of Commons Science and Technology Committee
HMG	Her Majesty's Government
ICCDR	International Conference on Carbon Dioxide Removal
ICEPT	Imperial College Centre for Energy Policy and Technology
IEA	International Energy Agency
IEA CCC	IEA Clean Coal Centre
IEAGHG	IEA Greenhouse Gas R&D Programme
IED	Industrial Emissions Directive
IGCC	integrated gasification combined cycle
IPAC-CO_2	International Performance Assessment Centre for Geologic Storage of Carbon Dioxide
IPCC	Intergovernmental Panel on Climate Change
JCOAL	Japan Coal Energy Centre

KEPCO	Kansai Electric Power Co.
KLIF	Norwegian Climate and Pollution Agency
kWh	kilowatt-hour
LTS	large technical system
m^3/day	cubic metre per day
MAC(s)	marginal carbon abatement cost(s)
MACC	marginal abatement cost curve
METI	Ministry of Economy, Trade and Industry, Japan
MHI	Mitsubishi Heavy Industry Ltd
MIT	Massachusetts Institute of Technology
MLP	multi-level perspective
MPE	Norwegian Ministry of Petroleum and Energy
MRCSP	Midwest Regional Carbon Sequestration Partnership
MW	megawatt
NDMA	N-Nitrosodimetylamin
N_2	nitrogen
NER	new entrants reserve
NETL	National Energy Technologies Laboratory
ng/m^3	nanogrammes per cubic metre
NGO	nongovernmental organisation
NILU	Norwegian Institute for Air Research
NIMBY	not in my back yard
NIPH	Norwegian Institute of Public Health
NMF	Norwegian Ministry of Finance
NMTI	Norwegian Ministry of Trade and Industry
NOAH	FoE Denmark
NOK	Norwegian krona
NO_x	nitrous oxides
NPD	Norwegian Petroleum Directorate
NSM	new social movement
NUMBY	not under my back yard
NZEC	Near Zero Emissions Coal project
OECD	Organisation for Economic Co-operation and Development
Ofgem	Office of the Gas and Electricity Markets
OSPAR	Oslo and Paris Convention for the Protection of the Marine Environments of the North-East Atlantic
PIU	UK Cabinet Office's Performance and Innovation Unit
ppmv	part per million by volume
PTRC	Petroleum Technology Research Centre
PV	photovoltaics
R&D	research and development
RCEP	Royal Commission on Environmental Pollution
RD&D	research, development and demonstration
RDM	robust decision-making

REACH	EU Regulatory Framework for the Management of Chemicals
SACROC	Scurry Area Canyon Reef Operators Committee
SBSTA	UNFCCC Subsidiary Body for Scientific and Technological Advice
SCCS	Scottish CCS
SCOT	social construction of technology
SP	Scottish Power
SRCCS	IPCC Special Report on CCS
STS	science and technology studies
t/day	tonne per day
TCM	Test Centre Mongstad
tCO_2	tonne CO_2
TIC	Technology and Innovation Centre
TIS	technical innovation systems
TNS	Taylor Nelson Sofres PLC
toe	tonne of oil equivalent
TSB	Technology Strategy Board
TWh	terawatt-hour
UAE	United Arab Emirate
UNFCCC	United Nations Framework Convention on Climate Change
WRI	World Resources Institute
WWF	Worldwide Fund for Nature
ZEP	Zero Emissions Platform
ZERO	Zero Emission Resource Organisation

1

INTRODUCTION

Nils Markusson, Simon Shackley and Benjamin Evar

Don't forget to breathe out! Right now, the future of carbon capture and storage (CCS) seems very much to be hanging in the balance. The suspense is high and tensions are surfacing. Lots of questions are being asked: if CCS promises so much as a response to climate change, then why are there such different views of it? If the technology is backed by resource-strong actors, why then does progress seem to be stalling? Should all effort be targeted on one option/project to get the technology up and going as soon as possible, and this way demonstrate a response to the urgent problem of climate mitigation, or should we adopt a more cautious approach? Can the technology be both robust and flexible enough to withstand the litmus test of trials, demonstration and deployment? Clearly, the future of CCS will be determined by things both apparently technical and social. This book is about the socio-technical and political dynamics of CCS. It will shed light on the related perceptions and representations, governance and decision-making and innovation processes.

If you are not familiar with carbon capture and storage, you will find a short description of what it is in Chapter 2. There, you will also find a concise description of the key policy, regulatory and economic issues surrounding CCS and you may wish to read that chapter to familiarise yourself with the background information necessary to make sense of the debates covered in this book.

So why should you read a book about the social science of CCS?

CCS has risen meteorically among climate change mitigation options and over the last decade it has come to be widely accepted as a key technology option[1] to decarbonise the energy system. It features in the climate policy documents of many countries as a cornerstone of their mitigation strategies and is widely explained and promoted as part of a palette of technologies with which to paint decarbonisation scenarios. Academia has been central in developing CCS knowledge and visions and in legitimising the technology and giving it credibility. The 2005 Special Report on CCS from the Intergovernmental Panel on Climate Change (IPCC 2005) represents a key event in this process. Whatever one's beliefs and opinions about CCS, it is clear that it has played, and will continue for some time to play, a very important role in our deliberations on climate change and how to mitigate it. So if you are interested in the climate change problem, and especially in what we do about it, then this is a book for you.

Climate change is one of the most pressing challenges for the future development of the energy system, alongside concerns of supply security, affordability, competitiveness and choice of governance structures. Among technical decarbonisation options, CCS offers the unique prospect of reconciling substantial climate change mitigation with the continued use of fossil fuels. This would preserve the economic value of the existing fossil fuel regime, with its capital assets, skills and institutions – a fact that helps in explaining the wide appeal of CCS and its rapid rise on climate and energy policy agendas. As a consequence, the technology has gained salience in corporate as well as public sector energy strategies. If you are interested in the future of the energy system, this book will speak to that topic as well.

CCS is of potentially huge importance and of immediate relevance to a wide range of people across the climate change and energy domains. This book sets out to shed light on the social nature of its development, and to help guide the thinking and actions of everyone with an interest in the technology, be it private interest or professional, from a supportive or sceptical position and with or without training in the social sciences or indeed in the science and technology of CCS.

The realities of technology development have yet to catch up with the grand visions formulated within the CCS arena of supporters. Whilst mature component technologies and small-scale pilots exist, the world has yet to see a full-scale, integrated system in operation on power plants.[2] During the last few years, the CCS community has been up against political hesitation to meet the cost of large-scale demonstrations. There also remain technical, scientific, economic, financial, regulatory and public acceptance uncertainties, but the technology is on the verge of being tested at demonstration scale. This makes for interesting times as high expectations meet practical experience and we should expect surprising developments to unfold over the coming years. This makes the timing excellent for a book that looks into the activities and processes of technology development and deployment, scientific uncertainty, policymaking, public understanding and other issues that social science is apt at exploring and explaining.

One aspect that hopefully makes this book attractive is that it shows how social science can contribute to our understanding of CCS beyond the analytical perspectives available from engineering and natural science, which have tended to dominate discussions around the technology to date. There is a growing body of social science, broadly defined, research on CCS. One focus of this literature is based on economic modelling, exploring scenarios of deployment and assessments of the impact and cost efficiency of CCS and other climate mitigation options. Another major focus has been work on understanding the perceptions of, and opinions on, CCS of publics and other actors. Beyond this, there are important strands of research on policy, economics and finance and on innovation processes. Social science research on CCS thus spans a wide range of disciplines, methodological approaches and epistemological assumptions.

There have been efforts to integrate this growing body of research with perspectives in natural science and engineering – not least the book *Carbon Capture and its Storage: An Integrated Assessment*, edited by Shackley and Gough (2006). This work presented an introduction to CCS from geological, engineering and social science perspectives and provided case-studies of CCS deployment in the UK. A similar work released in 2007 and edited by Wilson and Gerard – *Carbon Capture and Sequestration: Integrating Technology, Monitoring and Regulation* – was written primarily from an American perspective and dealt more directly with the issue of permitting and liability. There have also been substantial outputs from integrative social science projects. The book *Caching the Carbon: The Politics and Policy of Carbon Capture and Storage*, edited by Meadowcroft and Langhelle (2009a) provided accounts of CCS development and debate in a range of countries and valuable insights based on a comparative analysis of policies and politics. Most recently, a special issue of the journal *Global Environmental Change*, edited by Bäckstrand *et al.* (2011) brought together contributions from social science researchers, also with a strong focus on policy and politics.

This book builds on this emerging literature and is the first book to present a wider range of social science perspectives on CCS, beyond analyses of policy, politics and regulation. It will demonstrate the contributions that social science can make to a wide range of topics, from actors' varying perceptions of the technology, to complex governance challenges, to the peculiar nature of CCS innovation processes.

From the point of view of social science practitioners, CCS offers intriguing possibilities for new analysis and learning. Climate change is one of the more complex problems facing humanity, a type of problem that has been called 'wicked' and 'messy' (Verweij and Thompson 2006). One important reason that it is so difficult to address and resolve is that our societies, with their underpinning capitalist economic engine, are arguably built and organised around the use of fossil fuels. Evolutionary economists have therefore begun to reinterpret the history of industrialisation as a process in which human societies have captured the energy stored in fossil fuel carriers (coal, oil and gas reserves) in ever more efficient means, such that structural and transformative economic, technological and socio-behavioural change not only has been manifested, but become locked-in (see Box 1.1) (Arthur 1989). In other words, our dependence on fossil fuels has not only enabled, but also defined and limited our current socio-industrial systems and their future development.

BOX I.1 CARBON LOCK-IN

Researchers studying how technologies evolve have developed the concept of 'technological lock-in', meaning the way in which technologies, social practices, regulatory regimes, standards, skills and management systems co-evolve and hence become highly interconnected over time (Arthur 1989, 1994). If a technology wins out amongst a range of options as the main choice in a particular technological sub-domain, it is not necessarily selected because it is automatically 'superior': timing, strategy and historic circumstance, as much as optimality, can determine the winner (Utterback 1994). But once widespread selection has taken place, apparently inferior designs can become locked-in through a path-dependent process (Unruh 2000: 820). Lock-in means that any substantial change to a socio-technical regime will require the overcoming of the inertia inherent in the status quo. A whole set of consecutive changes will be needed in other parts of the associated system (such as skill-sets, regulations and standards, companies with appropriate capabilities, a high standard of performance by the challenging technology, together with a body of consumers that understands and knows – or is prepared to invest time and effort in learning – how to use it).

Unruh (2000, 2002) has coined the term 'carbon lock-in', which we define here as: 'the interaction between fossil fuel technologies, infrastructure, institutions and social practices such that strong socio-technical, economic practices and behavioural incentives favour continued use of carbon intensive technologies'. Examples of carbon lock-in include: expansion of electricity networks powered by centralised fossil fuel fired generating plants; and the petrol-fuelled personal car transport system.

Studies of CCS and lock-in suggest that CCS (assuming it develops according to current assumptions) would reinforce carbon lock-in (Unruh and Carillo-Hermosilla 2006; Markusson and Haszeldine 2010; Vergragt *et al.* 2011; Shackley and Thompson 2012). Indeed, a common argument for CCS is that it promises to conserve the investments made and social practices constructed around the burning of fossil fuels. However, CCS is as yet immature, and its future form cannot be known for certain. There exist competing visions, for example its application to biomass-based industrial processes, air capture etc.

The question, then, is whether we can assess the likelihood that CCS as a socio-technical entity in itself is being locked into a particular development path. Here opinions diverge. Vergragt *et al.* (2011) argue that biomass CCS, whilst most likely dependent upon fossil fuel CCS for developing storage infrastructure, is being marginalised and crowded out in R&D and policy visions by a strong focus on application on fossil fuels. Shackley and Thompson (2012), in contrast, distinguish between 'shallow' and 'deep' lock-in and emphasise the continuing flexibility of CCS and the uncertainty about its future.

A transition to a post-carbon society would affect most, perhaps all, aspects of society. Decarbonisation is clearly a threat to many established practices and institutions. It is arguably larger to those same institutions and interlocked systems than the impacts of climate change themselves, since global warming will impact on the global south more than the industrialised north, which is the source of the lion's share of the emissions. The key promise of CCS, to square our need of continued reliance on fossil fuels with climate change mitigation, therefore appeals not only to those industries that are directly dependent on fossil fuels, but a range of actors throughout society. This line of thought has been supported in economic assessments by institutions such as the International Energy Agency, which have estimated that the cost of climate change mitigation without CCS would rise by as much as 70 per cent (IEA 2009). CCS has thus offered an opportunity to reconcile previously opposing interests, and has, for example, played a role in bringing fossil fuel actors to the climate change negotiation table. This inevitably also leads to CCS embodying tensions between different interests. Meadowcroft and Langhelle (2009c) as well as Pollak *et al.* (2011) have identified an apparent paradox: that the actors expressing the most enthusiasm for CCS – government and industries with strong fossil fuel interests from either the use of production and export of fossil fuels – are not necessarily the same actors that have the highest carbon mitigation ambitions.

One consequence of CCS sitting on the fault line between the fossil regime and the climate change imperative is that CCS presents a great opportunity to study controversy, conflict and the politicisation of technology, especially as CCS migrates from the imaginaries of R&D to fully realised investments. The technology can work as a 'lab' for the social scientific study of technology politics. For the reader with an interest in the role of technology in society, this is another reason to keep reading this book. And, before social scientists are accused of indulgence and working with only an academic audience in mind, studying controversy and conflict is often incredibly useful to hard-nosed policymakers, industrialists and investment fund managers in the very practical sense of effecting change that is 'acceptable' (including the possibility of no change) and avoiding proposals that are likely to be unacceptable, and such costly mistakes. Conversely, social science insights may also be used by those opposing development, to guide their efforts. Furthermore, social science can provide invaluable insights into how research, development and demonstration programmes might be shaped and organised such that those issues 'uncovered' through controversy can be addressed and, perhaps, even resolved through alternative forms for decision-making.

We can explore this issue further. It is often stated that there is only one reason to develop and deploy CCS: climate change mitigation. This highlights the fact that CCS is an environmental technology whose envisioned function is to abate CO_2 emissions, rather than a technology that in itself produces low-carbon electric power like renewables or nuclear energy. But framing CCS this way downplays the other reasons why actors may show an interest in it. For example, for fossil actors, CCS represents a chance of survival, prosperity and competitiveness in a carbon-constrained future. In more specialised language, CCS – like other technologies – exhibits 'interpretative flexibility' (Pinch and Bijker 1984). Interpretative flexibility suggests that the meaning of the technology varies depending on whose perspective we adopt, and this may matter for what functional performance they seek from it and, eventually, for how it is developed and formed.

BOX 1.2 THE SOCIAL CONSTRUCTION OF TECHNOLOGY

In a classic study of technology development, Pinch and Bijker (1984) studied how different social groups in the nineteenth century perceived the early promises of cycling and what different requirements they therefore had for the development of the bicycle. Those interested in using bikes for sport prioritised designs emphasising speed, whereas other groups were more concerned about safety, preserving the reputations of women riders wearing skirts etc. The kind of design we see as a normal bike today is at least in part the result of an interactive process reflecting the interests of different social groups, to varying degrees.

The point here is that different social groups may see a technological artefact in different ways, and that this matters for how it comes to be developed. This suggests that technology does not just develop according to an internal logic, and moreover that knowledge about a technology is, at least in part, in the eye of the beholder. We may call this 'interpretative flexibility'.

A challenge, then, in developing technology is to construct and present it in a way that allows the forging of social support for the technology. It needs to attract enough interest from enough different actors, and without alienating anyone with the clout to stop it.

The role of CCS as a glue between the climate policy community and the fossil regime means that a heterogeneous coalition of actors with very different objectives in more or less qualified support of the technology has emerged (see Box 1.3). It could even be suggested that the survival, or perhaps revival, of climate change policy as a hot political issue to some degree depends upon the extent to which CCS turns out to be successful. Without a viable CCS option, the political costs of carbon mitigation may simply be too large for some powerful interests.

These different actors have different perceptions of the benefits and risks of the technology, which creates challenges for the governance and coordination of the overall CCS project as well as of specific projects. To develop policy in such a minefield, which of course also evolves over time in response to factors like fuel prices, the fortunes of climate policy negotiations, technology development and the like, is a serious challenge for the climate policy, energy policy and innovation policy domains.

BOX 1.3 ADVOCACY COALITIONS AND DISCOURSE COALITIONS

'Advocacy coalitions' (Jenkins-Smith and Sabatier 1994; Sabatier 1998) can be defined as sets of actors from different institutions who share an interest in an area of policy and who share key beliefs. Rather than assuming that actors will advocate toeing the line of their particular institutions, their preferences are assumed to be shaped in the coalition they form with others. Pollak *et al.* (2011) have applied this notion to analyse how the development of CCS policy in the US has come to be dominated by incumbent energy sector interests rather than climate policy actors.

The related notion of 'discourse coalitions' (Hajer 1995) highlights the role of science-based 'storylines' as the glue holding advocacy coalitions together. Hajer applied this notion to explain how a variety of actors supported the storyline of acid rain in the 1980s and supported policies aiming at reduced sulphur dioxide and nitrous oxides emissions. We argue here that the dominating storyline about CCS has worked to hold together the interests of the fossil regime and the climate policy community, although that coalition has perhaps now broken down to some extent.

Finally, CCS is a peculiar case for the study of technology innovation. The currently dominating theory of environmentally motivated system change – transitions theory (see Box 1.4) – tends to emphasise new technology invented and developed in protected spaces on the margin of dominating technical regimes, that later grows to challenge established practices. In contrast, CCS has emerged from *within* the fossil fuel regime as a defensive move to sustain it in the face of climate mitigation policy and it appears to date that the CCS technologies which look to be the most promising are emerging from the existing industry players, though game-changers from the sidelines are always a possibility.

BOX 1.4 TRANSITIONS THEORY, THE MULTI-LEVEL PERSPECTIVE AND CCS

A 'transition' can be defined as a major structural shift or transformation in a societal function such as transportation, housing or the energy system. Some examples of historical transitions – as studied using transitions theory – include the shifts from sailing ships to steam ships, from horse-drawn carriages to cars, from traditional factories to mass production, and in the context of energy, from coal to gas systems (Geels 2002, 2005b, 2006a; Correljé and Verbong 2004).

It should be noted that although these examples emphasise technological shifts, transitions are not only based around technological change. Transitions typically involve major changes in infrastructure, social norms and practices, regulations and other elements (Geels 2002). A transition typically requires that these different elements change alongside each other, to form a new socio-technical system.

The Multi-Level Perspective (MLP) (see Figure 1.1) on transitions maintains that radical innovation comes from the margins – from the niche hinterland of dominant socio-technical regimes.

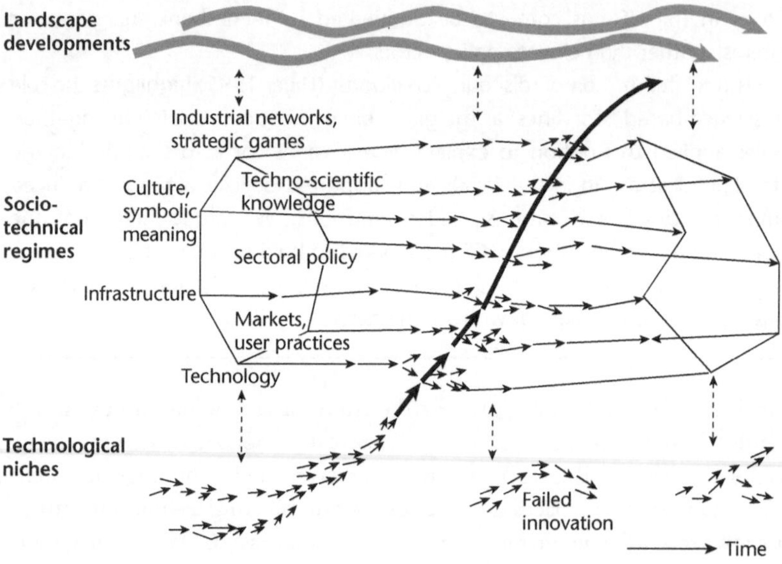

Figure 1.1: The Multi-Level Perspective (MLP) on sustainability transitions (source: Geels 2002).

The low-carbon transition technologies might be things like micro-grids populated by distributed power generation (PV, wind, biogas etc.), or larger grids using concentrated solar power (CSP) from North Africa, offshore wind, hydrogen fuel cells, algal bio-reactors, maybe even nuclear fusion, and so on. All these ideas have been widely debated with some fairly radical ideas being presented over the past few decades. But what is CCS? An interloper from within the socio-technical regime itself! It comes from the same companies that already produce power and chemical plants or components thereof such as Mitsubishi, Doosan, Siemens, Alstom, Fluor etc. CCS is promoted, developed and planned for by the incumbents in the electricity generating sector (existing utilities) and from the oil and gas companies that develop and manage hydrocarbon reservoirs. There are, admittedly, a very few smaller companies 'on the edges' which are active in developing CO_2 capture technologies and who knows whether a 'game-changing' and 'disruptive technology' will come crashing in from the sidelines at some point?

One interpretation of CCS is that it is mostly a defensive technology developed by the existing system incumbents to ensure that they are able to respond to existing and future demands for CO_2 emission reduction. CCS means that they have a clear future role in a carbon-constrained world. Its development allows system incumbents to hedge their bets. If governments are prepared to find a way to pay for CCS, then companies will (hopefully) be in a position to develop it. If politicians renege on their strong commitment to CO_2 reduction targets and intentions, however, the existing companies will not have lost out that much if they divert some resources to CCS RD&D, especially since this expenditure can be recovered from the end-consumer or through the windfall profits of the EU ETS. Many of the same companies are, of course, investing in a range of renewables and nuclear fission as part of this risk-spreading, hedging, early market penetration strategy.

Arguments framed in terms of 'either renewables or CCS' as pushed by, e.g. Greenpeace (Greenpeace International 2008) and by CCS critics such as Tsouris *et al.* (2010), underplay the importance to large energy multinationals of risk-reduction through portfolio management that are regulated in different national contexts. Simply comparing 'marginal carbon abatement costs' (MACs) of CCS and observing that there are cheaper MACs arising from the use of renewables does not begin to address the real complexity that arises from making all these technologies work in practice in systems largely not designed for them, nor the effects of technological learning in bringing down the MACs over time. In summary, CCS has been most commonly conceptualised as a non-disruptive 'end-of-pipe' technology which allows continuity in a situation of high fossil fuel lock-in within industrialised countries (Unruh and Carrillo-Hermosilla 2006).

An alternative interpretation of CCS is that it is a potentially more radical type of innovation which is not simply about continuing with centralised grid provision of electricity but rather an element in new configurations of gasification, power and heat generation and CCS technologies which can occur on a range of scales (Shackley and Green 2007). This vision is more speculative, based as it is on bold and disruptive technological innovation which comes partly from within the dominant regime but only by selected risk-taking companies. In the past few years, the vision has been looking increasingly tentative, due to challenges facing the major proponents and projects extolling this particular vision (such as BP and Statoil at their Carson, Abu Dhabi and Mongstad projects).

This origin in the fossil fuel regime and the privileged position CCS therefore enjoys, drawing on the well-established practices, knowledge and resources of this regime, also helps explain its perhaps surprisingly rapid emergence as a favoured policy option. It is frequently said that all the technologies needed are already in use, and that all that is needed is to integrate them and deploy them at scale in response to the urgent need to mitigate CO_2 emissions. This is arguably a serious underestimation of the technical challenges involved, however, as quickly becomes evident in more nuanced discussions. There has, in short, been limited time and practical experience so far to discover and develop all the technical and scientific knowledge needed for CCS. The result is a 'double exposure' of a technology that is in need of further development and at the same time already mature.

Integrating social perspectives on CCS

Hopefully we have now convinced you that this book could be interesting in several ways and raised your curiosity about its content. A key strength of the book is the contributions from a range of leading social science CCS researchers. In addition to this, the book seeks to develop an integrative approach that identifies the links as well as tensions between the contributions, and which aims towards an overall synthesis of the research presented.

To that end, we need to first take a step back and reflect on different ways of understanding and analysing technology and innovation. In addition, this will help explain the actual and potential roles of social science in the development of knowledge about CCS in particular and technology more generally.

The way we think about technology tends to be dominated by popular ways of understanding technology and at odds with the specialised perspectives social scientists have developed. It can be helpful to see this as a contrast between a traditional perspective – influenced not least by engineering – and a social scientific one. Although this contrast may, in some ways, exaggerate the differences, and certainly conflates a wide range of views among social scientists, engineers and others, it nevertheless depicts a real tendency of divergent thinking.

The traditional perspective tends to focus on technology as artefacts and their associated physical properties and environmental impacts, and to see this as the exclusive domain of engineers (and scientists). Digging only a little deeper quickly reveals that also the economics of the technologies matter profoundly to the practices of developing and using technologies as do socio-economic impacts and repercussions. Before actual deployment, economists and social scientists are therefore regularly sought out to assess what engineers term the 'non-technical impacts'. Identifying further dimensions to technology, such as politics, organisation and cultural meanings, is typically less easy and made all the harder by the dominant view of technology which does not include these aspects.

Moreover, in a traditional view, technology design and development is a matter of rational, technical-problem solving. At worst, technology is seen merely as a simple extension of knowledge gained from scientific inquiry. This perspective is problematic in several ways. First, it excludes from view the wide range of knowledges that technology development in practice draws on. They include the practical experience of engineers and technology developers, but also the experiences, skills and knowledge of customers, users, critics, competitors and others. Second, it tends to obscure the fact that technology design and development involves making choices, that there is more than one rationality involved in defining and solving design problems and that, in fact, technologies often could have come to be developed quite differently from how they were.

The traditional view instead typically depicts technology as driven by an internal logic, with both problem formulations and their solutions emerging from the depths of engineering practice as 'hard' technical facts. In reality, technology is most often developed through interaction among a range of actors, and its development benefits from, or is at least shaped by, their input.

From this follows all too easily the notion that technology development is a separate phenomenon from the evolution of wider society. If technology is seen as a matter of technical facts, determined by a delimited expert cadre, following an impersonal, internal logic and based in objective scientific facts, it becomes hard to see how technology is influenced by the wider society (apart from the fact that society created a 'demand' for a given technology to be developed in the first place).

Instead, the relationship with society is envisioned as a one-way flow, in which technology impacts on society rather than the other way around. Public debates are permeated by these types of narratives about how technologies will affect society – as a source of bounty and societal improvement or as a sinister threat and the bringer of needless risks. Either way, society can only choose to adapt to, or possibly ban, the technology. The option of reshaping technology is not on the table. At its most refined, the traditional view acknowledges that technology dynamics is driven by the interaction of technical imperatives and the demands of the market, but recognition of wider influences such as cultural, social or political factors is relatively rare.

In the traditional view of technology and how to study it, there is a rather limited role for social scientists, and only for particular kinds of social science. When technology is seen as only a set of artefacts, with purely physical properties, there is little need to understand their cultural meanings or societal roles. As a worst case, everything technology-related is seen to be the domain of engineers, with no need for social scientists to bring their expertise to bear on matters. If technology development is seen to follow an internal logic, even debates on policy and management options regarding funding, regulation etc. are sometimes considered better left to engineers and scientists. More typically, there is acknowledgement of the usefulness of some social sciences. In the context of CCS, the most well-received social science is arguably economics, used for modelling the energy system, whereas qualitative and more overtly political approaches are more rarely in demand.

At most, social scientists can be seen to have a legitimate role in analysing how technology 'impacts' on society, and how it is that the general public does not appreciate the gifts brought to them by technology developers. This frequently takes the form of social scientists being brought in to study public acceptance; typically not as an integrated part of R&D efforts at early stages of development, but as an add-on, end-of-pipe activity seen as a way of managing the reaction of publics to new technology once developed. At times, this leads to further social science engagement, but this is not necessarily a typical outcome.

In the same vein, social scientists may also be given the somewhat broader role of analysing so-called 'ELSI' projects, studying ethical, legal and social impacts of a technology. Again, this often takes the form of an activity carefully kept apart from 'actual' technology development, and as an afterthought rather than proactively contributing to problem definition.

Against this traditional view, we can articulate a view of technology more suitable for social scientific analysis. A more socialised conceptualisation of technology means that we have to acknowledge that designing and using technologies requires not only physical artefacts but also people and institutions, and that technologies have physical, but also economic, cultural, legal, political etc. properties.

These aspects or elements of a technology make up more or less coherent and well-functioning socio-technical systems or configurations, which in turn are embedded in wider societal contexts. In a process of technology development – an innovation process – these different elements interact. In principle all elements can interact, and co-evolve, with each other, although in practice some interactions will be more important that others, depending on circumstances.

In particular this means that not only will technology development impact on society, but also that societal developments will shape technology (Mackenzie and Wajcman 1999). Much work has been done in the social sciences to elaborate these kinds of dynamics.

A classic example of how social dynamics may shape technology comes from the studies of the development of the keyboard layouts. The dominating layout in Anglo-Saxon countries and beyond is the so-called QWERTY configuration. This configuration was chosen at a very early stage since it enabled salesmen to demonstrate typing the word 'typewriter' using the keys of only one row for impressive typing displays. Other layouts have since been shown to be much preferable for more realistic typing behaviour, but by then the design was so well entrenched in terms of dedicated production capacity, typing skills etc. that it has remained with us since (David 1985).

The people involved in the innovation process can be a wide range of actors, certainly including technology developers and engineers, but also users, operators, customers, competitors, regulators, marketing staff, media etc. These different categories of actors may bring their different ways of thinking – their different rationalities – to the process, and their different perceptions of the technology. An example is the way that the risks of CCS are discussed in different contexts (see Box 1.5).

BOX I.5 DIFFERENT REPRESENTATIONS: INEVITABLE VARIATION OR SOMETHING MORE IMPORTANT?

It is 13.00 on a wet Edinburgh afternoon in spring 2011 and a visiting geologist is just starting a seminar on an evaluation of the leakage risks arising from storing CO_2 in a deep saline reservoir. The visitor describes the properties, depth and impermeability of the cap-rock (which sits above the reservoir and provides an important part of the reassurance that the CO_2 will not simply leak out). She then goes on to describe the other rock formations which occur in the so-called 'over-burden' – the rock between the cap and the surface. But what if, by some unlikely event, the CO_2 did manage to escape these various containment levels? Where would it go? The 'what if' analysis led the geologist to explore the potential CO_2 leakage pathways and one possibility is that the CO_2 could get into a highly permeable aquifer at a higher level than the cap-rock and start moving laterally across several kilometres. In this way, the CO_2 could arrive at a fault or abandoned oil well in the rock and could escape to the surface. Being a land-based CCS project, the storage site is located below agricultural land, biodiverse nature reserves and human habitations, so that a leak could mean health, safety and environmental impacts. The scenario is possible though very unlikely but to this audience of mostly geologists it can be openly discussed and evaluated. Scientists frequently consider remote possibilities as part of their everyday deliberations, so they are used to considering extremely unlikely events while not regarding them as anything other than 'extremely unlikely'. More generally, in discussions over coffee and lunch breaks at CCS meetings, geologists frequently express the view that some lateral movement of CO_2 outside the boundaries of the 'storage complex', and even some eventual leakage to surface or ocean floor bottom, is very likely – but that it is not a problem because it would be such a small amount relative to the huge CO_2 volumes being stored. And, what's more, the bystander also learns that the extent of leakage could depend in part upon the degree (e.g. resolution) to which the leaked CO_2 is actually monitored across the very large surface area over which leakage could occur.

Meanwhile, throughout 2008, in the corridors and debating chambers of the European Parliament and European Commission, lawmakers in Brussels worked hard to put the finishing touches on a new 'Directive of the European Parliament and of the Council on the geological storage of carbon dioxide' (EC 2009). To be granted a CO_2 storage permit, the developer has to demonstrate that there is 'no significant risk of leakage'. The legislators appear to be implicitly acknowledging that some risk of leakage persists. The CO_2 storage site can only be handed over to the state for 'good keeping' in the long term, however, if the stored CO_2 has been demonstrated to be 'completely and permanently contained'. Hughes (2009: 1) notes that this later requirement 'demands a higher level of containment security than the "no significant risk" required for the permitting of a site'. From the developer's perspective, it is not clear when a site might be handed over to the state, even while operational costs would continue to be incurred with no revenue post site-closure. Other such 'indeterminate risks' are introduced elsewhere in the Directive and, from reading this document, the non-geologist might suppose that the risk of CO_2 leakage can indeed be zero.

This difference in representation of CO_2 leakage risk in a university seminar and during coffee breaks at scientific conferences from 'possible in a what-if-scenario', the likelihood of which cannot be stated, to 'zero-risk' in a body of European legislation that frames CCS regulation in 27 European countries, can be regarded as an inevitable and non-surprising consequence of different audiences needing different things – it's a case of 'horses for courses' it might be said. But there is another school of thought which maintains that the coexistence of such representations of the 'same thing' – that of 'potential risk if x, y and z do or don't happen' and of 'zero-risk', and of the tensions thereby created by the apparent contradiction – is quite fundamental to modernity, to today's society. Such tensions generate what Bruno Latour calls 'hybrids' (Latour 1993), which then come back to haunt us, defying as they do ready categorisation and generating all sorts of unanticipated effects. Even if one does not wish to take the argument so far, there is a pragmatic sense in which conflictual representations are worthy of study and bringing to the light of day. The formal legalistic narrative of the CCS Directive can result in high expectations on the part of policymakers and the public which cannot possibly be met by scientific assessment. This disparity could easily hold back CCS projects on the basis that they turn out to be – shock horror – 'rogue reservoirs' and not zero-risk after all.

These actors also bring a wide range of knowledges, skills and kinds of expertise to innovation processes. Much of this knowledge is informal and tacit, not necessarily based in science and sometimes lacking the legitimacy that only science and engineering 'experts' enjoy in the traditional, technocratic view of innovation. Clearly 'innovation' describes a wide range of processes (see Box 1.6).

In this perspective, technologies are the emergent outcomes of interactions between actors and artefacts, costs etc. These interactions take place in wider societal contexts providing them with resources as well as presenting constraints for the actors. Technologies are emergent in the sense that they can be as much an outcome of contingent events as of plans and strategies. As complex systems, socio-technical systems are therefore never entirely predictable. A good example of this is the so called dash-for-gas in the UK in the 1990s, which came as a surprise to most initiated observers of the energy system, and emerged as a confluence of a range of factors, including both technological, political and economic ones (Winskel 2002b). This also means that technologies are not entirely controllable or governable. Alongside the dash-for-gas in the UK, the nuclear industry ground to a halt, in spite of favourable political will – albeit trumped by a very strong commitment to market liberalisation, which came to favour gas (Winskel 2002a).

BOX 1.6 INNOVATION CONCEPTS

At the most general level, 'innovation' can be defined as a process by which a new technology is developed and used as part of wider processes of socio-technical system change and stability. But as argued here, innovation is a catch-all phrase for many different processes, where the everyday understanding of innovation as a linear process of research, technology development and subsequent deployment is only one among many different possibilities, and only ever a partial picture of reality.

Studies of technology have developed a wide range of concepts and models for analysing such processes, reflecting this diversity (for a useful overview, see Russell and Williams 2002). As a contrast to the linear model posited here as a default notion (and made to play the role of straw man, although the belief in the notion as a universal truth is fairly widely spread), we could mention the notion of 'social learning' which has been used to understand the learning and accommodation needed on behalf of a wide range of actors for (some) technologies to function. For CCS, arguably, not only engineers and suppliers need to understand the technology, but also equipment operators, policymakers, capital providers, local communities etc.

Crucially, this range of actors involves both those who will come into hands-on contact with the technology, as well as those who will know about it in the abstract. The technology needs to accommodate both users as well as this wider range interests.

The social science perspective of technology outlined above cannot give full justice to the variation within social science, in terms of the assumptions made, the theoretical commitments held, or the concepts used. This book will seek to integrate insights from a range of specific social science positions. We approach this with a healthy respect for the challenges of combining, for example, realist research traditions assuming that reality can more or less straightforwardly be observed 'out there', common in, for example, economics, with constructivist traditions stressing the active and creative role of the researcher in interpreting any phenomenon studied, strongly represented in, for example, anthropology.

The perspective set out here is an attempt at building a platform for meetings across social sciences and between social scientists, other academics and practitioners. And, as the contrast between traditional and social science perspectives does not have to be as stark as it can at times be and as depicted here, we hope that the book can also speak to wider audiences.

The focus and structure of the book

To further facilitate this integration of perspectives, we will here introduce three over-arching themes of the book, with broad interest. The first theme is that of 'perception and representation'. By representation, we mean the ways in which different actors express their understanding of a technology, in this case CCS. The actors may represent the technology more or less reflexively and strategically, to achieve different objectives, and are quite capable of varying their message to fit different situations and contexts. These contributions are useful not least for giving insights into how different actors come to see CCS. But the contributions can also help us understand the processes wherein such representations impact on the actions of those engaging with CCS. This theme picks up on a broad concern with knowledge, its production and its use in the social science study of technology.

A second theme is that of 'governance'. The term is here used to denote a wide view of the challenges that the rapid evolution of CCS present to politicians, policymakers and industry managers. This chapter focuses on the possible effects of policy choices on complex technology systems, how to make informed decisions about uncertain futures, and the impact of the technology community on the sphere of policy and funding.

The final theme is that of 'innovation', by which we mean the process of development and uptake of a technology. The focus is here shifted away from knowledge production, to the interaction and alignment of heterogeneous elements and actors in the process of establishing a new socio-technical system. This theme is useful for suggesting what mechanisms are involved in making technology work on the ground. It can give insights into how different technical and social elements of CCS interrelate: for example, how public understanding may contribute to shaping apparently technical choices made in CCS projects.

The contributions of the book reflect in some ways the status quo of social science research on CCS. To date, a key focus has been on public perceptions and engagement – a theme well covered not least in the first part. As mentioned above, issues around policy and politics have also been explored in a growing body of work, and this is reflected here as well, particularly in the second part. The theme of innovation has been less well studied, and the book's third part makes a contribution here. This is not to say that the book can fully represent all the diversity of work carried out in the social sciences broadly defined, but it does provide an overview of key themes, and it includes a range of perspectives and approaches.

The book includes mainly contributions from qualitative social science, only very briefly reviewing some results from, for example, economic modelling and learning curve analysis. In a field dominated by science and engineering and the quantitative social sciences, this book sets out to make the case for qualitative work. We are also aware, as editors living in the UK, of being prone to see the world from the vantage point of that country. For this reason, we are pleased that the contributions cover CCS developments in the EU, the US, Norway and Japan.

Before we go into the first part of the book – that on perceptions and representations – we will in the next chapter be given an introduction to CCS technology. This introduction will take you through a brief history of the emergence of the technology, explain the main technology components and varieties, as well as provide an overview on the related economics and law.

Notes

1 A CCS system design utilises a range of technologies to realise capture, transport and storage functions (cf. Chapter 2). We will here nevertheless refer to CCS as one technology, for several reasons. First, to conform to common language use on this topic and, second, since the integration of CCS into one, functioning technical system is exactly what is at stake at the present time.

2 As opposed to the (relatively) straightforward cases where CO_2 is already removed from natural gas before shipping the gas to customers; there are only three large-scale operations of this type globally.

2

AN INTRODUCTION TO KEY DEVELOPMENTS AND CONCEPTS IN CCS

History, technology, economics and law

Benjamin Evar, Chiara Armeni and Vivian Scott

This chapter introduces several aspects in geology, engineering, economics, policy and regulation that have implications for the governance of CCS on global, regional and national levels. It also provides a short history of developments in the field from the 1970s until the present day. The issues covered here set the stage for discussions in subsequent chapters and introduce essential background material to the general reader.

What is CCS?

CCS is defined as the combination of carbon dioxide (CO_2) capture from flue gas or other process emissions arising from power generation or industrial operations, with compression and transportation of that CO_2 to a suitable geological formation for permanent disposal. While storage of CO_2 for subsequent use may be possible, this is not the aim of the technology system, which has been conceptualized as an option to reduce atmospheric emissions from the energy and carbon-intensive industries.

Several governments around the world, including Member States of the EU, Norway, the United States, Australia and Japan, have established state-sponsored RD&D programmes dedicated to CCS over the past decade, and energy ministries have expressed hopes that these activities will support widespread deployment by the early 2020s. The International Energy Agency (IEA) has estimated that for CCS to contribute significantly to climate change mitigation, 100 demonstration projects will need to be deployed by this time, and scaled up to more than 3,000 projects by mid-century to capture around 15 per cent of worldwide CO_2 emissions. This will require an investment in the order of trillions of dollars (IEA 2009) together with a very fast scale-up of capacity in supply industries.

The build-up in CCS capacity may likewise provide a window of time to develop more cost-effective renewable technologies (Brown *et al.* 2007), and yield significant savings in the costs of climate change mitigation (Stern *et al.* 2006). Calculations from the IEA suggest that these cost reductions compared with a non-CCS scenario may be as large as 70 per cent by 2050, and that to achieve such savings as much as 10 gigatonnes (Gt) of CO_2 will need to be captured annually at this time, 5.5 Gt from the power sector alone and the rest in carbon intensive industries (IEA 2009). That is the equivalent of more than 2,500 average sized coal-fired power plants. Forecasting in energy generation and distribution costs has historically been notoriously inaccurate (Smil 2000; Utgikar and Scott 2006), and exact cost savings and development trajectories are therefore associated with a high degree of uncertainty until experience has been gained from constructing demonstration plants.

A short history of CCS

This section will present a brief overview of the history of developments in CCS technology, including early work in academia and industry, up to and including recent plans for large numbers of demonstration plants.

From the broadest perspective, the development of CCS as a climate mitigation technology from conception to the present day has gone through the general (and to some degree overlapping) stages set out in Table 2.1.

TABLE 2.1: Stages of CCS technology development, 1970 to present day

Period	*Key characteristics*
1970s and 1980s	Concept creation and early (isolated) academic research
1990s	The mitigation of fossil fuel emissions becomes a major political concern and leads to early research and development on CCS; incentivised by isolated progressive climate policies; first demonstration projects commence
2000–2006	Detailed studies of CCS as a potentially large integrated system, initially at national, and then broadening to international, scales; CCS part of official future energy and climate agendas and ambitions, as expressed in government policy
2007–2010	Publicly funded demonstration programmes created to attempt to prove and accelerate technology development and deployment and many small-scale pilot projects established
2010–present	Initial demonstration funding awards allocated and first demonstration projects under consideration with detailed front end engineering design; policy expectation that multiple CCS demonstration projects will be in operation around the world sometime after 2015

The developments that have taken form at these stages will be elaborated on the following page.

The early days

The concept of capturing CO_2 from large point sources and storing it in isolation from the atmosphere was first proposed as a climate mitigation method by Marchetti (1977), albeit with more emphasis on storage in the ocean water column than in subsurface geological reservoirs. While emissions abatement and climate change mitigation using CO_2 capture and storage was a novel concept, it is worth noting that neither the capture of CO_2 from large point sources, nor the injection of CO_2 into the subsurface were unknown practices at the time. Natural gas sweetening operations to separate co-produced CO_2 (vented to the atmosphere) from methane was developed in the 1930s (Bottoms 1930) and the first commercial injection of CO_2 (extracted from a natural reservoir) into a depleting oil field to enhance oil recovery (EOR), commenced in 1972 at the SACROC unit of the Kelly-Snyder Field in Texas, USA (Donaldson *et al.* 1989).

Limited academic research on CCS followed through the 1980s (e.g. Hoffert *et al.* 1979), and an early project to store CO_2 in the subsurface was initiated in 1986 (Shackley and Gough 2006). CCS was brought further into the mainstream of academic research and government interest in the 1990s, when the mitigation of energy-related CO_2 emissions from the combustion of fossil fuels became a major subject of scientific and political attention. The IEA took an early role in stimulating international research through its Greenhouse Gas Research and Development Programme, and activities were funded through government and industry platforms such as the CO_2 Capture Project collaboration between oil and gas operators, launched in 2000.

The first demonstration projects

The first demonstration CCS project had, however, been developed some years earlier. In response to political debate attempting to reconcile a progressive green domestic and international agenda with its role as a major exporter of oil and gas, Norway was at the forefront of CO_2 mitigation, and introduced the world's first tax on CO_2 emissions from off-shore industrial installations in 1991 (Sumner *et al.* 2009). In 1996, this resulted in the operation of the Statoil Sleipner gas processing and CO_2 injection platform in the Norwegian North Sea, the world's first large-scale project capturing and storing CO_2 for the explicit purpose of climate mitigation (Statoil 2009a). CO_2 co-produced with methane is here separated using amine absorption in post-combustion, and injected into the Utsira sandstone saline formation above the producing gas fields. The CO_2 emissions tax meant that CO_2 storage was economically attractive (the capture costs having already been incurred in gas processing during the preparation of natural gas).

The Sleipner platform has operated continuously, capturing and injecting around a million tons of CO_2 per year, with planning underway to connect the facility to additional gas fields in the region. While the CO_2 has migrated upwards in the reservoir faster than initially expected, monitoring to date does not show any signs of leakage (Arts *et al.* 2004; Boait *et al.* 2011).

Three further large demonstration projects linked to gas processing facilities followed, with a total of four projects in operation globally. The Weyburn Midale project is currently the 'world's largest', storing 3 million tons of CO_2 per year and commenced operations in 2000 (PTRC 2011). In 2004, the In Salah gas processing facility in Algeria – operated by BP with partners Sonantrach and Statoil – commenced storage operations, re-injecting 1.2 million tons of CO_2 annually into the Krechba gas reservoir (In Salah CO_2 2010). Lastly, the Snøhvit gas processing and CO_2 storage facility in Arctic Norway commenced operations in 2008 (Statoil 2009b). CO_2 is here scrubbed from the gas and injected into a saline formation near the producing fields. As with Sleipner, this project was developed in response to the Norwegian tax on CO_2 emissions.

For all of these projects, their novelty lies in the explicit storage (i.e. injection and monitoring) of CO_2 for climate mitigation, rather than simply the separation of CO_2. It is worth noting that in the US, several large natural gas processing facilities, e.g. Exxon Mobil's Labarge facility in Wyoming, sell scrubbed CO_2 for use in enhanced oil recovery (EOR) operations, but these are not operated or monitored as projects for permanent CO_2 storage.

Because CO_2 is removed during natural gas processing, the incremental costs of a CCS project are much lower relative to power plants, per kilowatt hour (kWh), and this explains why three of the existing projects are of this type. The exception is the Weyburn Midale project (which utilises CO_2 from a coal gasification plant), but this is an EOR project, hence it has an additional revenue stream.

Recent growth in RD&D

Since the mid-2000s, research, development and demonstration activities have grown and spread, not least in response to and in anticipation of policy and financing initiatives. This has included the development and operation of a number of mostly commercially led small-scale pilot capture projects (often with a degree of public financial support) and a lesser number of small-scale injection and monitoring research projects (Russell *et al.* 2011; GCCSI 2011b; ZEP 2011a). Many of these pilot capture projects were located at sites that could possibly serve as larger demonstration facilities, to test and develop capture technology *in situ* on flue streams from coal-fired power plants. In contrast, storage test injections generally have made use of CO_2 from natural reservoirs, rather than CO_2 captured from power plant emissions (for obvious cost reasons). Only one pilot project, the Lacq oxy-fuel gas power plant pilot in south-western France, performs full-chain CO_2 capture, transport and storage (ZEP 2011a).

In addition to the four operational demonstration projects detailed above (Sleipner, Weyburn Midale, In Salah and Snøhvit), only three forthcoming demonstration projects have made final investment decisions to go forward with construction. These are the Gorgon gas processing facility on Barrow Island in

Western Australia, the Boundary Dam coal-fired power plant in Saskatchewan, Canada, and the Kemper County coal gasification power plant in Mississippi, USA.

CO_2 capture and transport

CO_2 capture systems fall into one of three different categories: post-combustion, pre-combustion and oxy-fuel combustion. Post-combustion technologies are the most advanced family of capture applications available and are being planned for most demonstration projects funded by the EU Commission. In post-combustion, CO_2 is captured from a power plant's flue gas after regular combustion, using amines, chilled ammonia solutions or other CO_2 separation chemicals (Whitton 2009). A main advantage of this approach is that it provides the possibility of retrofitting capture units to existing power plants, which have sufficient space nearby and which can transport CO_2 to a permanent geological store or to a temporary storage hub.

Pre-combustion capture works by first reacting a fuel such as natural gas with air to produce a synthesis gas composed of carbon monoxide and hydrogen. The carbon monoxide is reacted with steam to produce CO_2 and additional hydrogen whereafter the CO_2 is separated in an absorption process. This leaves a separate hydrogen-rich stream, which can be used as an input fuel elsewhere, e.g. in a hydrogen-based transport infrastructure. The resulting high concentration of CO_2 produced in pre-combustion capture can be more easily and cheaply separated for storage than is expected to be possible with post-combustion applications (IPCC 2005).

Finally, oxy-fuel capture works by burning coal or gas with purified oxygen supplied by membranes or chemical looping, to produce a flue gas that consists mainly of high concentrations of CO_2 and water vapour together with excess oxygen (IPCC 2005).

For all these options there will be an associated energy penalty, reflecting the increased power needed to operate the CO_2 capture plant. A summary report recently published by the IEA (Finkenrath 2011) notes that the net efficiency penalty can be expected to have a range between 7.5–10.5 percentage points, equivalent to a relative net efficiency penalty range between 15–25 per cent, depending on the type of power plant and average efficiencies.

After capture, the CO_2 gas stream is compressed to around 10 to 100 bar and transported either via pipeline or ship to a geological storage location. Existing pipeline infrastructure used for natural gas and EOR may in some places be used to transport the CO_2 to offshore storage fields (IPCC 2005). In the EU, much of the discussion around CO_2 transportation has focused on the feasibility of constructing oversized pipeline networks managed by a government body or contracted out to a private operator, with right of access granted to a number of private power plant operators and carbon-intensive industries. This would allow the pipeline network to accommodate increasing amounts of CO_2 as the price of carbon increases in the future, or as mandates on emissions performance are placed on power plant capacities or carbon-intensive industries.

CO_2 storage

Storage of CO_2 is mainly envisioned to take place in depleted oil and gas fields in the near-term, largely because such structures are relatively well mapped and have already been proven to contain hydrocarbons for thousands of years. In the longer term, most experts agree that larger deep saline aquifers, such as the Utsira formation beneath the Sleipner production well, will increasingly be necessary to allow injection of the substantial amounts of CO_2 envisioned (SCCS 2011). The injected CO_2 is stored in the pore spaces of permeable rocks such as sandstone, displacing and, to some extent, slowly dissolving in saline water. Injection takes place at a minimum depth of 800 metres, at which point the CO_2 typically exhibits a sharp increase in density, and a corresponding decrease in volume. With this increased density the CO_2 can remain stored as a supercritical fluid, a phase that shares characteristics of both a fluid and a gas. Because the CO_2 will be buoyant relative to saline water at this depth, it is important that the sandstone formation is overlaid by an impermeable mudstone cap rock. This way, if faults or fractures in the sandstone facilitate the upward migration of the CO_2, a final permanent trap is ensured by the dense mudstone cap rock. It is possible to inject the CO_2 to depths where it is no longer buoyant relative to saline water, but as the depth increases so does the cost of injection and the difficulty of monitoring sites. Other storage options are available including coal seams and disused mines and caverns, but such sites may not ensure sufficient integrity against leakage. Storage at shallower depths is also possible, but usually onshore where groundwater is stored. This therefore poses a risk of local groundwater contamination and also increased chance of public concern over other environmental and health risks (Shackley and Gough 2006).

Injected CO_2 is trapped within geological structures via four physical and chemical reactions that work over different timescales. In the period immediately following injection, CO_2 will be trapped primarily by virtue of the stratigraphic and structural characteristics of the geology. Faults and rock fractures may function as traps under some circumstances and in others facilitate the upwards migration of CO_2. The presence of an impermeable cap rock is therefore important to ensure that the CO_2 does not migrate to the surface. Hydrodynamic trapping can occur in saline formations that do not have a closed trap, where saline fluids migrate slowly towards the surface over very long distances. As CO_2 rises towards the top of the formation, it slowly dissolves in the water, and may become trapped as residual CO_2 saturation. In the long term, migration towards the surface may take place over millions of years after dissolution. Geochemical trapping of CO_2 takes places through a number of processes. First, through solubility trapping as CO_2 rises to the surface and dissolves in the saline water, thereby eliminating the buoyant forces. Second, the CO_2 will form ions as the rock dissolves. Finally, as the injected CO_2

reacts with the geology over thousands of years, some of it may be trapped as stable carbonate minerals (IPCC 2005).

Cost estimates

In a review of ten cost studies carried out for the IPCC's Special Report on CCS, Rubin, Chen *et al.* (2007) summarise a range of near-term CO_2 avoidance costs for coal- and gas-fired power plants. These findings only account for the cost of capture, by far the most expensive part of CCS operations, and range between \$29–51/tCO_2 (€21–37/tCO_2) for coal-fired power plants and between \$37–74/tCO_2 (€27–54/tCO_2) for gas-fired power plants. In their own cost model they include transport and storage costs, which add an additional cost increase of 4–10 per cent on top of the cost of capture, resulting in an average full-chain avoidance cost of \$61/tCO_2 (€44/tCO_2) for coal-fired power plants and \$72/tCO_2 (€52/tCO_2) for gas-fired power plants. A higher set of price ranges for CO_2 capture are reflected in an IEA cost summary report from 2011 (Finkenrath 2011). The CO_2 avoidance cost for capture in coal-fired power plants here ranged between \$43–58/tCO_2 (€31–43/tCO_2), and for gas-fired power plants was estimated to be \$80/tCO_2 (€58/tCO_2). The EU Zero Emissions Platform (ZEP) also recently completed an updated cost study, this one based specifically on CCS for the European power sector using data from the ZEP member projects (ZEP 2011b). The CO_2 avoidance cost ranges for full-chain CCS in this study were based on estimates for early commercial experience and concluded that the cost range for coal-fired power plants was between €40–50/tCO_2 (\$55–69/tCO_2) and between €80–110/ tCO_2 (\$110–151/tCO_2) for gas-fired power plants.

The significant price difference between applications for coal- and gas-fired plants in these studies reflects a relatively high degree of natural gas price fluctuations, as well as a higher per unit capture cost in the relatively low CO_2 concentrations of natural gas combustion. With limited real world experience with CCS to support such assumptions, the robustness of cost models generally has been shown to be relatively poor. Even minor adjustments to input assumptions have been shown to result in substantial changes to total project costs, and associated long-term deployment prospects (de Coninck, Stephens *et al.* 2009).

Prospective analyses have also considered the potential for long-term cost reductions. In one study Rubin, Yeh *et al.* (2007) used learning curves from a number of technologies to show that early costs could decrease by as much as 40 per cent after the first 100 gigawatts (GW) of installed capacity have become operational. If the IEA's hopes for CCS hold, this could happen as soon as 2020. However, the history of technology development suggests that poorly coordinated development worldwide may result in little knowledge sharing and slow cost reductions (Neij 1997). A substantial number of demonstration facilities may therefore not be sufficient, without appropriate channels for sharing best practice (van Alphen, Hekkert *et al.* 2010). Costs have moreover risen for some technologies like flue gas desulphurization after early large-scale integrated demonstration projects were constructed, due to unforeseen complications (Rubin, Hounshell *et al.* 2004) as well as limited competition among technology suppliers and the presence of a single dominant buyer (Rai *et al.* 2009).

Incentivising and financing CCS

Financing wide-scale deployment has so far been imagined in the form of a mix of private and public funds from an increased carbon price in the EU Emissions Trading Scheme (ETS), government-specific funding for CCS (such as feed-in tariffs, levies and direct demonstration and deployment subsidies), financial transfers to developing countries under the Clean Development Mechanism (CDM), and carbon taxes. Obstacles exist for all of these to be implemented and to support large-scale CCS development.

In the case of the EU ETS, which has long been seen as the primary funding mechanism for stimulating RD&D in and deployment of low-carbon technologies, an adequate price signal has yet to emerge. A lack of credibility in the ETS may also create perverse incentives for private developers to invest in coal as the price of carbon is discounted, in turn further eroding the market's credibility (Blyth *et al.* 2007).

In anticipation of the EU ETS price failing to meet the near-term CO_2 avoidance cost for CCS in power plants, the EU Commission has announced plans to build 12 demonstration projects by 2015 with public financing from the Zero Emissions Platform (ZEP) (*c.*€1 billion), the European Energy Programme for Recovery (EEPR) (*c.*€2.2. billion), and 300 million ETS allowances under the New Entrants Reserve (NER, valued at *c.*€2.5 billion at the start of 2012). Some of these funds (the ZEP and EEPR) have already been allocated to projects while the Commission plans to allocate NER funds some time in 2012. Receipt of funding usually comes with a requirement that developers share knowledge with one another and find additional financing elsewhere. Limited allocations under the EEPR (€180 million per project) has however meant that some successful entrants have declined to go ahead with projects due to a shortfall in total funds. According to the Global CCS Institute, a total of $21 billion of public funds are expected to be allocated worldwide in the next couple of years to support CCS RD&D (GCCSI 2011b). However, the allocation process in the EU has been delayed since first announced and worldwide several projects are facing reduced public financing commitments.

After years of negotiations, CCS has now been accepted as a mitigation technology eligible for financing under the CDM's project activities. This opens the door for non-Annex 1 countries with relatively large fossil fuel usage – China and India in particular – to invest the necessary resources to install CCS on power plants, supported by the technology specific funds. It is, however, unclear what a project application for CDM certification will look like at this stage and if the path towards funding support will be favourable to prospective developers.

Finally, carbon taxes may be combined with revenues from a carbon market in some jurisdictions, but only Norway has so far implemented a tax that is high enough to make CCS technology development cost-competitive with unabated emissions of CO_2. This is, however, restricted to offshore oil and gas extraction, rather than the entire power sector.

As an alternative to financing mechanisms, a CO_2 emissions performance standard (EPS) could be passed on an individual country basis or for a region such as the EU. Such an EPS was first introduced in California in 2007 where the allowable average emissions are just below 500 gCO_2/kWh. This is lower than conventional pulverised coal-fired plants, but higher than most combined cycle gas turbine (CCGT) plants, and therefore does not effectively mandate use of CCS in the near-term.

Similar legislation has since been introduced in Oregon, Washington and Montana where it complements the voluntary carbon market, the Western Climate Initiative. In the EU, modeling work has suggested that the introduction of an EPS limited at 350 gCO_2/kWh for plants above 100 megawatts (MW) capacity from 2010 onwards, could reduce power plant emissions by a third (Wartman *et al.* 2009). Legal analysis has concluded that such a standard would be admissible as part of existing Directives (Client Earth 2008). However, the EU Parliament has announced that an EU-wide EPS will not be passed. It is therefore left to individual Member States to decide whether they want to include a standard as part of a general incentive package for CCS.

CCS legal developments under international law

Global legal efforts to govern and support CCS deployment have initially focused upon the removal of unwitting legal barriers, particularly with regards to CO_2 storage, and subsequently upon the design of dedicated legal and policy frameworks to facilitate capture, transport and storage. Currently, developments in many jurisdictions are mainly driven by the implementation and consolidation of such legal and policy decisions (Macrory 2011).

Early legal questions were concerned with the legality of CCS under international marine legislation. In the absence of a specific provision, the 1996 Protocol to the 1972 London Convention on the Prevention of Marine Pollution by Dumping of Wastes and Other Matter (the London Protocol) and the 1992 OSPAR Convention on the Protection of the Marine Environment of the North-East Atlantic were seen as involuntarily prohibiting CO_2 storage under the sub-seabed (Hendricks *et al.* 2005; Havercroft and Purdy 2007). Over a relatively short time, agreement was reached for amending such treaties (2006 amendments to Annex I of the London Protocol and Annex II-III of the OSPAR Convention, London Protocol (2006), OSPAR Commission (2007a, 2007b, 2007c)) to enable offshore storage of CO_2, under specified conditions (Armeni 2011). In 2009, an important amendment to Article 6 of the London Protocol was also adopted to enable the transboundary export of CO_2 (London Protocol 2009). Despite the significance of these amendments, some issues remain. While the amendment to the OSPAR Convention is about to enter into force the CO_2 transboundary export amendment to the London Protocol has only been ratified by Norway, and there is a high probability that implementation will take a long time. This results in such prohibition still being in force, and limits investor confidence in undertaking financial and operational investment decisions towards CCS.

CCS legislation in the European Union, Australia and North America

These international hurdles have, however, not prevented CCS from gaining a primary role within national climate change strategies – mainly within the European Union and its Member States, in Australia and in North America. The adoption of the 2009 EU CCS Directive (EC 2009) represents the 'first comprehensive legal framework for the management of environmental risks related to CCS worldwide' (Doppelhammer 2011: 93), placing the EU in a leadership position on the regulation of the technology. The Directive mainly governs onshore and offshore CO_2 storage (from exploration to post-closure), but some provisions also cover capture and transport. It establishes a permitting regime and criteria for enabling storage activities in the EU, with special attention to site selection, risk management and monitoring. In case of leakage, the Directive also establishes that the operator is responsible for corrective measures, a surrender of CO_2 allowances under the EU ETS (EC 2003) and preventive and remedial actions under the Environmental Liability Directive (EC 2004). It includes an obligation to provide financial security for stored CO_2 and the transfer of long-term responsibility for site safety to a competent national authority, if the site operator fulfils specified conditions.

The Directive also establishes a carbon capture readiness requirement for all combustion plants with a rated electrical output of at least 300 MW. This means that consent to build new combustion plants is subject to an assessment of the technical and economic feasibility of retrofitting CCS in the future, when the technology is proven. If such assessment is satisfactory, sufficient space must be left available at the site for CCS equipment to be installed. Finally, the Directive removes legal barriers to CCS deployment embedded in existing EU laws (including the Water Framework Directive, the Integrated Pollution Prevention and Control Directive, the Waste Framework Directive and the Environmental Impact Assessment Directive) to avoid double regulation.

All EU Member States were obliged to adopt national legislation implementing the CCS Directive by June 2011, but as of February 2012, some Member States have yet to comply (e.g. Germany, where the parliament's Chamber representing the 16 Federal States – *Bundesrat* – rejected the proposed law in 2011). The European Commission has since issued four Guidance Documents to support Member States with the interpretation and implementation process (EC 2011a).

Australia has also been a pioneer in the area of CCS regulation, which has developed at various levels of government. An overarching Commonwealth framework regulating offshore CCS (Australian Department of Resources, Energy and Tourism 2011) combined with implementing regulations and state legislation

– both onshore in Victoria (Victorian Government 2008), Queensland (Government of Queensland 2009) and South Australia (Government of South Australia 2011), and offshore in Victoria and Western Australia (Government of Western Australia 2008) – is currently in force. However, different approaches and inconsistencies can be observed in the various frameworks (Gibbs 2011). While the Commonwealth Offshore Act, the Victorian Offshore Act and the South Australia onshore legislation are based upon existing petroleum legislation, both the Victoria and Queensland onshore legislations constitute autonomous regimes. In one case (Western Australia), project-specific legislation has even been enacted merely to regulate the Gorgon Project. Different approaches to regulation have emerged in particular provisions, such as those governing long-term liabilities, which can either be transferred to the Commonwealth/State or remain with the operator (as in Victoria and Queensland).

The state-of-the-art in CCS law and policy is also quite diverse in North America, which has chiefly been influenced by the experience with EOR. In the absence of a concrete US federal government initiative, the Federal Environmental Protection Agency (EPA) has led regulatory activity for CCS. In 2010, the EPA finalised both the requirements for CO_2 storage under the Safe Drinking Water Act's Underground Injection Control (UIC) Program, including the development of a new class VI of injection wells (EPA 2010a), and the reporting requirements under the Greenhouse Gas Reporting Program for facilities that inject CO_2 underground, with or without permanent storage (EPA 2010b). At State level, extensive legal activity has produced a patchwork of CCS legislation initiatives. These activities have largely not been coordinated and legal concerns related to transboundary migration and differentiated long-term liability frameworks may therefore need to be further addressed in the near future.

In Canada, the province of Alberta has recently taken on a national leadership role for the policy and regulation of CCS. In 2010, Alberta adopted a series of amendments to existing legislation to clarify the regulatory framework for CCS in its territory (Alberta Minister of Energy 2010). The Act vests the government with the ownership of the pore space for CO_2 storage and authority to issue storage licences and leases. Upon certified site closure, ownership of stored CO_2 is vested in the Crown, which assumes all long-term liabilities (including claims in tort), when the claim is the result of activities carried out as part of an agreement with the government for injecting CO_2 and the operator has complied with all other relevant regulations. A financial contribution to a post-closure stewardship fund is also required to enable the province's government to cover the costs associated with the transfer (liability as well as monitoring and management of facilities). Recently, the Alberta government has also adopted regulations on pore space tenure rights (Alberta Minister of Energy 2011); undertaken a wide regulatory framework assessment for CCS in order to issue regulations, bringing together international expertise; and proposed the introduction of an EPS and a CCS readiness requirement for new coal-fired power stations. Despite an initial paucity of legislative activities when the EU was developing a bespoke framework with the

CCS Directive, Alberta appears to have waited in regulating CCS in order to benefit from best practice from early movers. Legal developments can also be observed in the province of Saskatchewan, where amendments to existing legislation have recently been passed to regulate CO_2 transport and storage.

CCS prospects in emerging economies

The potentialities and prospects of CCS have also started to be discussed in several emerging economies (Kyoto Protocol non-Annex I countries). The International Energy Agency (IEA) has estimated that, in order to maintain a commitment to reduce emissions by 50 per cent by 2050 compared to 2005 levels (IEA 2008), 50 CCS projects will be needed by 2020 in non-OECD countries (IEA 2009). In recent years, several non-OECD countries have shown significant interest in the technology and some have already started an assessment of the capacity of their legal regime to incorporate CCS activities, leading to a definition of such countries as 'second generation CCS lawmakers' (GCCSI 2011b: ix). China has been particularly interested in CCS deployment and has established financial arrangements, international RD&D partnerships and capacity collaborations to support six major demonstration projects. However, a legal framework is yet to be developed.

South Africa has set up an ambitious roadmap for commercial deployment by 2025 and is committed to producing a draft legislative and regulatory framework in the next couple of years. South Africa is also considering the introduction of a carbon tax and has established a carbon capture ready requirement for the proposed 5,400 MW coal-fired Kusile power station (Beck *et al.* 2011). Conversely, India has been more reluctant to commit to CCS, mainly because of cost considerations, competing priorities for development and poverty eradication, and the risks and uncertainties associated with CCS (Kapila *et al.* 2009; Rajamani 2011). However, steps have been taken in the context of international RD&D collaborations, capacity building and financial support.

Brazil has likewise been cautious on CCS and has actively opposed the inclusion of CCS within the CDM. This was due, inter alia, to concerns associated with the monitoring, environmental and health impact, liability and, more broadly, the risk that allowing CCS projects within the CDM 'would divert from the central idea of the CDM which is to promote long-term benefits in the direction of low carbon economy towards creating subsidies to enhance fossil fuel production' (SBSTA 2010: 8). Despite such opposition, some initiatives on CCS are being developed in the country (international R&D collaboration and capacity building initiatives). Important CCS initiatives have also been launched in Korea, Indonesia and Malaysia (GCCSI 2011b).

Key legal issues in regulating CCS

The resolutions to specific legal definitions have been pivotal in establishing legal and regulatory frameworks for CCS. Particularly important is the qualification of CO_2 as waste or as a commodity. This question has been partially resolved by excluding permanently stored CO_2 from the list of waste, where disposal is prohibited under international marine dumping legislation and from the scope of EU waste legislation. Recently, the US EPA has also proposed to exclude CO_2 from the legislation governing hazardous waste, provided that storage complies with rules established under the Safe Drinking Water Act. However, some interpretative issues remain with respect to the treatment of CO_2 under the 1989 Basel convention on the control of transboundary movements of hazardous waste, and the 1991 Bamako convention on the control of transboundary movements of hazardous waste into and within Africa, neither of which mention CO_2 specifically.

Another important issue is related to storage permanence and the consequences of leakage. Leakage from a CO_2 store ('non-permanence') could result in damage to the environment, human health and physical property, as well as in wider economic losses from price effects on land. Leakage will also have implications for climate damage, as CCS mitigation efforts would be compromised by any CO_2 released into the atmosphere. Provisions to ensure permanence and to regulate the consequences of leakage are at the core of CCS legislation. Adequate site selection, risk assessment, risk management and monitoring are therefore crucial elements of regulatory frameworks. At the same time, a comprehensive short- and long-term liability framework is essential to ensure that the consequences of non-permanence are adequately repaired and compensated.

The IEA has analysed several legal issues and best practices discussed in this chapter, across a number of jurisdictions, with a view towards supporting development of a bespoke legal and regulatory framework (IEA 2010a). However, it is important to note that individual jurisdictions are likely to modify such tailor-made frameworks to fit their interests and concerns.

PART I

Perceptions and representations

3

INTRODUCTION TO PART I

Perceptions and representations

Nils Markusson and Simon Shackley

Topic/scope

The theme of perceptions and representations centres on social science research on different actors' – both lay people's and experts' – perceptions and understandings of CCS technology, that is, how they interpret the technology and construct representations of it. These perceptions and representations matter for decision-making and for our understanding of the actions of the actors involved with CCS, as their actions will be shaped by how they perceive the technology, its potential benefits and drawbacks.

One of the areas that has received most attention so far from social scientists studying CCS is that of public understanding and acceptance. The focus has been on not least the general public and its knowledge of and opinions about the technology. This research has shown, for example, that CCS is more acceptable if presented as one among several mitigation options. Another important focus has been on people in local communities affected by CCS projects, showing that local circumstances vary and that the reception of CCS proposals depends on local context and history.

Representations research covers a broader range of actors than just individual citizens. Not least, there has been interest in collective actors and organised interests. Environmental NGOs have received special attention for their role in articulating as well as forming wider public opinion. Also, the role of media has been studied as an arena for CCS debates or as an actor in itself making decisions on how to portray the technology.

Moreover, research in this tradition has evolved from simplistic assumptions about strict distinction between expert actors with reliable knowledge and lay actors who have opinions rather than knowledge, to more subtle models. This has opened up for the social science study of how expert knowledge is produced. The research theme of representations includes studies of all kinds of experts in science, engineering and other disciplines and in academic, industrial, policy and other settings.

Questions addressed in this research theme include:

- What shapes different actors' perceptions of CCS?
- What are the effects of different actions or interventions on actors' perceptions?
- What kinds of impacts can public understanding have, apart from just stopping projects?
- How do the representations of CCS produced by experts affect the actions of themselves and others?

Relevance to non-academics

Research on representations of CCS matters practically in several ways. It matters most obviously, perhaps, in terms of understanding and engaging with public responses to concrete CCS projects proposals. The most high-profile case of public protest stopping a project to date is the Barendrecht project in the Netherlands (Vergragt 2009; Feenstra *et al.* 2010)

The main point to make here is that public acceptance can matter profoundly – both as a show-stopper and in shaping the designs and plans for projects – and that social science research can help us understand how this might happen. The knowledge and opinions of the general public with respect to CCS is also important as it can come to impact on policymaking. Below, we will review in more detail what the results are from research on public understanding and acceptance so far (see also Appendix 3.1 to this chapter for a tabulated summary of results from selected publications).

Clearly, many actors strive to contribute to how CCS is represented in the public sphere and to influence the perceptions of CCS in circulation. Social science research can help understanding the processes through which that can happen. A note of caution is needed here. Social phenomena are rarely predictable, at least not fully and in any straightforward manner. This limits the scope for producing any simple 'how to' recommendations for managing public perception and an awareness of contingency and unexpected outcomes are necessary. Having said that, social science can provide insights into what may or may not work, which can help guide action at the level of specific projects (Hammond and Shackley 2010) as well as at the level of wider societal debate. Moreover, the results can of course also be useful to a variety of actors with different motives, be it to build support for the technology or mount resistance.

It is important to note that what is at stake here is not just a binary of actors being for or against the technology, but to understand the role of representations in shaping the evolution of a technology. The articulation of visions, plans and scenarios, and the claims made about their benefits and drawbacks, help guide the actors, and ultimately, therefore, contribute to shape the emerging technology and its related institutional arrangements. Social science can also contribute to understanding such wider processes and assess what works well or not, from different viewpoints including efficiency – is the message getting out there? – as well as democracy: is it a transparent process? In the part of the book on governance, we will return to these issues.

This part seeks to address these kinds of uses. Practical questions that will be addressed include:

- How may the design of CCS projects in general, and public outreach activities in particular, affect public acceptance?
- What framings of the message may cause what reactions?
- How can actors engaged with CCS learn more about their (our) roles in shaping perceptions of CCS?

Literature review (see also glossary in Appendix 3.2 to this chapter)

We know that public understanding of CCS matters. Public support for CCS can make or break CCS projects. There are examples of both public protests leading to aborted projects and local publics being supportive of CCS plans (Feenstra *et al.* 2010; Markusson *et al.* 2011). Social science can contribute to explaining how this happens.

CCS actors are aware of the importance of publics and concerned about its impacts, but not always very good at dealing with this issue. Shackley *et al.* (2009) concluded that efforts at communicating CCS to the public had in general not been well coordinated or effective. The authors proposed a more proactive, interactive approach, and the provision of more accessible materials.

It is clear that acceptance of the technology, and its impact on policy and projects, is contingent on the local and national context (Markusson *et al.* 2011). How technology is governed varies across places, cultures and political systems. More technocratic, less participative systems are less sensitive to public concerns. Acceptance of CCS is also likely to depend on preceding experiences of interaction with the relevant actors – as employers, as developers of other projects, etc. – and the resulting levels of trust or distrust (Huijts *et al.* 2007). Trust is based on perceptions of both the competence and the intentions of the actors backing CCS projects (Terwel *et al.* 2009). Broad coalitions of actors, including a range of competences, as well as representing different interests and concerns, may stand a better chance of getting their plans accepted. Support from environmental NGOs can increase trust (Huijts *et al.* 2007).

In comparison with local communities potentially affected by CCS projects, studies of the population at large tend to show low levels of awareness of CCS (Wallquist *et al.* 2009; Malone *et al.* 2010). There are also results showing that public perceptions of the technology are rather fickle (de Best-Waldhober *et al.* 2009), suggesting that they are not based on thorough reflection. The future of general public perception of CCS is therefore uncertain. Very different scenarios are possible, including continued lack of interest, polarised views à la nuclear technology and more temperate opinions.

The linchpin argument for CCS is that it is the only technology that squares the circle of continued reliance on fossil fuels with climate change mitigation. Hansson and Bryngelsson (2009) have mapped the representations of CCS by experts, as expressed in public domain reports, scenario exercises etc. and found that a more complex range of arguments were marshalled in favour of CCS. They identified three main storylines, summarised in the following:

- *CCS can sustain the modern lifestyle and is compatible with current social structures.* There is plenty of coal left, and humankind will not be able to abstain from using it. CCS is a modern solution, guaranteeing our high-energy lifestyles. CCS is the only solution, since all other solutions are likely to fail.
- *CCS provides a bridge to a sustainable energy system.* CCS will buy time for the further development of renewables. CCS will not crowd out other technologies, but competition on the market will make sure that many solutions are developed, all of which are needed. Some experts thought that CCS may in the longer term break the fossil lock-in, and bridge over to a hydrogen economy.
- *CCS is about solidarity with the developing world and about peace.* Rich countries should help developing ones by funding the development of CCS. Without CCS, global warming impacts may lead to large-scale global migration and wars. CCS contributes to security of supply and economic growth making society more resilient.

The authors (Hansson and Bryngelsson 2009) also found evidence of a systematic optimistic bias among experts. This is not surprising, and part of the promotion needed to bootstrap innovation processes. Such optimism can be problematic in some ways though. Sometimes the dynamics of technology promotion develops into cycles of hype and disappointments (van Lente and Rip 1998), which means a bumpy ride for those seeking to develop the technology. In the case of CCS, this would at least imply delayed development. Optimistic bias among experts could also mean that the advice provided to policymakers is misleading, potentially underplaying the problems and risks of the technology, although of course the advice could also be too pessimistic (cf. Hansson, Chapter 5, this book; Shackley and Evar, Chapter 9, this book).

The counterpart of the CCS visions and promises described above are the main concerns expressed about CCS by experts, publics and other stakeholders, notably environmental NGOs. The concerns that have been voiced have been constructed around not least the following frames:

- *Carbon storage is unreliable and risky.* CCS has a risk of CO_2 leaking from storage sites, and may have impacts on both human health and the environment.

- *CCS will be costly and will not be ready on time.* CCS comes with a severe energy penalty and adopting CCS will add to the cost of power production prohibitively. It will take too long time to develop and deploy CCS for it to contribute to meeting climate mitigation targets.
- *CCS-abated fossil fuel plants will out-compete renewables.* CCS will crowd out efforts to develop renewable energy sources and improve energy efficiency, and lock us in further into an energy system reliant on fossil fuels.
- *CCS is a barrier to profound societal change.* The compatibility of CCS with existing societal arrangements has sometimes been seen as a disadvantage, hindering a transition to a society with a radically lower resource use.

(adapted from: van Alphen *et al.* 2007; Wallquist *et al.* 2009)

The presentation above of the main arguments is a caricature and should not necessarily be read as implying that there is a polarised debate. The reality is more nuanced than that. For example, environmental NGOs express very different positions on CCS – ranging from strong opposition to enthusiastic support (Corry and Riesch, Chapter 6, this book). Surveys of CCS stakeholders have found support for the technology, but that the support is conditional, and that the stakeholders have an awareness of the risks (Shackley *et al.* 2009; van Alphen *et al.* 2007; Johnsson *et al.* 2010). It should also be noted, again, that both support and scepticism is shaped and coloured by the local context with its specific circumstances and history.

We know a few things about what designs and technology varieties may be more or less acceptable. Experiences in, for example, Germany and the Netherlands suggest that onshore storage may be seen as risky, and a common expectation is that offshore storage may be more acceptable. In the case of offshore storage, it may be that the siting of pipelines is the bigger obstacle (Shackley *et al.* 2009).

Provision of information about climate change and about CCS may increase acceptability of CCS, but this is not clear (Huijts *et al.* 2007; Malone *et al.* 2010). Whilst ignorance does play a role, lack of knowledge is not enough to explain the lack of stable, strong public support. What information is provided matters for CCS perceptions, but also, crucially, how the information is provided and by whom (Malone *et al.* 2010).

The acceptability of CCS also varies with how it is presented and framed. Shackley *et al.* (2009) argue that CCS should not be promoted as a magic bullet. The technology is more acceptable if seen as part of a wider set of mitigation options, including renewables and energy efficiency.

Finally, it is perhaps worth noting here that the link between representations and action is complex. Even strongly articulated opposition to the technology may not hinder deployment, depending on the societal context. Conversely, industry acceptance of the technology may not in itself resolve issues around, for example, who will foot the bill. A tendency has been observed in policy domains (by Langhelle and Meadowcroft 2009, comparing across countries, and by Pollak *et al.* 2009, comparing across US states) to develop CCS policies rather than emissions control policies when dependent on fossil fuels. In the continued absence of strong climate change policy commitments, we may expect more talk than action. Further issues around the governance of CCS will be discussion in the second part of the book.

Key theoretical perspectives

Several disciplines, including psychology, communication studies and sociology, have contributed to the study of CCS representations. These disciplines and the associated theoretical perspectives come with quite different – partly conflicting, partly overlapping – concerns, assumptions and methods.

In psychology, the main focus is on individuals, as part of large populations or more concrete groups. The research is about the mechanisms through which individuals' perceptions and attitudes are formed, including information provision and interaction with other individuals. See for example Polson and Curtis (2010) for the case of expert assessment of CCS risks and de Best-Waldhober *et al.* (2009) who studied the Dutch general population.

The sociology of knowledge provides a different perspective on the construction of knowledge, shifting the focus further away from individuals towards collective social processes, and emphasising more strongly the impact of societal contexts on knowledge production. Research in this vein has shown how even the 'hard facts' of the natural sciences are the products of social processes (Bloor 1991). Through extension, this research has shown how also knowledge about technologies is open for interpretation (Pinch and Bijker 1992).

Sociological research shows that the boundary between the knowledges of experts and lay people may be less clear than is often assumed (Wynne 1996). This research has provided a useful reference point for further work into public understanding of science and technology, emphasising the contingent and context-dependent nature of the knowledge of both experts and lay actors.

The sociology of technology has provided further insights into how representations of technologies may matter for technology development and deployment. Articulation of expectations about technological advances and their future development helps attract interest and resources to the efforts in developing the technologies and may therefore act as self-fulfilling prophesies (Michael 2000). This strand of research can help explaining how representations do not just matter for opinions and decisions to attract or reject a technology, but act in more subtle ways to shape the actions of the actors involved and the form the technology eventually takes.

The methodological choices in this field are plentiful, and the normative lessons taught vary, but through a – crude – simplification, we could say that:

- psychological approaches tend to collect data through surveys or experiments and rely in statistical analysis; and that

- sociology of knowledge type approaches rely much more on qualitative data from documents, interviews and focus groups, and emphasise the need for interaction and dialogue.

The contributions of the chapters in this part

The first chapter in this part, by Judith A. Bradbury, applies socio-cultural theory to the issues of public understanding of CCS. Drawing on case studies of community responses to CCS project proposals, she notes that effective public engagement needs to start from an understanding of the communities' social characteristics, including people's broader policy views and life experiences, rather than the technology as such. A key recommendation emerging from this study is the need for integrating public engagement work at a strategic level in project organisations, and to utilise social science expertise.

Anders Hansson turns the sociological gaze on the CCS experts and how they (we) represent CCS in scenarios: reports as well as acceptance research. He observes a tendency on behalf of experts towards optimistic portrayals of CCS futures and of downplaying their reservations in key communications. Hansson applies results from the field of expectations research, and notes that whilst there is a need for any new technology to seem promising, this is problematic in terms of communication with lay persons. Proposed solutions include explicit methodological reflexivity, holistic analysis of uncertainties and new forums for learning about emerging technologies.

Finally, Olaf Corry and Hauke Riesch study how environmental NGOs have come to have widely diverging positions on CCS. They apply problem-definition literature to the topic of CCS, to make the point that decision-making is a messy process which depends on the framing of both the problem and the solution. And since there is always scope for varying framings, there is scope for disagreement. The chapter makes the diversity among environmental NGOs more intelligible, by showing how their positions vary depending on whether they see emissions or underlying fossil fuel use and industrial arrangements as the problem to solve, and whether they see CCS either narrowly as a technology in an engineering sense or more broadly as a socio-technical arrangement.

Appendix 3.1 Summary of results from selected public perception research

TABLE 3.1: Findings from anticipative basic research (i.e. undertaken without reference to a particular real-world case)

Issue	*Notes*	*Examples/references*
Large-scale leakage of CO_2 from geological storage sites	Concerns about safety of storage – lack of understanding of how liquids or supercritical fluids occur in rock formations Concerns over explosive release of pressurised CO_2 'bubble' or 'balloon' Defeats the purpose of CCS Health and safety risks arising to local populations Ecological impacts of leakage Induced seismicity/earthquakes Onshore storage sites viewed less favourably because of risks of leakage and induced seismicity Impacts on ground water	Palmgren *et al.* (2004) Shackley *et al.* (2005) Wallquist *et al.* (2009) Huijts (2003), Huijts *et al.* (2007) Gough *et al.* (2002) Oltra *et al.* (2010) Sharp (2008a, 2008b) Sharp *et al.* (2009) Terwel *et al.* (2009) Ashworth, Mayhew *et al.* (2007)
Leakage of CO_2 from pipelines and low-level leakage from storage sites	Risk of explosions Health and safety risks Local ecological impacts of leakage on plants, animals and ecosystems	
Avoids tackling 'real issue' of moving towards a genuinely sustainable future energy system	Short-termism End-of-pipe technology: better to reduce emissions than to store Allows unsustainable use of fossil fuels to continue into the future Lets us 'off the hook' Could deter investment in, and policy attention directed towards, renewable energy development Could deter action on energy efficiency and reducing energy demand Green light to other underground storage of other waste streams Could it result in a larger rebound effect?	
Feasibility and costs	Expensive Who is going to pay? Not tested/technology not ready yet Is there enough storage capacity? Requires a massive infrastructure that does not exist today	
A use for CO_2?	Is there any way that CO_2 can be utilised such that it comes to have a value? Preference for making use of something that costs a lot of money to remove	

Issue	*Notes*	*Examples/references*
Trust and confidence	Industry should not be allowed to profit from this Motivations of industry and politicians involved are not trusted Where arguments promoted by industry are felt to be incongruent with that organisation's own motives, trust decreased Range of view about trustworthiness of companies, governments, NGOs and scientists Liability/responsibility: who is responsible, especially in the long-term?	
Strategic issues	CCS as a 'bridging mechanism' while other low-C technologies developed Relies on government subsidies There are better options for carbon reduction than CCS	
Moral issues	Are we 'playing God'?	

Source: Hammond and Shackley 2010.

TABLE 3.2: Key public concerns regarding CCS arising from applied research (surveys, focus groups, etc.: i.e. undertaken in the context of a real-world proposed project)

Issue	*Notes*	*Examples/references*
Local benefits	Importance of direct economic or other types of benefit to the local community – e.g. job creation, money into the area, prestige, advanced technology development, centre of excellence, a visitor centre promoting local tourism etc. Maintaining core industries and skills and allowing continued use of resources such as coal Potential use of CO_2 in local industries or in Enhanced Oil Recovery, etc. Energy diversification?	FutureGen work in the USA Hund and Judd (2008) Bielicki and Stephens (2008) Bradbury *et al.* (2009)
Local issues and possible concerns	Impacts on water consumption/availability Production of saline water Costs of power Area disturbances – end up living in an industrial park Land-use rights Decommissioning Monitoring	

Issue	*Notes*	*Examples/references*
Property rights	Appropriate compensation packages Neighbouring farmers/landowners wanting compensation for CO_2 storage and possible loss of mineral rights Property prices affected? Negative experience in the past, e.g. with pipelines adversely affecting agricultural land	
Role of information provision	Some evidence that concerns over CCS are positively correlated with understanding In one case, respondents had general concerns about CCS which became better articulated as more information was received In one case, respondents' support for CCS did increase over a one-day seminar, but only as part of a portfolio of approaches Physical models of CCS helped to increase understanding Local media positive about job creation and ability to sustain coal industry Deficiency of expert information	
Stakeholder engagement	Active engagement with a range of local and regional stakeholders, opinion-formers and community-leaders appears to have helped in building local acceptance of projects	
Trust and fairness	Lack of trust in government and the private sector Poor experience of oil and gas sectors in the past Desire for transparency, participation and redress should something go wrong Sense of empowerment would increase acceptability (empowerment being defined as ability to mitigate community-defined risks and to ensure that just procedures would be adopted) Concern that CCS plants will be dumped in poor and voiceless communities The key risk could be that of being neglected and ignored in the event of project failure Fear of being in a 'sacrifice zone' and being treated as 'guinea pigs' Lack of confidence in monitoring regime and lack of clarity on who would listen to concerns	

Source: Hammond and Shackley 2010.

Appendix 3.2 Glossary

The term	*What it means*
Public(s)	The public can mean different things in this research. First, there is the distinction between the general population (in, say, a country) whose opinions may matter for policy, and the local population with an interest in, for example, the siting of a capture plant. Second, the term is often used in the plural: publics in this kind of research, to remind us that 'the public' is not homogeneous, but often better understood as composed of various categories, groups, networks etc. with different knowledge, preferences and frames of reference, affecting how opinions are formed.
Stakeholders	This term denotes any actor who has a stake in something, for example, the fate of CCS. Who is seen to have a stake is a rather subjective thing. In the context of CCS projects, it is often understood as external actors, but not necessarily. (In management literature, a distinction is made between stakeholders and shareholders of a company.) In the context of CCS as a socio-technical system, a stakeholder means roughly anyone involved with the technology or who at least has an opinion about it.
NIMBY	NIMBY is an acronym for Not In My Back Yard, and refers to the phenomenon of actors supporting CCS (or other technologies or initiatives) in general, but not when there are plans for sites close to where they live. The term is in common use, but has been avoided here since it has been discredited in parts of the literature as being patronising as well as misleading in terms of explaining local resistance, which can happen for many reasons other than the crass egotism suggested by the NIMBY term. In the context of carbon storage (and other things buried in the ground), the term Not Under My Back Yard, NUMBY, is also used.
Lay – expert	In everyday language this is a straightforward distinction, where experts are recognised as having expertise based on objective and universal knowledge, whereas lay people do not. Social science has challenged this distinction, showing that there are several possible bases for knowledge about any one thing, including practical experience specific to local contexts. The point is not to say that everything is relative and that anyone's opinion should count as much. The point is that to understand many of the things discussed in this part it is necessary to go beyond the simple expert-lay model and recognise different degrees and kinds of expertise.
Knowledge	The conceptualisation of knowledge varies, from the view of objective facts established by scientific methods, to broader (and sometimes conflicting) views recognising the variety of valid and effective rationalities across different social groups and that knowledge is context dependent.

	Knowledge is not merely impersonal information, but something that is understood, interpreted and communicated by people. 'Learning' is, therefore, a much more complex process than just the transmission of facts.
Framing	Knowledge is always about interpretation of information guided by specific frames of reference. There is always (more or less) scope for choosing different frames for the interpretation of any given piece of information or fact. The choice of frame alters the meaning of the information, shifts the emphasis and makes different aspects visible or invisible. The term 'framing' denotes this choice of reference frame. Frames may be more or less explicit or implicit in any representation, and they may be more or less consciously and strategically chosen. Framing clearly matters for how to communicate about CCS, for example, seeing CCS as a stand-alone solution or as part of a portfolio of mitigation options, as discussed above.
Performativity	There are bootstrapping-type mechanisms at play in many (possibly all) social activities. It is useful to think here about self-fulfilling prophesies, for example when governments seek to talk up the state of the economy. A similar dynamic is at play in technology development, when visions of the functionality and wider benefits of a technology help attract resources as well as guide development efforts. Claims made about (the future of) the technology may thus be self-fulfilling – also called 'performative' – and arguably this is a necessary feature of technology development. Conversely, claims about the difficulty of making CCS happen may contribute to making that true. The processes involved may be complex, but the concept helps in understanding how representations of CCS shape action and ultimately the technology itself. Related concepts used include: 'expectations' and 'promises'.
Representation	This term is used here in addition to perception. We take perception to denote any actor's understanding of and view on the technology, its status of development, performance, risks etc. In contrast, the term 'representation' emphasises more that actors are actively taking part in the construction of knowledge about the technology, and create and communicate claims about it.
Interpretative flexibility	This term denotes the way on which different groups can have very different understandings of a technology, its properties, benefits and drawbacks.

4

PUBLIC UNDERSTANDING OF AND ENGAGEMENT WITH CCS

Judith A. Bradbury

Introduction

Resolution of the technical and economic questions surrounding carbon dioxide capture and storage (CCS) cannot be achieved without actual, real-world demonstration. The reality of demonstrating CCS at a particular site and in a particular community requires developers to understand and work with its members to gain their permission (also referred to as acceptance), be it tacit or explicit. This permission, as highlighted in several recent controversies in Europe and the US, is not a given. Thus, while developing the technology is essential to bringing down costs, as well as verifying scientific predictions and understanding long-term implications, difficulty in gaining acceptance can potentially add to overall costs and hence, impede adoption of the technology.

A growing number of research studies have examined public attitudes and understanding of CCS in general. Additionally, experience in developing and implementing demonstration projects is adding to our understanding of some of the key factors that may affect community response and potentially effective engagement. Some of this social knowledge, which is part of the broader learning process discussed in Chapter 13 of this book, by Markusson, Ishii, and Stephens, is being documented in 'lessons learned' reports by practitioners in the field.

The focus of this chapter is on public understanding and community engagement in deployment of a specific CCS project, where the potential impacts are concrete rather than abstract in nature. The discussion is limited to projects in the US and is presented in four sections, following this introduction. The following section provides a theoretical background and overview of the current status of applied social science knowledge about community perceptions and engagement. What are the key research findings on CCS siting and what do we know about the elements of an effective public engagement approach? Subsequent sections, which examine six examples of CCS deployment, primarily in the Midwestern US, identify key factors that affected community understanding and response and the potential for effective engagement. The chapter concludes with a summary of the key findings and suggests several implications that emerge from the combination of these empirical data and the broader theoretical framework.

Current social science knowledge: theory and guidance

Recent findings on public response to CCS can be interpreted within a broader theoretical framework that offers valuable insights into public perspectives of technology. This section outlines the theoretical background underlying the analysis presented in this chapter and summarizes recent applied research and guidance on public understanding of CCS.

Theoretical background

The work of social theorists such as Otway, Rayner, and Wynne, summarized below, provides the background for understanding public perspectives, as well as a constructive basis for addressing them. Building on work in the sociology of science and the early sociocultural work of Douglas and Wildavsky (1982), these theorists have emphasized the social processes of knowledge production and utilization and the social processes within which opinions about a particular issue are formed. Thus, people (including both developers and potentially affected community residents) do not evaluate technology such as CCS in isolation from their broader life experiences and social affiliations. Rather, they base their evaluation on an existing cultural frame of reference – their values, interests, previous experiences and ways of interpreting and responding to the world.

This literature highlights differences in social and cultural frameworks that influence what is important to people, how they define problems and issues, the aspects of the situation on which they focus, the type of evidence they seek, whose opinion they believe and their preferred process for decision-making. Such differences extend beyond the expert/layperson distinction to form what Schwarz and Thompson (1990) have described as a consistent package of biases that constitute differing political cultures by which humans interpret and respond to the world. Understanding these cultural frameworks, which are likely to occur both within and among communities, can provide important insights into public responses to CCS.

The literature also emphasizes that public concerns about a technology are likely to extend beyond technical risk and safety issues and that technology decisions need to incorporate the full societal dimensions of the policy issues at stake – a point emphasized by Funtowicz and Ravetz (1985) as being of particular relevance for policy problems where factual and evaluative dimensions are high. Otway and von Winterfeld (1982) and Otway (1992), for example, pointed out that opposition to or support of technologies may be caused by factors that go beyond risk and other characteristics of the technology to include social and economic issues quite separate from the risk of the technology per se.

The broader issues that arise in technology deployment include issues of equity and trust. Rayner and Cantor (1987) have emphasized that since decisions about project implementation are not just technical but also involve decisions concerning the level, acceptability and distribution of impacts, the acceptability of a risk is essentially an ethical question of "How *fair* is safe enough?" Significantly also, Wynne has shown that knowledge is conditional and that validity is dependent on conditions embedded in a particular knowledge model being fulfilled in actuality. He argues that experts' trust in supporting institutions may not necessarily be shared by the public and that this is key to understanding public perception of the risk of a technology (Wynne 1991, 1996).

The bottom line derived from this school of thought is that if we want to understand public perceptions of CCS, we should first start with the social context of the perceiver rather than characteristics of the technology.

Applied CCS research and guidance

Reports from researchers and practitioners presented at the annual international Greenhouse Gas Technology conferences over the past decade have contributed to development of an international knowledge base on CCS and the public. In addition, the past two years have witnessed publication of several comprehensive documents on research and practical guidance covering European, US and Australian CCS experiences that provide a solid basis of social science knowledge on public understanding and engagement. They include case study research conducted by Hammond and Shackley (2010) and by Ashworth *et al.* (2010b) for the Global CCS Institute (GCCSI); a review and analysis of the NearCO_2 public participation practices (Desbarats *et al.* 2010); a *Best Practices* manual published by the US Department of Energy (DOE) National Energy Technology Laboratory (NETL 2010); a GCCSI Toolkit, which was a companion to the case study research by Ashworth *et al.* (2010c); and *Guidelines for Community Engagement* published by the WRI (2010) that was prepared for regulators and local decision-makers, as well as developers.

In their *Review of Research Findings, CCS Project Experiences, Tools, Resources and Best Practices*, Hammond and Shackley (2010) analyze 13 case studies in Europe and the US and provide a comprehensive overview of current social science knowledge that forms a basis for understanding and engaging with residents in CCS communities.[1] They identify several patterns distinguishing successful from failed CCS projects, including: trust in developers; a history of extractive and fossil fuels industry in the area; population density; early, informal engagement; matching expectations of stakeholders to the engagement and decision-making process; and transparency and timeliness in responding to the public. Also identified were the types of concerns raised by local publics. The storage component of CCS (especially on-shore storage) is perceived as something new, unknown and potentially risky, and, as may be expected, the overriding public concerns are related to safety and risk to personal health, homes, and land values. Concerns that the authors term secondary were voiced over lack of belief in the efficacy of CCS to combat climate change, especially in comparison to other sustainable energy options; liability; decreased plant efficiency; and perpetuation of the fossil fuel industry. They also reported that an understanding of climate change, energy planning, and the carbon cycle are "essential" to understanding the need for CCS (Hammond and Shackley 2010: 33) and noted the risk of CCS projects becoming "political footballs" which politicians support or oppose in order to gain popularity (ibid.); the impact of policy issues external to the local community; and the likelihood that concerns, statements, and activities can be amplified by the media and internet.

The authors also discuss briefly some of the pertinent basic and applied research findings related to public perceptions of CCS. These include level of awareness (low and typically unstable – see especially Daamen and de Best-Waldhober 2008); the limitations of surveys in delineating reactions in the abstract to poorly understood technologies (Malone *et al.* 2010); mixed findings concerning the reaction to provision of more information (see Upham and Roberts 2010, for a succinct overview of research into public perceptions, most specifically the relationship between information and attitude development and change); and the importance of perceptions of fairness, equity, sense of empowerment, and trust in institutions (Bradbury *et al.* 2009; see also Wong-Parodi and Ray 2009). They emphasize that the public is composed of multiple publics and that siting issues are more likely to focus on concrete, local impacts than on CCS as an abstraction dealing with the abstract problem of climate change.

Desbarats *et al.* (2010), reporting on the first phase of the NearCO_2 Project, provide a systematic analysis of lessons learned from the communication strategies used in eight studies of CCS and analogous developments. Findings of particular note include: the need to understand the existing local context, by conducting preparatory social characterization to identify factors to be taken into account in planning communication; the need to recognize that people tend to experience "place attachment" that will encourage defending one's community against perceived threats such as CCS; that people expect procedural justice and to have their concerns taken seriously; good quality information and a mix of contexts (formal vs. informal, for example) contribute to effective engagement; and involving experts who are likely to be perceived as independent may engender trust. They conclude that regulations and local considerations may have a significant impact on public perception of projects and complicate the developer's communication and consultation efforts and that numerous factors may influence the effectiveness of communication and consultation at the local level. These include the timing of consultation, the type of material being communicated, and the potential for two-way dialogue. One significant conclusion is that:

> The effectiveness of communication and consultation is not amenable to simple inference. Not only is there no simple relationship between communication,

> consultation and public perception, but inferring the nature of this relationship in real-world settings is rarely straightforward, due to a wide variety of additional, simultaneous issues.
>
> (Desbarats *et al.* 2010: 42)

Case study research sponsored by GCCSI (Ashworth *et al.* 2010b), examined five projects in Europe, Australia and the US, focusing on lessons learned for best practices in communication and outreach. The report concluded that best practices in two-way communication and outreach are not sufficient, alone, to ensure successful project deployment but that factors contributing to success include: alignment among key government and development team members; integration of two-way communication and outreach as a critical component from the beginning; and adaptation of planning and management to the project's social context. The five projects were evaluated on the basis of factors derived from the case studies as a guide for future projects to use in assessing their social contexts. The companion GCCSI toolkit *Communication/Engagement Toolkit for CCS Projects* (Ashworth *et al.* 2010c) offers hands-on guidance, basic worksheets and resources for the practitioner.

Best Practices for Public Outreach and Education for Carbon Storage Projects (NETL 2010), prepared for NETL by the outreach leads of the seven DOE Regional Carbon Sequestration Partnerships, outlines ten best practices, based on the partnerships' early project experience. They note that effective outreach will not guarantee public acceptance but can contribute to schedule adherence, cost controls, and community goodwill. First and foremost, the manual emphasizes the need to integrate public outreach with project management and establish a strong outreach team that includes (and aligns) both technical and communication/outreach staff from the various organizations involved in the project. A series of appendices, providing detailed examples of how to plan and manage specific activities, includes a planning and management matrix, communications plan, press release, fact sheet, and poster; and planning a site visit, conducting social characterization, and conducting a focus group.

In combination, these documents provide a solid foundation for understanding the likely factors contributing to success, the types of issues that will arise and a wealth of detailed, hands-on guidance for the practitioner on the why, what, when, how, and with whom of planning and implementing effective public engagement. Yet they also highlight the dynamic nature of the social environment and the multitude of variables to be considered. *Significantly, all documents emphasize that engagement should begin early and demonstrate transparency.* Additional key findings include:

- Patterns that have been found to distinguish successful from failed CCS projects, include: trust in developers; a history of extractive and fossil fuels industry in the area; population density; early, informal engagement; matching expectations of stakeholders to the engagement and decision-making process; and transparency and timeliness in responding to stakeholders.

- Engagement is not simply an add-on activity but should be an integral component of the overall management strategy from the outset – including coordination or "alignment" among key government and project team players and inclusion of social factors in informing project design and implementation.
- An understanding of climate change, energy planning and the carbon cycle is "essential" to understanding the need for CCS.
- There is no one way to conduct engagement: all communities differ. Knowing your community and adapting your engagement strategy to the social characteristics of that community is fundamental to effective engagement.
- Engagement does not guarantee success.

Some insights from US experience

This and the following sections analyze six recent CCS projects in the US. The analysis draws on the author's experience in implementing public engagement (outreach and education) at four sites in the Midwest for the Midwest Regional Carbon Sequestration Partnership (MRCSP) funded by the US DOE and also on findings from three, separately conducted research projects. Only publicly available data are used to substantiate the discussion, throughout.[2]

Although this analysis focuses on US experience, it indicates that findings are very similar to those in the studies summarized in the previous section. Cultural variations (for example, in views on climate change) are noted where applicable.

The MRCSP projects include three small-scale Validation Phase CCS projects in different types of regional geology and one large Development Phase project. Research findings include focus group research conducted in collaboration with two other regional partnerships (Bradbury *et al.* 2009) and the FutureGen 1[3] and Carson projects that were included in the comparison of five international case studies discussed in the previous section (Ashworth *et al.* 2010b). Supporting data are also drawn from the study *Communicating the Risks of CCS* prepared by Bradbury *et al.* (2011) for the Commonwealth Scientific and Industrial Research Organisation (CSIRO). The study was based on the experiences of five companies implementing CCS projects in the US and Canada: the Illinois Basin-Decatur Project, Mountaineer, Quest, Weyburn, and several projects conducted by Southern Company including Plant Barry and Plant Daniel.

This section describes the six projects that are the primary focus of analysis and their social context, while the following section focuses on differences in patterns of public response among the sites. What were the issues of public concern and what distinguishes successful from unsuccessful siting attempts? How do these findings compare with others' findings and what can we learn from them that will help in future CCS siting and associated public engagement planning?

Project overviews

Of the six projects examined in this chapter, four were successfully sited and two were withdrawn. The three small MRCSP Validation Phase projects were completed within a four-year period ending in September 2009, while a larger-scale planned Development Phase project was withdrawn in 2009 for business reasons well before the permitting stage and is currently being reconfigured at another site. The Carson project, announced in 2006, was also withdrawn prior to permitting for business reasons and is being replaced elsewhere in California. The FutureGen 1 site in Illinois was selected in 2006 with strong support from state and local residents and is included in the successful siting category, although it was later reconfigured into a different project by the DOE, primarily for budget reasons. None of the projects experienced controversy over pipeline construction. Although FutureGen 1 and the Carson project would have required off-site pipeline construction, the willingness of nearby landowners to allow transportation access was one of the criteria for FutureGen 1 selection, while plans were not sufficiently developed at Carson prior to cancellation for such controversy to emerge.

The three Validation Phase projects were small-scale demonstration projects undertaken by the MRCSP under the second phase of the US DOE's Regional Carbon Sequestration Partnership Program. The partnerships are associations of private companies, comprising research and utility companies; educational and environmental organizations; and government agencies, including state geologic surveys. First established in 2003, the partnerships were designed to test the technical potential, economic viability, and public acceptability of carbon sequestration in each of seven regions, nationwide. In an initial phase, the Characterization Phase, the MRCSP had focused on characterizing the carbon dioxide sources and sequestration potential in the nine-State MRCSP region. The second phase, the four-year Validation Phase, focused on actual, small-scale demonstrations of the technology (and also on terrestrial sequestration) at specific sites in different types of geological formations in eastern Ohio, Kentucky and Michigan (see www.mrcsp.org).

The third, Development Phase of the DOE Regional Partnership Program involved injection of a larger amount of CO_2. The MRCSP's larger-scale project, originally planned for Greenville, in western Ohio, planned to inject over one million tonnes of CO_2 over a four-year period. The entire project, including characterization and monitoring was scheduled to take ten years.

The proposed Carson facility was a commercial project located in the Los Angeles Basin, California. It was to be a first-of-a-kind 500 MW hydrogen power plant using petcoke from nearby oil refineries and storing 90 per cent of its CO_2 which was expected to be piped approximately ten miles to existing oilfields for enhanced oil recovery.

FutureGen 1, developed as a public-private partnership between DOE and a non-profit consortium of utilities and coal companies (the Alliance), was to be sited in Mattoon, Illinois. The project's location was the result of competitive bidding from interested States and communities. It was intended to be an international model research facility comprising a 275 MW integrated gasification combined cycle (IGCC) power plant that would store 90 per cent of its carbon dioxide in a saline formation.

Some key project features are apparent from this summary description:

- First, one half of the projects (the three Validation Phase projects) were very small scale and of short duration. The other half were much larger and of longer duration.
- Second, three projects (Greenville, Carson, and FutureGen 1) were new facilities.
- Third, the project focus was on facility siting only.
- Fourth, Carson differed from the other projects discussed in this chapter in that the primary focus of attention was on the innovative power plant rather than on carbon storage.
- And finally, the FutureGen 1 project was unique in that its location was self-selected and the winner of an intense and highly publicized national siting competition. As will be discussed in the fourth section, below, this process of self-selection was a major factor in local and state response to the facility.

Site social characteristics

The social characteristics of each site varied, although most were located in areas with small populations and only one (Carson) was in a highly populated, urban setting.

All three Validation Phase MRCSP projects were somewhat distant from nearby small population centers. Two were located on power plant sites owned by utility partners in Kentucky and Eastern Ohio, respectively, and one at an existing oil and gas field in northern Michigan. The nearby populations were predominantly white. Of the three projects, the Kentucky project near Rabbit Hatch, Kentucky was the closest to a large metropolitan area – Cincinnati, approximately 40 miles away. However, the immediate area surrounding the power plant was rural and all CCS activities were confined to the plant site. The closest population centers were just across the Ohio River in the State of Indiana (population approximately 2,500); and the Boone County seat with a population of almost 11,000 at a distance of approximately 11–12 miles. The Eastern Ohio project was sited at one of a number of coal-fired power plants located along the Ohio River and strung out between small population centers. The nearest towns were 1.4 miles away with a population of approximately 10,000 and 4.5 miles away with a population of just over 3,600. The third Validation Phase project was undertaken in northern Michigan near a year-round recreation and tourist center with a population of between 3,000 and 4,000. It was sited at an existing oil and gas field used by one of the partners, used existing oil pipelines, and obtained CO_2 from a nearby gas processing plant which was owned by another partner (CO_2 is a byproduct of gas processing). The Kentucky and Ohio power plants had both been in operation for many years

– since 1981 for the former and the 1940s for the latter. However, CCS activities were new to local residents. In Michigan, by contrast, area residents were familiar with enhanced oil recovery and natural gas processing plants and much of the infrastructure needed for the demonstration such as compressors, pipelines, and injection system was already present at the site.

The proposed MRCSP larger scale, Development Phase project was located at the site of an ethanol plant in an industrial park on the outskirts of Greenville, Ohio, although the project footprint extended beyond land owned by the company. The city's 2000 population was approximately 13,300 and primarily white. It is the largest city and county seat of rural Darke County, population approximately 53,300, where most residents are employed in agriculture (especially corn and soybean production, and hog raising). Greenville historically has served as the main metropolitan center of the county, providing services to county residents (Ohio Historical Society, www.ohiohistorycentral.org). Unlike the population in rural Darke County which has been decreasing in recent years, the Greenville population has increased slightly.

The Carson project was to be located in a highly industrialized area in the Los Angeles Basin, California. It included a site for the proposed new power plant in an industrial area adjacent to oil refineries in the city of Carson, as well as a storage site in Long Beach. The city has a very diverse, high-density population of almost 90,000. The population in the Long Beach area where the CO_2 was to be piped was larger (over 461,000), also very diverse and of high density.

Mattoon, in Coles County, Illinois, the site for the proposed FutureGen 1 project, has a population of approximately 18,000, primarily white; the county population, overall, is approximately 52,000. The town's history is linked to that of local railroads, with manufacturing and industry developing in the twentieth century (www.colestoday.com/aboutarea/localhistory.htm). Interviews with local residents indicated that the area has been losing manufacturing jobs (Hund and Greenberg in Ashworth *et al.* 2010b Appendix 3); job growth rate between 2000 and 2008 was negative at 12.9 per cent. Economic development was an important priority.

In summary:

- First, with the exception of Carson, all projects were located in or near relatively small, predominantly white, population areas. Only Carson was in a large, diverse urban setting.
- Second, with the exception of Greenville and the possible exception of Carson, all were located in areas where the project host had a history of economic and social ties to the area and well-established, community relationships. Although the Carson developer had ties with the State political establishment, the size of the population limited the extent to which close personal relationships could be developed with community members.
- And finally, with the sole exception of Greenville, all were located in areas familiar with coal use, oil production or manufacturing. Greenville differed in that it was the center of a historically agricultural county.

Discernible patterns in successful CCS siting

Although members in all project communities expressed concerns about CCS technology, only two of the six projects encountered controversy. What were some of the factors involved? This section identifies six patterns that distinguish successful from unsuccessful projects:

- Types of issues raised.
- Site selection process.
- Economic and social fit.
- The nature of the relationships with the community on the part of the host site and developer.
- The impact of broader policy issues on perceptions of CCS.
- Public engagement activities.

The analysis is summarized in matrix form in Table 4.1.

Issues raised

A qualitative analysis of the issues raised by the public at the six project sites indicates that, in general, they were similar to those raised in other venues and reported in the literature cited previously, although as discussed below, the Carson project was somewhat different in that the primary issues of contention arose over the power plant siting rather than the CO_2 storage.

Although almost all of the concerns about CCS technology that were noted in the literature were raised in some form at the US sites under discussion, distinguishing features of the two cancelled projects were that (1) issues of concern expanded into opposition – opposition, which in the Greenville case, became widespread and (2) linkage to broader policy issues and, in Greenville, social issues related to relationships, accountability, and involvement in project decisions figured prominently in the rationale for opposition.

As in Hammond and Shackley's cases, nearby residents at all US sites viewed the potential for impact in very concrete terms. Health and safety risks that could directly affect their lives were clearly of greatest concern. In some cases also, CCS discussions became politicized and were amplified by the media and internet. CCS-specific issues included potential leakage, especially a catastrophic leakage (the Lake Nyos case typically was cited as an example); earthquakes and induced seismicity; impacts on water supplies; the energy burden of CCS which would require using additional electricity; and impacts on housing and land values, especially for residents living close to the site. CCS-specific issues also included the formation of carbonic acid that could eat through or corrode concrete and metals. Other concerns raised that were characterized by Hammond and Shackley as "secondary" were related to policy and social issues. These included: liability; skepticism about the existence of climate change; the efficacy of CCS to combat

TABLE 4.1: Summary project analysis

Project/Factor	*Validation Phase #1*	*Validation Phase #2*	*Validation Phase #3*	*Development Phase*	*FutureGen 1*	*Carson*
Site selection process	Volunteered by partners to provide regional geologic diversity	Volunteered by partners to provide regional geologic diversity	Volunteered by partners to provide regional geologic diversity	Ability to supply required amount of CO_2	Self-selection and competition	Proposed plant located near to petcoke supply from nearby oil refineries
Economic and social fit	Perceived economic benefit outweighed risks	Perceived economic benefit outweighed risks	Perceived economic benefit and familiarity with oil/gas production activities	No perceived economic benefit but substantial perceived risks	Economic benefits perceived as far outweighing risks	• Power plant: economic benefits countered by pollution concerns • Storage site: uncertain
Relationships and social issues	Well-established in the community, with credibility built up over past actions	Well-established in the community, with credibility built up over past actions	Familiarity with local oil and gas company activities, with added credibility of utility and university partners	• Little opportunity for host site and developer to establish long-term, credible community relationships • Major concerns: liability/accountability, trust and input into decision-making	• Joint State/local effort with well-established relations with economic development agencies • Liability issues resolved by State Legislature	Large, diverse urban setting poses a potentially greater challenge than in small communities for establishing relationships

Project/Factor	*Validation Phase #1*	*Validation Phase #2*	*Validation Phase #3*	*Development Phase*	*FutureGen 1*	*Carson*
Policy issues	None apparent	None apparent	None apparent	• Lack of belief in climate change and/or efficacy of CCS compared with green alternatives • Opposition to continued coal use	Strongly supportive of CCS and continued use of coal	• Environmental justice concerns re increased pollution in already polluted area • Overlapped with hotly debated energy policy and use of coal
Public engagement	Proactive, aligned and coordinated with host site and regional partners	Proactive, aligned and coordinated with host site and regional partners	Proactive, aligned and coordinated with host site and regional partners	Reactive, weak alignment and lacking multiple supportive partners	Very proactive, multiple level effort, aligned with state and local agencies	Initial, proactive focus on state and local officials; subsequently expanded

climate change, especially in comparison to other sustainable energy options; opposition to a technology that would perpetuate the fossil fuel industry; lack of trust in the developer; and frustration that residents apparently had no say in project decisions.

However, while most members in communities hosting the three Validation Phase projects and the FutureGen 1 project were reassured on these points,[4] others, especially the Greenville community, were not. In both the FutureGen 1 and Greenville projects, the media played an important role in amplifying issues – in the former case in favor of the project, in the latter in opposing the project, with local political leaders lining up accordingly. Significantly also, leading opponents of the Greenville project were skilled in the use of the internet to link to opponents in other CCS communities regionally and even globally. However, active support for their cause was primarily limited to grass roots organizations in the neighboring state of Indiana and the larger, national environmental organizations did not become involved.

Greenville was notable in that the types of concern most stridently expressed in addition to those related to health and safety, were the broader policy and social issues – and these were frequently interwoven with CCS-specific concerns. As discussed in subsequent sections, some residents were very vocal in their lack of belief in climate change; some criticized CCS as a waste of tax dollars; and others expressed the belief that "green" alternatives would suffice to mitigate any potential problems. Also very apparent in that community was the belief that the project was not a good fit for their type of community and the expression of social concerns related to relationships, the decision-making process and accountability (Who is liable? Who will take care of us when something goes wrong? Are we guinea pigs for an unknown technology?) (see, especially, www.marketplace.org/topics/sustainability/greenville-opposes-burying-co2 and www.citizensagainstco2sequestration.blogspot.com).

Although the issues raised in Carson also included the health and safety concerns cited above, the project differed in that concerns stemmed primarily from the power plant siting and the increase in conventional pollutants and associated environmental justice impacts. The project occurred at a time of heated debates in the California State Legislature over the future of energy projects which also affected public discussion of Carson, even though the project was not directly involved. Initially, the project seemed likely to gain broad acceptance: it was strongly and publicly supported by the State Governor and offered significant economic benefits to the local city as well as providing carbon-reduced energy. But the power plant also was to be sited in an area with pre-existing air quality problems that already were a major issue for environmental justice groups and those opposed to continued use of coal – and these issues became pre-eminent. Whether or not the project would in fact have been successfully permitted became moot since it was cancelled for business reasons after the emergence of concerns about the technical and commercial viability of the proposed storage area (Bradbury and Wade in Ashworth *et al.* 2010b: Appendix 2).

Site selection process

The fundamental drivers in project selection were the geologic potential of a site and, for the larger projects, proximity to a CO_2 source. Within these two parameters, however, the particular way in which individual sites were selected had significant implications for the effectiveness of the engagement process and future success of the project. The selection process served to place the Validation Phase projects and FutureGen 1 at an advantage from the outset. The former three projects all benefitted from the assistance provided by locally well-established and committed MRCSP partners who had volunteered to serve as hosts. FutureGen 1 benefitted especially from the self-selection process in combination with inclusion of socially based criteria in the request for proposals that encouraged identification of a site that was appropriate from a public, as well as a geologic perspective.

The MRCSP Validation Phase sites were selected to represent different types of geologic formation representative of the region. Since the capture component of CCS is not yet far advanced, the two power plant locations planned to truck in the small amount of CO_2 that would be needed; the Michigan site took advantage of the ready access to CO_2 already being piped through existing infrastructure and, in fact, was able to inject approximately 60,000 tonnes in a two-phased approach. Significantly, from the social perspective, the volunteer project hosts were already committed MRCSP partners, knowledgeable about the DOE program, including the public outreach requirements, as well as being economically and socially well-established in, and knowledgeable about, their communities.

In selecting a site for the larger, Development Phase project, however, the partnership looked beyond its member partners to select an ethanol plant under construction in Greenville, western Ohio, on the technical basis of its producing a pure source of CO_2, as well as offering potentially promising geology. Unlike the previous tests, selection in this case was constrained by the key DOE funding requirement, applicable to all regional partnerships, that at least one million tonnes of CO_2 be injected over a four-year period. Although many coal-fired power plants are located in the MRCSP region and over 75 per cent of the region's electricity production comes from coal, the lag between capture and storage capabilities meant that none of these plants and none of the active utility program partners was in a position to supply the requisite amount of CO_2. As a result, the project location lacked some of the built-in advantages of the early test sites in terms of economic and social fit and linkage to a well-established host.

For the Carson project, the focus of selection was the location of the power plant – and also the focus of opposition that subsequently emerged. Its joint developers, BP and Edison Mission Energy, selected the location because of its proximity to needed supplies of petcoke from the oil refineries and also from recycled industrial wastewater. Storage was tentatively planned, but not finalized, for the large Wilmington oil field in nearby Long Beach that was viewed as a likely location for using the CO_2 to enhance production.

Selection of FutureGen 1 was a very open and highly publicized process, based on competitive bidding from interested States and communities. Twelve proposals were submitted which were subsequently narrowed down to four communities from the two States of Texas and Illinois, and finally selection of the winner, Mattoon in Illinois. The project was actively sought and every effort made to demonstrate that Mattoon was an appropriate site, both geologically and from a public perspective. Public acceptance was specifically mentioned: "A primary goal of FutureGen is to build industrial and public acceptance for future zero emission, coal fuelled plants of similar design characteristics" (FutureGen evaluation, cited by Hund and Greenberg in Ashworth *et al.* 2010b: Appendix 3, p. 11; see also p. 14 of the 2006 FutureGen 1 evaluation report[5]). Significantly also, the willingness of landowners to allow monitoring on their land and access to transportation lines and land use were included in the more than 100 criteria used in selecting a suitable site.

Economic and social fit

For all four successful sites, there was a notably strong economic interest in promoting CCS over the longer term, even if the size of the project was not sufficiently large to generate many jobs and immediate economic benefits. For these communities, potential benefits were seen as outweighing potential risks.

At the three Validation Phase and FutureGen 1 project sites, people were familiar with coal, oil, and gas production or manufacturing, and the continued use of fossil fuels was important to the State and regional economy. Two Validation Phase projects were sited at power plants that were a well-established part of the community and where CCS could potentially enable future coal-fired electricity generation with the associated benefits of continued local employment and contributions to the tax base. The third was in an oil and gas producing area where residents, familiar with activities such as oil drilling and seismic surveys, could see an economic benefit.

In Illinois, the State's interest in promoting the coal industry combined with the jobs and economic benefits that would accrue to the state and local area (including a promise of over US$2 billion of investment for the selected community) made FutureGen 1 a very attractive and very actively pursued proposition, especially in an area that had been losing manufacturing jobs. Illinois State and local development agencies had been very active in recruiting industry in general and FutureGen 1 in particular. Local economic development officials typically played an active role in ensuring that there would be a good fit between a proposed project and the host community, as described by Hund and Greenberg:

> Acting on behalf of the community, the local development office uses agreed criteria and instinct to determine whether to pursue projects. It considers the balance between community benefits, challenges, resource allocation (i.e., water and/or land resources, community fit, environmental impact (i.e., pollution

potential), economic growth potential and other factors to decide if a project is good for a community.

(Hund and Greenberg in Ashworth *et al.* 2010b: Appendix 3, p. 14)

The proposed project in Greenville, in contrast, was located in an agricultural area which was culturally very different from the eastern section of the State where one of the Validation Phase projects was located. In this community, the long-term economic benefit to the local area from coal and the immediate benefit from the limited number of jobs that the project would provide were not apparent – particularly for a politically conservative area where the reality of climate change was disputed and/or other alternatives for mitigating carbon emissions were advocated. The risks to landowners, water supplies, and the lack of assured compensation should unforeseen problems arise were seen as far outweighing any possible abstract benefits from addressing climate change (even if it existed).

Carson was somewhat different in that the storage aspect of the project received less attention than the innovative power plant. Although the power plant would have brought economic benefit to the city, it was its contribution to air pollution in an already polluted area that extended beyond the city that was the primary cause of controversy rather than the storage component of the project.

Local relationships: trust, accountability, and decision-making

Comparison of the six cases, in combination with data obtained in the risk communication study conducted by Bradbury *et al.* for CSIRO (2011), highlights differences in local relationships between successful and unsuccessful projects. These social factors – relationships, accountability, and input into decision-making – reinforced the economic contributions of a host site to form a foundation for acceptance of a CCS facility.

The cases examined in this chapter demonstrate that projects benefit when they are able to build on existing good relationships. As noted previously, Wynne's early research pointed to trust in the managing institutions as being a key factor in perceptions of risk. Additional research specific to CCS also emphasizes trust and accountability as important factors in facilitating acceptability of new technological developments. For example, in their focus group studies conducted in three different regional partnerships, Bradbury *et al.* found that "In all cases, trust in authority and concerns about the fairness of CCS implementation procedures were the most strongly expressed concerns" (Bradbury *et al.* 2009: 4668; see also Huijts 2003; Hammond and Shackley 2010; Desbarats *et al.* 2010). The cases examined here expand those findings by indicating that a significant element is trust that has been built up on the basis of previous relationships with the community. Host sites that have a demonstrated history of credibility in previous interactions can be expected to be similarly committed in future dealings to taking the community into consideration – and thus, be accountable.

The element of trust is especially evident in the Validation Phase and FutureGen 1 projects as compared with Greenville. For the Validation Phase projects, neighbors at two sites had a longstanding relationship with the host power company which was an important contributor to the economy and community life and at the third, with local oil and gas drilling companies. In these projects, the project developer was able to benefit from coordination with the host sites and partners in the local areas and was therefore less likely to be seen as an unknown and possibly untrustworthy outsider. The significance of this element was highlighted in the CSIRO risk communication study, where more than one interviewee pointed to the importance of:

- Credibility built up through past actions.
- Being part of the community in which the project occurs.
- Communication about CCS being part of ongoing communication with employees and communities about power plant operations.
- By extension, trust in one partner can act to boost the credibility of other team members.

As one interviewee stated:

> We're speaking from experience, we're learning by doing .... [We] build credibility in the community by doing the things that we're doing – to have the credibility to back up what we're saying because we are living and experiencing it.
>
> (Bradbury *et al.* 2011: 11, 12)

Another reported that "We have a good thing going to build public confidence in CCS because of our bigger picture attributes and things that we've done in the past and over time" (ibid.: 12).

Also:

> We're already ingrained in the community. I always say this to everybody because I've found this to be the most important thing with our risk communications is [the community people] are our employees ... that work at these plants, you know there's hundreds of people that work at [our company], they all live in the local area around the plant. So we start internally addressing these people and the risks and that is both internal and external communications at the same time. We're addressing multiple stakeholders at one time because the employees are the local community around the plant (ibid.: 11).

And:

> One of the benefits that we have with some of our projects is that we're doing them at existing [facilities] where we've had ongoing community relations and risk communications in place already about operations .... So we know the

people who need to be engaged and we've been engaging them over a period of time.

(ibid.: 16)

Further, in the words of another: "We brought in experts from … other sources and I think, because we trusted them, that made the community trust them more" (ibid.: 12).

The significance of these factors is further confirmed by the FutureGen 1 project. Although the project was new and could not draw on an existing facility's relationships, as for the Validation Phase projects, relationships with project developers benefitted from established and credible relationships with state and local economic development agencies and also the Illinois State Geological Survey, all of whom worked closely with the Alliance. An additional, very important reinforcing factor was that the FutureGen 1 Request for Proposals asked respondents to indicate how liability could be addressed and virtually all final proposals, including those from Illinois, obtained state legislation accepting liability for the project (though not generally for all CCS projects). Residents also had a genuine opportunity for input into project decisions – not only in the original siting but also in project design features such as changing the facility entrance or conducting seismic surveys at a convenient time for farmers.[6]

In contrast, the Greenville project suffered from a lack of long-established relationships that could have provided a basis for evaluating the trustworthiness and accountability of the host company and developer in taking the community's interests into consideration while deploying a new technology. At that site, the ethanol plant was very newly constructed and there had been no opportunity to establish long-term relationships, either with the host site or the developer; indeed, relationships with neighbors had been negatively affected by depletion of their water wells by draw-downs for the plant following its start-up just a few months before the CCS project was announced.

Social issues, including accountability, relationships, and involvement in decision-making are clearly discernible in the Greenville media reports throughout and these social factors were frequently interwoven with, and exacerbated, concerns about the technology per se. Long-term liability coverage for host sites was not provided under the DOE Partnership Program and no special State legislation providing for liability had been passed in the State of Ohio, as in Illinois for FutureGen 1. Further, in Greenville, the storage and plume area extended beyond the land owned by the host; consequently, liability, including related questions about what would happen and who would be accountable if something went wrong, was a major issue from the outset. This issue, which simmered for over a year, reached its height over plans for seismic surveys which encountered opposition because of "too many unanswered questions regarding liability and responsibility issues from possible tile, road, property and other damages" (Vehre 2009a).

Concerns were frequently expressed also that the decision was a "done deal" in which local residents had no say and many questioned why the community should trust the "sequestration experiment" when "of course those who stand to profit, as well as those who would test, approve the method, and compress and pump the gas; [sic] do not live in the area that would be greatly damaged" (Surber 2009). The intermingling of the various social concerns is apparent in one resident's critique:

> This is not your town. This is not [the developer's] town … the plan sound[s] like a foregone conclusion …. There appears to be little tangible benefit to Greenville from the ethanol plant and now the city is becoming a guinea pig for the planned carbon sequestration experiment. Who is going to be around in a 100 years to check this?
>
> (Shaner 2008)

Policy Issues

The broader policy environment may add a complicating factor to discussion of the appropriateness of CCS technology in a particular community. None of the six projects discussed in this chapter suffered from targeting by external groups opposed to CCS; however, linkage of CCS to broader policy issues can be seen in the contrast between Carson and Greenville on the one hand, and FutureGen 1 on the other. Policy opposition was very evident in the two former projects, although the specific issues of concern differed.

Issues raised in Carson were focused primarily on broader environmental justice concerns and the potential for increased pollution from the power plant component. In addition, public discussion of the entire project was subsequently impacted and complicated by the broader debate in the California State Legislature over CCS regulations and energy policy, including opposition to coal-fired electricity generation. As concluded by Bradbury and Wade: "As concern about air pollution rose, it became possible for activists to raise unanswered questions about the site of the plant and the storage, contributing to a withering of active support" (Bradbury and Wade in Ashworth *et al.* 2010b: Appendix 2, p. 22).

Perceptions of CCS in Greenville appeared to be strongly influenced by positions on broader policy issues. As reported widely in the local media, specific policy issues included:

- Lack of belief in climate change. Although no surveys were conducted in the Validation Phase project communities to quantify attitudes toward climate change and an associated need for CCS, the general impression was that climate change was not of particular significance in residents' lives – and thus, not a motivator for either accepting or rejecting a need for CCS. However, the experience in Greenville suggests that a strongly held belief that climate change is *not* occurring does affect attitudes toward CCS. Some of the most vocal opponents in Greenville strongly disputed the scientific basis for climate change, arguing that the CCS program was based on the "lie" of climate change (Bliss 2009).

- Conservative political views and the belief that CCS is a "government boondoggle" and a waste of taxpayers' money. Politicization of CCS is very evident in national politics in the US and may be especially problematic for the deployment of the technology.[7] (For pertinent discussion of US opinions on climate change, see McCright and Dunlap 2011; Borick *et al.* 2011; Reiner *et al.* 2006).
- Opposition to the continued mining and use of coal. Some of the most active opponents of the facility were strongly opposed to coal. One leading opponent stated, "We should not forget that coal has a dark side in the toll it takes on our heath, our environment and our communities" (Teaford 2009).
- Commitment to the use of renewables. Opposition to coal was frequently accompanied by a belief that, if climate change is indeed occurring, renewables will provide the solution. For example, "CO_2 sequestration is an expensive and impractical waste of tax dollars that could be better spent for solar and other alternative energies" (Vehre 2009b) and, as queried in a brochure distributed by the opposition group prior to a public meeting held in June 2009, "Why sequester it, when it can be mixed with algae to produce oil or biodiesel to help reduce America's dependence on foreign oil?" (www.citizensagainstco2sequestration.blogspot.com)

In contrast, state and local FutureGen 1 supporters did not dispute the policy basis and need for continued use for coal: the project, which was actively sought, was viewed as a good fit and residents expressed civic pride in hosting such an internationally recognized, innovative research project.

Public engagement activities

The factors discussed in previous subsections provide the background against which the projects' public engagement process was conducted. All of these factors affected the nature and effectiveness of that process.

Each project implemented a more extensive public information and engagement process than was required by regulation although, with the exception of FutureGen 1, none undertook engagement in the sense of planned, informed two-way discussion as to whether the project was appropriate for their community. The projects differed primarily in the timing and nature of their engagement process and the degree to which it was aligned with project partners and integrated with the overall management strategy. The engagement process associated with successfully completed projects built effectively on the foundation established by the selection process, social and economic fit, trusting relationships, and congruence with policy perspectives. It was marked by the following features:

- Early initiation.
- A proactive approach.
- Integration of communication with technical activities.

- Alignment and coordination among the various members of the project team, including collaboration with members knowledgeable about the local community.[8]

In addition, although not explicitly stated, among successful projects there was at least implicit recognition from the outset of the importance of addressing and committing resources to the non-technical risks as part of a comprehensive project management approach. This underlying orientation was succinctly expressed by one interviewee in the CSIRO risk communication study:

> I think we as a company over the past number of years have had experiences with communities that tell us that our non-technical risk is actually more of a challenge than the technical risk. … So absolutely committing to managing your non-technical risks right up-front early on a project is absolutely essential to any project's success. …we [the host company] have seen things go south. And so, we're acutely aware of the downside of not managing stakeholder engagement. But you know it could still go south. It can go south so easily.
>
> Definitely recognizing that non-technical risk can delay or stop a project, committing to adequate resources for it and having a plan – a mitigative plan that covers all the aspects – the stakeholder engagement, the regulatory applications, making the right assumptions early on. I mean, those are all best practices absolutely. But they don't guarantee you anything other than you hope that if something happens, you can jump on it quickly and respond and that you have a level of trust or respect that you'd be able to get it back under control quickly.
>
> (Bradbury *et al.* 2011: 17)

A snapshot of the engagement (outreach) activities undertaken by each of the three Validation Phase projects is available on the MRCSP website. However, the full report for this phase (also called Phase II) is not publicly available as of this writing. Activities shown on the website include, for each project, a simply-written and illustrated series of updates on technical activities. Also included are photographs and overviews of informal public and employee meetings conducted for the projects (including copies of technical presentations, poster exhibits, videos, and fact sheets available at each meeting). Separate tabs on the site provide fact sheets on specific sites and also general CCS information, as well as links to a variety of resources and an interactive feature that enabled readers to submit and receive responses to any questions. Videos available include an in-house video on seismic testing and a video on well-drilling at one site that was created by the Annenberg foundation, in consultation with the MRCSP team. Unlike the Michigan site, which was located in an oil and gas drilling area, local publics in Kentucky and Ohio communities had little familiarity with these types of activity.

Particular features of the Validation Phase projects were the integration of engagement with technical activities, alignment of government and project partners and the benefit resulting from coordination among established local partners. The engagement process was initiated informally and early, especially at the two power plant sites where seismic surveys, unfamiliar to most residents, were to be conducted. Some of the preliminary work conducted prior to the onset of technical activities to form the foundation for future relationships at each Validation Phase project site is described in the MRCSP report for the previous phase (Phase I) which is available on the website. This foundation was further strengthened by collaboration in planning and implementation with partners at the host sites.

At the two power plant sites, the host utility companies played a leading role in communicating with their employees, local officials, and nearby residents, as part of their regular, ongoing communication about plant activities. This process was initiated early and in advance of any technical activities such as the seismic surveys. The importance of early, "no-surprises" communication, using existing communication channels, was also stressed by interviewees in the CSIRO risk communication study. One person reported that:

> When we were awarded the site characterization project … I would go in and they'd have all the plant employees come in and we would give a presentation on what we're doing. And be very open with them, and say, OK you are all going to be seeing some drill rigs coming onsite, you're going to be seeing some trucks hauling CO_2, you're going to be seeing contractors and stuff like that. So we try to address stakeholders early on when things are happening and so we tend not to have anybody not understanding what is being done – you know they'll ask a few questions, but then they tend to view it as not a big deal, they're doing it for our benefit and we're listening to them and that sounds good and let's go back to our jobs.
>
> (Bradbury *et al.* 2011: 12)

And another said:

> We're always trying to be at the forefront when new information comes out. If we're going to do a modification to the power plant, install new equipment or change the way we operate things, we always try to lead into that with a lot of public outreach so that people are comfortable with what is going to happen before it happens.
>
> (ibid.)

In Michigan, project team members included, in addition to the host partner, both technical and outreach staff from the partnership lead and a utility and local university partner. The DOE *Best Practices* manual outlines the generic planning strategy used to ensure integration of technical and public engagement activities at the site (in Appendix 2) and describes the coordinated project team engagement approach:

> The team provided diverse perspectives upon which the project could draw, including both technical understanding of planned activities, valuable knowledge about local culture and politics, an existing network of media and local contacts and effective ways to communicate with local residents.
>
> (NETL 2010: 17)

As shown in the media reports on the web site, local reporting about the Validation Phase projects was neutral or positive. Some reports were especially informative – for example the *Bay City Times* provided an informative and well-illustrated overview of CCS (Lounsbury 2007).

Among all projects, FutureGen 1 was the most notable for the openness, extent, multiple and proactive nature of its engagement process and the alignment, coordination, and very active part played by project team members. However, FutureGen 1 was also unique in that early public engagement was essentially an integral part of the self-selection process which, by its very nature, ensured participation of communities most committed to CCS. In interviews conducted by Hund and Greenberg, respondents "highlighted choice as an important dimension ... they viewed the project as a competition that their community was vying to win" (Hund and Greenberg in Ashworth *et al.* 2010b: Appendix 3, p. 14).

The engagement process was conducted in parallel by the national-level Alliance and also by the FutureGen for Illinois Project Team. The latter team was led by the Illinois Department of Commerce and Economic Opportunity, supported by the Illinois State Geological Survey, the competing two Illinois communities (Mattoon, which was actually selected, and Tuscola), community economic development teams, industry partners, consultants, and State and local politicians. As reported by a local community leader, "The community engagement and education process was critical in generating the support the original FutureGen project enjoyed in Mattoon" (Griffin, in WRI 2010: 45).

Although other members of the community were not necessarily as immediately supportive as their leaders, they were engaged early, even as proposals were being written and submitted and had an opportunity to listen to, and learn from, questions that were submitted by various groups. As described in the GCCSI and World Resources Institute (WRI) documents, a variety of materials including posters and models were made readily available, and engagement took place in a variety of forums and formats with different community organizations and groups, frequently in local schools, and with different sponsors. The local media were active in disseminating information and airing public opinions.

The contrast between the Validation Phase Projects and FutureGen 1 on the one hand and Greenville on the other is striking. Activities related to the project in Greenville are no longer shown on the website; however, a partial record can be glimpsed from local media reports. These media reports were overwhelmingly negative from the outset. The engagement process was less proactive and less timely than at the Validation Phase sites. In part, this was because little opportunity was available for foundation-building prior to site announcement, as had been the situation for the earlier projects. In part also, because the host site owners were not MRCSP partners and were new in the local community (the ethanol plant had just begun operations), engagement planning was unable to benefit from the proactive, combined efforts of a broad-based team that was known to community members and was knowledgeable about local culture and politics, as well as being familiar with the regional partnership program and its emphasis on outreach. When site selection was announced unexpectedly in May, 2008 by the DOE at a national CCS conference, the team was placed in a reactive position of decide-announce-defend. By the time of the public meeting, which was not held until August, residents had many unanswered concerns about the project (Shaner 2008). The delayed engagement strategy, which never recovered from its reactive nature, added to the other contextual barriers to successful siting, as illustrated in negative, local media reports and amplified in a website that was later established by local opponents.

In Carson, communications staff engaged and sought support from State and local officials, as well as some community organizations, prior to announcement of the project. The announcement by the California Governor, with fanfare and high visibility, highlighted development of the innovative power plant and the leadership role of California in reducing greenhouse gases. Early, pre-announcement engagement activities were focused primarily on state and local officials. Local community organizations were also engaged, including some environmental justice groups; however, the representatives from other environmental justice groups who later spoke out against the project were not engaged early. Some concern was expressed by those interviewed for the GCCSI case study that

> the team had "not done itself any good" by making a "grand announcement" before knowing where the CO_2 would be stored and before evaluating the extent of potential opposition to any new, air-polluting facility by selecting a site which was "notorious" for the presence of environmental justice activism
> (Bradbury and Wade in Ashworth *et al.* 2010b: Appendix 2, p. 19)

Summary and implications

Issues and patterns of community response

Several patterns emerge from the analysis of projects included in this chapter.

- The patterns affecting community response as summarized in Hammond and Shackley's studies were largely replicated in the US cases analyzed in this chapter.

  Key patterns in successful siting that can be seen in both sets of cases were: trust in developers; a history of extractive and fossil fuels industry in the area (associated in the US cases with recognition of the long-term benefit of CCS); early, informal engagement; and transparency and timeliness in responding to the public. An additional factor identified in the cases examined here, and also in the cases included in the study by Ashworth *et al.* (2010b) is alignment and coordination among project partners. Hammond and Shackley also identified increased population density as a barrier to successful siting, but this was not found to be a distinguishing factor in this study. However, this may be attributed primarily to only one project (Carson) being in a diverse urban setting and it is not possible from the available data to isolate this as a primary factor. As Hammond and Shackley noted, where there are less people, there is less likelihood of opposition snowballing and also fewer people (with diverse perspectives) to engage with, resulting in a more manageable situation. There is also a potentially greater opportunity to build personal relationships.

Issues of concern were also similar among both sets of studies, although a particular US contextual factor is evidenced in lack of belief in climate change. Primary issues at all sites were related to health and safety. Secondary concerns were related to policy and social issues: belief in the efficacy of CCS to combat climate change, especially in comparison to other sustainable energy options; liability; decreased plant efficiency; perpetuation of the fossil fuel industry; and understanding the need for climate change were also identified as factors in both sets of studies. Concerns about liability and lack of belief in climate change were especially important factors in Greenville.

- Success is more likely when a social and economic fit exists between the project and community and the potential benefits are viewed by its members as outweighing potential risks.
- Social factors such as relationships, accountability, and decision making may be interwoven with technical concerns. Of these, well-established community relationships provide a solid foundation for credibility, trust, and accountability that facilitates the potential for technical progress. Relationships may be with an existing facility or with other organizations, such as trusted state and local organizations that, in the FutureGen 1 case, worked with the development team.
- The analysis confirms the findings from previous studies that engagement does not guarantee success. Understanding and building relationships with one's community is paramount but multiple variables can affect the deployment process. External factors in addition to community and project characteristics may play an influential role. In particular, as illustrated in Greenville and Carson, perceptions of CCS were negatively affected by broader policy issues. In the former case, support for CCS was affected by policy positions characteristic of both the political right (strong disbelief in climate change and opposition to government spending) and of the political left (sustainable, "green" solutions for climate change over perpetuation of fossil fuels). In the latter case, controversy emerged primarily from concerns related to the potential for increased pollution from the proposed power plant in an area known for environmental justice concerns, as well as from opposition to perpetuation of the coal industry.

- Engagement is more likely to be effective if it is able to draw on a foundation of economic ties and trusted relationships and is:
  - Undertaken as an integral part of a coordinated management strategy in which the team players are aligned and are knowledgeable about local culture.
  - Characterized by a "no-surprises" approach that is initiated early and conducted proactively, rather than reactively. (Although the optimum timing for initiating broad engagement may vary, depending on the community.)
  - Part of a self-selection process.
  - Part of a longer-term process that extends beyond site selection to active involvement in decisions about plant design.

Implications

Interpretation of the findings about community understanding and engagement can be facilitated by linkage to the broader theoretical framework offered by sociocultural theory, outlined previously. Several implications emerge from the combination of this theoretical perspective and the empirical evidence developed to date.

First, in seeking to deploy CCS, one can expect to encounter differences in social and cultural frameworks that will need to be accommodated. These frameworks will influence what is important to people, how they define problems and issues, the aspects of the situation on which they focus, the type of evidence they seek, whose opinion they believe, and their preferred process for decision-making. Differences in frameworks are likely to occur *both among and within communities*. In some communities, residents may be more willing to host a CCS project: for reasons such as those identified here, they may see the project as consistent with their desired way of life. In other communities, the social cards may be stacked against constructive public engagement in a potential project. Understanding the factors that may potentially affect willingness to even initiate a dialogue and/or indicate challenges that will require an exceptional commitment in terms of compensating benefits and additional communication staff and financial resources can provide an important guide to developers.

Second, people's views about CCS do not occur in a vacuum, but are linked to their broader policy views. Key policy issues demonstrated by the case studies in this chapter that can affect support for CCS include lack of belief that climate change is occurring, opposition to the use of fossil fuels and advocacy of renewables as the sole solution for addressing climate change. In the US, lack of belief in climate change is also likely to be associated with politically conservative views that advocate limited government expenditures.

Third, public concerns about CCS are likely to extend beyond technical safety issues such as potential leakage or induced seismicity to include social issues such as trust, accountability, and involvement in the decision-making process. Trusting relationships, in particular the credibility of the host and project team and managing institutions in safeguarding the community's interests, are key. Building trust may be much more challenging for new facilities or in larger, more diverse communities than in the smaller communities included in this analysis; nevertheless, these broader management issues, which are likely to underlie community response, are a priority to address. As Bradbury *et al.* concluded in their comparative CCS focus group research:

> Resolution of safety issues such as those related to potential leakage, seismicity and long-term containment are, and will continue to be, essential to successful deployment of the technology, but … *management of these risks* is the critical factor for public acceptance.
>
> (Bradbury *et al.* 2009: 4671)

Finally, if CCS technology is to progress, site selection and deployment will require more than meeting technical and geological criteria. Currently, despite growing recognition of the need to address the social aspects of deployment, CCS projects are typically selected on the basis of technical criteria alone. While selection of a site with the "right" geologic structure and CO_2 supply is clearly essential, the value of social science knowledge is not always recognized. Too often, such knowledge may be viewed by technical staff as an add-on resource that can help "fix" public perceptions after the real work is done.

The self-selection process used in the FutureGen 1 project was very effective in ensuring that CCS was not simply acceptable, but also actively sought by the receiving community. However, as demonstrated by the withdrawal of Mattoon support for a refigured FutureGen project that limited the facility solely to storage, the self-selection process was insufficient per se to ensure positive community response (for a full explanation see Griffin in WRI 2010: 14–15).

But such a process, if carefully structured, may nevertheless offer a valuable approach to marrying technical and social needs. The large government subsidies and complex, lengthy process that would inevitably exclude communities less able to compete, may at first blush seem to render the approach infeasible – but perhaps the feasibility may appear less problematic when compared with the experience of a highly publicized failure such as Barendrecht.

A productive, more generally applicable approach may be to incorporate social science knowledge into the site selection, deployment and overall project management and budgeting process. Such an approach would explicitly recognize that the biggest challenges to successful siting may stem from non-technical factors and would plan accordingly. Management plans would include allocation of sufficient funding and staff; alignment and coordination among key government and project team players; and inclusion of social factors both in selecting a site and also informing project design and implementation as the project progresses.

Social characterization, a term that has been introduced to describe the process of collecting and incorporating stakeholders' views, may provide a key to successful siting (Greenberg and Wade 2011; see also Desbarats *et al.* 2010), provided potential environmental justice concerns can be avoided. At a minimum, much like the geological screening that precedes site selection, social characterization could facilitate identification of communities that are potentially the most receptive locations or where additional effort may be needed to address social challenges – and this could be especially important for future CCS development during the current, initial period when attempts are being made to demonstrate the technical and economic viability of the technology. At best, it would assist also in collaboration with a community in project development, recognizing that the social data that are collected represent sentient human actors and that, from an ethical as well as a practical perspective, acknowledged community representatives should be involved in the process of identifying issues, information needs and potential benefits to be maximized from the community's perspective. Such collaboration would truly represent public engagement in the sense of informed, two-way discussion as to whether a project is appropriate in a particular locality and context.

Notes

1 Considerable overlap exists among the terms most frequently used to describe the process of informing and involving the public – including outreach, communication, and engagement. The term outreach is generally regarded as a one-way process of providing information, while communication may variously be described as one-way or two-way. The US Department of Energy, for example, uses the term "outreach and education" for the involvement process of the Regional Carbon Sequestration Partnerships although, in actual practice, the process may involve two-way discussions. For the purposes of the discussion in this chapter, the author does not distinguish among the terms used by the various researchers and agency programs, albeit recognizing that the ideal to be attained is Hammond and Shackley's description of engagement as "informed, two-way discussion as to whether a project is appropriate in a particular locality and context" (Hammond and Shackley 2010: 4).

2 Unless otherwise noted, data for this and the following sections are taken from census data available on the web, the MRCSP website (www.mrcsp.org), media reports, or from individual case studies of the Carson and FutureGen projects included in the studies by Ashworth *et al.* (2010b).

3 FutureGen 1 refers throughout to the original version of the facility which was the subject of the case study included in Ashworth *et al.* (2010b). The facility, sponsored by the US Department of Energy, was to be a model for the integration of power generation and CCS located in Mattoon, Illinois. As discussed in Chapter 13 in this book, by Markusson, Ishii, and Stephens, the concept has now been restructured and a new Illinois location selected.

4 The exception was a lone landowner who challenged and delayed issuance of the permit for the Michigan validation project; however, he did not receive community support and his case was dismissed.

5 The 2006 FutureGen 1 evaluation report is available at: www.fossil.energy.gov/programs/powersystems/futuregen/fg_proposals_evaluation_report.pdf.

6 The importance of incorporating neighbors' input into project design was also emphasized by a respondent in the CSIRO risk communication study where the developer listened to, and changed pipeline routing in response to, affected landowners' concerns (Bradbury *et al.* 2011).

7 One participant in the risk communication study prepared for CSIRO reported that, rather than focusing on climate change, his team orients the discussion to the need to be a leader in helping the technology to control CO_2 in advance of possible climate change legislation that could affect the plant's ability to provide future employment (Bradbury *et al.* 2011: 16).

8 Lack of alignment among key government and project development players was one of the striking characteristics of the failed Barendrecht project, as reported by Feenstra, Mikunda, and Brunsting (in Ashworth *et al.* 2010b: Appendix 1). The number and variety of organizations typically involved in a complex research project may pose a particular challenge for internal communication and constrain effective, external communication and engagement with the public.

5

COLONIZING THE FUTURE

The case of CCS

Anders Hansson

Introduction

In the CCS debate, a common claim is that humankind stands at a historic fork in the road: the two options are either to continue releasing CO_2 into the atmosphere or to bury it in the crust of the earth. Posing the choice in this way marginalizes other mitigation options. Arguments for the necessity of CCS are often expressed with strong conviction and claims of consensus in political documents, research reports and the mass media; the expectations of the technology are obviously high. Statements claiming that coal power and CCS are compatible with sustainable development, while still contested, are no longer unusual. What was formerly deemed a problem, the combustion of coal, has recently been framed as a possible solution that enables even more stringent stabilization targets if included in the mitigation portfolio. However, as no CCS demonstration coal plants exist yet, CCS development still entails significant scientific, technical and economic uncertainty (Hansson 2008).

CCS is in a developmental stage in which technology-related activity is on the increase, and in which visions and ideas about the technology's implications, significance and future are being constructed. Uncertainty is inherent in any claims about future technological development and expectations are socially constructed by central actors such as scientists, experts and other influential system builders. Borup *et al.* (2006: 286) broadly define technological expectations as 'real-time representations of future technological situations and capabilities … [as] wishful enactments of a desired future.' Creating expectations is important, according to Brown and Michael (2003), especially when practicality and value have yet to be demonstrated and when support must be mobilized. This is particularly the case for CCS since, though the technology has not been applied at scale, expectations regarding the technology's potential, as expressed in scientific reports, policy documents, and the mass media, are nevertheless high (Hansson 2008; Hawkins *et al.* 2009; Herzog 2011; Markusson *et al.* 2011; Shackley and Evar, Chapter 9, this book).

In this chapter, I will describe and analyse the expectations of CCS embedded in various practices and briefly analyse how these practices are articulated and constructed. The boundaries between these practices are indistinct – expectations are co-constructed. The practice examined here is the construction of scientific scenarios; I examine its links to public acceptance research and policymaking in order to enable a tentative analysis of performativity[1] and co-construction. The main reason for choosing the scenario literature is that such studies can play a key role in developing shared visions and making sense of the future; they are also involved in creating expectations of emerging technologies and in mobilizing the resources necessary for their implementation (cf. McDowall and Eames 2006). The chapter will conclude with a few tentative suggestions regarding how to manage expectations and improve decision-making.

Sociology of expectations

Two distinct branches of research focus on expectations. The first one has its roots in economics and the rational expectations literature of the 1960s. The second is constructivist in outlook and emerged several decades later in the science and technology studies (STS) field. The former applies a realist perspective and aims to analyse the differences – and incongruence – between expectations and the real world. For example, activities related more to expectations than to the real world eventually create hype and consequently result in a crash. One assumption, when applying the realist perspective, is that there is a calculable difference between expectations and the real world and that this difference can be adjusted and minimized. This assumption rests on a belief that a technology's promise can be evaluated *before* it has been implemented and tested or, in cases in which new technologies are being constructed, that claims about, for example, a technology's progress, can be logically distinguished from its actual progress (Borup *et al.* 2006). However, calculating the future is highly problematic, so the perspective of this chapter is inspired by the second branch of research, more specifically, a sociology of expectations with a social constructivist stance.

According to Brown and Michael (2003), there is another significant distinction between a STS perspective on expectations and more traditional future research. While the latter looks *into* the future, an STS perspective instead looks *at* the future, meaning that it is not predictive. The analytical focus is the social construction of the future (Brown and Michael 2003). It is also acknowledged that by looking *into* the future, the future may be colonized – the activity itself is thus performative, mobilized in the present in a self-fulfilling or self-negating way (McDowall and Eames 2006). This performativity is also briefly examined in this chapter. Brown and Michael (2003: 5) claim that 'the past and the future … are constantly created and recreated in the present.' The sociology of expectations aims to offer a framework for understanding how expectations are constructed among key actors (Brown and Michael 2003); this perspective is briefly explained below.

As Borup *et al.* (2006: 285) point out, 'future-oriented abstractions are among the most important objects of enquiry for scholars and analysts of innovation. Such expectations can be seen to be fundamentally "generative", they guide activities, provide structure and legitimization, attract interest and foster investment.' Hence, expectations drive both technical and scientific activities and also extend to the macro (i.e. national policymaking), meso (i.e. innovation networks) and micro (i.e. engineering projects and the work of individual scientists) levels. According to Borup *et al.* (2006), expectations mediate between these scales and change in response to new problems or conditions. Expectations are developed and co-constructed in scientific and political activities, making them both the cause and consequence of these activities. The importance of expectations for technological development and its interlinking of a range of actors and technologies can be explained by the perceived increased importance of societal development. Scientists are increasingly expected to establish relationships with various societal stakeholders; consequently, policymakers and the business community are constantly confronted with predictions of the promise of nascent technologies.

One characteristic of early expectations is that they often are technologically deterministic, downplaying other considerations such as cultural, societal and organizational factors. The past is also often selectively downplayed when these expectations emphasize newness. These processes may partly be explained by the history of disappointment in a variety of technological fields, such as energy, stem cells and biotech. However, past failures are often explained and rationalized by claims that they occurred due to unique or rare circumstances not present today (Borup *et al.* 2006: 289). Borup *et al.* (ibid.) claim that expectations evolve over time and may be cyclic, oscillating between hype and disappointment. Shackley and Evar (Chapter 9, this book) analyse this cyclicality and the case of CCS by applying 'the issue attention cycle' concept. Among other things, they claim that, while public attention to CCS has generally been mounting since 2008, an increasing lack of confidence in CCS has emerged.

Expectations not only change over time, they also vary between groups and contexts. People closely involved in a technological development process may be contradictory in their expressed expectations: when making public statements, they often promote a confident view of promise being fulfilled, but when among peers, their stance is more cautious and reflexive. There is clearly a high degree of interpretative flexibility across communities depending on the social context (cf. McDowall and Eames 2006; Hansson and Bryngelsson 2009). Methodologically, expectations may be analysed by examining, for example, metaphors and future-oriented claims, determinism, problem framing and a technology's connection to positive values in relevant documents.

Brown and Michael (2003) argue that it is important to consider the contexts of expectations. They claim that statements regarding expectations and uncertainties will vary, intentionally or unintentionally, according to the audience addressed.

Accordingly, I present several examples from contexts with different main audiences, examples that enable the analysis of potential inconsistencies regarding how the expectations are constructed. The examples relate to the three levels, i.e. the micro (i.e. individual scientists and scientist networks), meso (i.e. communicating expectations/information to the public), and macro (i.e. high-level policymaking) levels, that Borup *et al.* (2006) claim play a key role in mobilizing resources to create expectations and visions.

CCS technologies mainly exist in laboratories, small-scale pilot plants and the imaginations of developers and promoters. Nonetheless, a large volume of CCS papers and documentation has been produced by a wide range of actors, including scholars, engineers, business representatives, politicians, journalists and ENGOs. The common theme running through many of these texts is that they present discourses about expectations. In the 1970s and 1980s, there was growing interest in quantitative predictions. According to Utgikar and Scott (2006), most predictions related to energy technology development made in the 1970s regarding conditions in the early twenty-first century have proven to be overoptimistic. Furthermore, there seems to be no correlation between the prediction technique used in the forecasting exercise and the accuracy of the predictions. Hence, quantitative prediction has gradually been replaced or complemented by more qualitative methods, such as scenarios and Delphi exercises, though they are still part of the same prediction and extrapolation paradigm. In the following section, I will offer some examples from scientific papers, more specifically, studies developing CCS scenarios.

Although my analysis is based on a limited number of texts and is not necessarily generalizable, it illustrates some of the methodologies, arguments and results that can easily be found in the peer-reviewed CCS literature. In the Special Report on Carbon Dioxide Capture and Storage (SRCCS) (IPCC 2005: 354–355), the IPCC reviews the findings of research papers, applying techno-economic modelling to estimate the future impact of CCS. The IPCC found that the average CCS share of total emissions mitigation globally this century across scenarios was ~55 per cent for the 450 ppmv scenarios and ~45 per cent for the 550 ppmv scenarios. Guided by these averages, the papers analysed in this section can be regarded as representative, in that they also project that CCS will have a very significant future impact. The methodologies and arguments I used here were also used in my previous, more thorough analysis of the CCS scenario literature covering 25 papers (Hansson 2008) and in the subsequent brief overview by Bryngelsson and Hansson (2009).

Illustrative examples

CCS in science and scenarios

Odenberger *et al.* (2008) and Odenberger and Johnsson (2010) investigate, by means of techno-economic modelling, the role of CCS technologies up to 2050 in a European context. They explicitly state that they are not making predictions, but are instead quantitatively analysing the possible roles of various technologies. By the end of the period, the results suggest that up to 50 per cent of Europe's electricity production may be provided by CCS-equipped power plants (equivalent to the entire electricity production in 2005, 3000 TWh). CCS is claimed to permit continued reliance on domestic fuel (i.e. coal) while still achieving national emission targets, consequently enhancing security of supply (these emission targets correspond to the European Commission's targets). CCS is framed as the only option that, under strict mitigation targets, enables continued coal or fossil fuel combustion while reducing the need for behavioural or societal changes. It is also claimed, with reference to the International Energy Agency (IEA), that fossil fuels will doubtless continue to account for the bulk of Europe's energy supply in the foreseeable future (Odenberger and Johnsson 2010), leaving few options but to develop CCS. For this reason, by referring to political targets and continued reliance on coal, CCS is already presented as inevitable in the introductions to these two papers. I have observed that this and similar structures are common in the CCS scenario literature (Hansson 2008). This 'inevitableness' framing is not unique to the CCS literature. For example, in McDowall and Eames (2006), expectations of hydrogen technology were also created by emphasizing the technology's inevitability, connection to energy independence, and eased behavioural change burden.

Odenberger *et al.* (2008) provide methodological information that casts the study's scenarios in another light: 'Finally, it should be noted that the cost estimates above are based on IEA estimates, which are lower than most recent industry estimates' (p. 436). Consequently, recalculation would likely indicate a significantly lower future impact of CCS, especially since initial costs are crucial when estimating future cost reductions. However, Odenberger *et al.* (2008) make no attempt to discuss the potential impact of using the more recent and higher IEA cost estimates or to estimate the impact on their results. In fact, the consequences of doing so is not even mentioned in either the study's summary or recommendations sections.

When explaining the methodology for projecting the future of CCS, several studies explicitly state that major barriers to CCS implementation are being ignored; alternately, if the barriers are mentioned, it is assumed that they will be solved with time, or they are commented on and then explicitly treated as external to the study's system boundaries (McFarland *et al.* 2004; Herzog 2011). These studies are coloured by technological determinism. This determinism is also exemplified by Stangeland (2007), who explicitly neglects economic and political barriers, factors that paradoxically are said to be significant elsewhere in the text. This methodology is rationalized as follows by the author:

> In most scenarios from the literature, the potential for CCS is limited because of significant economical [*sic*] and political barriers that can delay the deployment of new technologies. The aim of the present work is therefore to calculate the full potential of CCS assuming that there are only minor political and economical [*sic*] barriers to wide implementation of CCS.
>
> (Stangeland 2007: 419)

Despite the exclusion of significant factors and methodological concerns, Stangeland makes very general and far-reaching conclusions regarding the future of CCS. The paper ends with a conclusion regarding the future potential of CCS that is almost identical to Odenberger *et al.*'s (2008) and Odenberger and Johnsson's (2010): CCS will enable a 50 per cent reduction in CO_2 emissions in the EU by 2050 and a 33 per cent reduction worldwide (Stangeland 2007). Impediments to realizing 'full potential' are not featured in the abstract or conclusions, but are mentioned only in the body of the paper; they are thus likely to be easily overlooked, whether accidently or deliberately, when the study is consulted, for example, by policymakers or the mass media. Nevertheless, calculating a technology's *theoretical* full potential might be a legitimate scientific simplification tool in order to determine the technology's upper deployment boundaries.

According to previous research, over-optimism is often prevalent when examining the future potential of technologies. Utgikar and Scott (2006) claim that experts tend to reduce the complexity related to the development of new technologies, or assume that problems will be overcome with time. However, when the results derived using that methodology leave their original context, it is important to emphasize that they can only be achieved in an ideal world. Nevertheless, Stangeland's (2007) conclusion coincides with the basic assumption regarding the future of CCS expressed in the introduction to the European Commission's 2008 proposal for a directive on the geological storage of carbon dioxide (EC 2008a). This is just one example of the interrelatedness between the micro and macro levels and between the political and scientific contexts.

Some studies endeavour to evaluate the technical progress of CCS in more detail. As the hypothetical present costs of CCS are very high compared with those of available energy technology, the future of CCS in large part depends on significant relative cost reductions in the future. A common assumption is that the cost reductions and technical progress of CCS will be comparable to those of technologies considered to be similar. This phenomenon has been discussed by Brown and Michael (2003), who note that the progress of other successful technologies is commonly assumed to repeat itself. Riahi *et al.*'s (2004a: 923; see also Riahi *et al.* 2004b) study, which is central to the chapter 'Cost and Economic Potential' in SRCCS (IPCC 2005), assumes that the diffusion of CCS technology 'will lie well in the ranges reported in the literature for the diffusion of successful technologies'. Hence, technologies that were *not* successful are implicitly considered to be, as Borup *et al.* (2006) express it, peculiar or rare cases that should not be generalized. The so-called learning rate often cited in the CCS literature is ~13 per cent,[2] based on the record of desulphurization technologies in the 1970s and 1980s. This figure is definitely in the upper range of reported real-world learning rates for coal power-related technologies of ~3–13 per cent, versus the worst cases of non-existent or negative learning due to failed commercialization (Jamasb and Köhler 2007). Applying a learning rate of 13 per cent creates a 50 per cent larger CCS impact in a scenario for this century compared with a static learning scenario (Riahi *et al.* 2004a, 2004b; IPCC 2005).

It is very common to frame CCS as based on existing technologies (IPCC 2005; Herzog 2011). Emphasizing the familiarity of the individual technological components of CCS and citing their existence in mature industries (e.g. pipelines, desulphurization, enhanced oil recovery and acid gas injection) makes unproblematic implementation seem plausible. Hansson (2008) and Flannery (2011) point out that this is not the only analytical conclusion that can be reached. Markusson's *et al.* (2011) review of the social science literature on learning and demonstration regarding CCS also claims that up-scaling is not trivial and could well fail. There seem to be a few unexamined problems.

First, since the constituent technologies of CCS have already been in commercial use for decades, an alternative argument is that it is unlikely that the costs of CCS will continue to decrease significantly. This argument refers to investor experience of conventional technologies being implemented in a new context. Furthermore, the future costs of long-term CO_2 monitoring and managing the liability issues are also key uncertainties not included in the learning rate calculations. Any major cost reduction requires, according to Flannery (2011), a breakthrough based on new approaches, for example, a new type of power plant that minimizes the cost of electricity while capturing CO_2. In addition, economies of scale should not be taken for granted, since they are very sensitive to the business model used (which is still non-existent; see e.g. Herzog 2011) and to how such an unproven large and complex system is managed and integrated. The role of CCS as a mitigation technology is also often claimed to be larger as the stabilization targets get more stringent (cf. IPCC 2005); in other words, 450 ppmv enables significantly faster and stronger penetration of CCS than does 550 ppmv. On the other hand, 450 ppmv requires massive investments in all kinds of mitigation technologies and infrastructure, and consequently stronger upward pressure on capital costs and capacity shortages (Flannery 2011).

Nonetheless, one of the most frequently cited statements regarding the future of CCS is found in IPCC's special report, SRCCS. It is claimed there that CCS can account for 15–55 per cent of cumulative mitigation efforts worldwide up to 2100 if the stabilization target is 550 ppmv, and that the inclusion of CCS as a mitigation option reduces stabilization costs by 30 per cent globally. One of the main conclusions of SRCCS is that the mitigation potential becomes even greater if the stabilization target is more stringent. However, the more reflective parts of the IPCC report, stating that the projections are likely overestimated, are not emphasized:

> The actual use of CCS is likely to be lower than the estimates of economic potential indicated by these energy and economic models. As noted earlier, the results are typically based on an optimized least-cost analysis that does not adequately account for real-world barriers to technology development and deployment, such as environmental impact, lack of a clear legal or regulatory framework, the perceived investment risks of different technologies, and uncertainty as to how quickly the costs of CCS will be reduced through R&D

> and learning-by-doing. Models typically employ simplified assumptions regarding the costs for different applications and the rates at which future costs will be reduced.
>
> (IPCC 2005: 45, technical summary)[3]

Several other publications focusing on the future of CCS (Hamilton *et al.* 2009; McKinsey and Company 2008[4]) have analysed the impact of earlier scenarios in view of the sharp cost increases experienced between 2000 and 2008.[5] According to Herzog (2011), the estimates presented in the well-known 2007 MIT study of CCS, *The Future of Coal*, are already outdated due to cost fluctuations. The costs of conventional coal power plants increased by 80 per cent in only a short period after its publication. Thus, the preconditions for projecting scenarios of the future of CCS changed even during the formulation of many of the CCS scenarios existing today. Herzog (2011) discusses the difficulty of estimating the CCS deployment rate given that costs have risen so rapidly. To deal with the cost estimation problem, Herzog emphasizes the importance of building ten demonstration plants worldwide to gain an understanding of the real-world economics of the plants, generate data to inform the regulatory process, and demonstrate their safe operation in order to gain political support, public acceptance and experience integrating the various components of CCS. The challenges are described as significant – especially concerning costs and storage capacity – but they are framed as surmountable:

> When looking at the challenges facing CCS, it is easy to get discouraged. However, this is the reality of all technologies available to address climate change. Developing them to the point where they operate at the necessary scale is a monumental challenge. The correct response should be not to get discouraged, but to realize that vigorous action is needed to overcome the challenges and that this action needs to be at a significant scale.
>
> (Herzog 2011: 603)

Another crucial factor for the cost estimates is the calculated or estimated energy penalty. The energy penalty constitutes the largest cost in the CCS chain. According to Page *et al.* (2009), who are sceptical of CCS, a review of existing energy penalty literature reveals that most energy penalty values are derived from simulations and cost studies and that the specifications were based either on previous work or calculations. The authors state that the underlying data were generally not traceable in peer-reviewed literature or publicly available publications, so the transparency was very low. Page *et al.* (2009) analysed how the calculations were made and compared them with real-world experience and their own recalculations. They concluded that the values are difficult to validate, that most published data on energy penalties are too optimistic, and that the high costs make CCS implementation unlikely in the near term.

As Borup *et al.* (2006) have claimed, cultural, societal and organizational factors are often marginalized so as not to influence the outcomes of models. As demonstrated, however, the reports are somewhat reflective, and the researchers briefly note that the complexity is reduced in the models and that this is likely to influence the output in a way that increases the potential of CCS, making it seem more feasible. As a result, this may have a major impact on the expectations of CCS, for example, among the public. As Brown and Michael (2003) claim, there might be a certain degree of interpretative flexibility across communities, depending on the social context. Therefore, in the next section, I will discuss some aspects of how the complexity, reflexivity and promises of CCS are dealt with in public communications efforts and public acceptance research. Is there any inter-contextual inconsistency, as McDowall and Eames (2006) claim is common in other cases?

CCS and public communication and acceptance

As recently demonstrated, there are significant uncertainties regarding the estimation of future CCS implementation costs as well as a reflexive awareness in the research community – though it is still a dominant stated conviction that CCS is necessary or will enable significant cost and emissions reductions. In addition to the political and techno-economic uncertainties, a lack of public acceptance is often emphasized as one of the major CCS implementation barriers (IPCC 2005; Herzog 2011). In line with previous discussions, a common assumption is that the technical problems of CCS are manageable, so the major remaining hurdles are political and social, more specifically, a lack of public acceptance. Consequently, probably the most developed social science stream in CCS research is studies of social acceptance. In the following discussion, I deliberately exclude the ENGO community's various positions on CCS, even though ENGOs play a crucial role in political and public discussion, scientific debate and the mass media. It is worth mentioning that ENGOs are generally more sceptical of CCS than are other actor groups, such as industry, politicians and scientists, even though a number of ENGOs are actively promoting CCS development (Anderson and Chiavari 2009).

Bäckstrand *et al.* (2011) claim that the normative agenda in the social scientific CCS literature is primarily to remove barriers to CCS deployment by designing communication strategies or ensuring well-informed public debates. Furthermore, according to Corry and Riesch (Chapter 6, this book), acceptance research has focused mainly on supplying information about CCS technology. A widely held belief in the acceptance research community is that proactive public communication about CCS is required to promote the technology's acceptance:

> Establishing more positive attitudes towards the technology early in its life is likely to be far more beneficial than trying to overcome negative perceptions later on. Given the high number of people who know nothing about CCS, the earlier communication commences the better, as this presents an opportunity for positive attitude formation.
>
> (Ashworth *et al.* 2010a: 430)

Bradbury (Chapter 4, this book) provides a detailed presentation of the general findings of acceptance research. Laypeople in general have very low awareness of CCS or, even more commonly, have never heard of the technology, so their opinions are easily influenced, at least temporarily, by providing new information (Daamen *et al.* 2006). For example, the information supplied in questionnaires, in order to enhance respondent knowledge of CCS, is performative. Acceptance research is coloured by a conviction that the task is to inform the public about CCS, or at least of the necessity or importance of CCS. These affirmations are congruent with the CCS scenario literature:

> The more accurately informed individuals are about CCS, the more likely the technology is to have a smooth implementation. This will be essential for the acceptance of the continued use of fossil fuels around the world as we transition to a low-carbon economy and which, based on all forecasts, will be essential in addressing human poverty and raising the quality of life in many developing countries.
>
> (Ashworth *et al.* 2010a: 431)

Malone *et al.* (2010: 421) have critically reviewed a large number of CCS acceptance studies and make the following claim: 'Questions from many of [*sic*] CCS surveys reflect an unrealistic goal as well as unbalanced information. The goal appears to be a world in which technologies, including CCS, do no harm – or they will not be implemented.' This analysis may be exemplified by how Miller *et al.*'s (2007: 21) survey presents CCS to survey respondents as 'one of the most environmentally friendly ways to address the issue of increasing CO_2 emissions, yet there is a small risk of leakage'. CCS is apparently framed as an existing technology with no major problems, or with problems that can easily be solved in the near future. The scientific basis for public acceptance research, in other words, what is considered reliable information to communicate to survey respondents, largely corresponds to the results of scenario studies.

A problem often noted in acceptance research is that there is no reliable information regarding the future of CCS. In light of that admission, it is paradoxical that a common research focus is on how to provide and present *correct* information in an accessible, pedagogical form. In a review of CCS communication efforts, Ashworth *et al.* (2010a: 429) conclude: 'Several of the projects confirmed that discussions of CCS with accurate and objective information is likely to lead to more positive attitude formation.' However, information on CCS is not confined to estimates based on scenarios; there is also practical experience with pilot plants and individual CCS components. Hawkins *et al.* (2009) attempt to describe the general development of CCS in recent years (see also Evar and Shackley, Chapter 10, this book). They mention that the past two years have not been encouraging. Hawkins *et al.* refer to the initial euphoria surrounding several well-known CCS projects that have recently suffered from a series of setbacks and cancellations.[6] Nonetheless, they explicitly reject the idea that 'CCS has been off to a false start' (pp. 4404–4405). They claim that the problems so far experienced by CCS share one theme: economic challenges due to policy uncertainty and limited public support.[7] Because these problems are non-technical, they are described as manageable if policy measures and educational programmes are initiated to accelerate CCS deployment. Hence, the cancelled projects are framed as not affecting the actual technical potential of CCS – an affirmation that is strengthened by stating that CCS is technically mature by referring to companies' claims that they could even deploy the technology today. This rhetoric, downplaying the importance of past failures, is similar to what previous expectation research has witnessed (cf. Borup *et al.* 2006). The paper ends with the following conclusion, which touches on several of the above-mentioned themes, i.e. admitting severe problems, claiming that these problems are non-technical and therefore manageable, and against the background of CCS as inevitable, claiming that better CCS information and pro-CCS decisions are needed:

> Despite some setbacks and slow progress in recent years, CCS still holds considerable promise as a greenhouse gas mitigation technology. Tangled up in a web of reluctance for adoption, uninformed rhetoric, carbon policy limbo and unfavourable economic conditions, the technology in the U.S. and elsewhere has been viewed as hype by many observers. However, decisive policy action can change the tide and move CCS projects to the pipeline rapidly. Political will and decisiveness will be needed in order to institute these policy changes. It is only with this kind of will though, that the world will avoid the worst impacts of anthropogenic climate change.
>
> (Hawkins *et al.* 2009: 4409)

Although several communication efforts and acceptance studies emphasize the importance of two-way dialogue with the public, discussing both the pros and cons of CCS and its various risks, or even of deliberative democracy including, for example, consensus conferences and citizen juries (cf. Roberts and Mander 2011), the ultimate goal of these research activities is often simply to accumulate knowledge of how the public can be made to favour the technology more. This goal is justified by the claim that CCS is unavoidable if we seek to mitigate climate change – at least at reasonable cost. Roberts and Mander (2011: 6307), who apply what they call a deliberative bottom-up approach to assessing public perceptions of CCS, conclude in the paper's abstract that 'such approaches offer an effective way of communicating information and can significantly reduce opposition aiding implementation'. Affirming the technology's feasibility by referring to science, but not emphasizing the methodological reflexivity that occurs in science, could make the promise of the technology seem surer and heighten expectations among laypeople, politicians and other stakeholders. This is especially the case because science and scientists, according to Roberts and Mander (2011), are considered by laypeople to be among the most reliable sources of information.

Another key arena in the co-construction of expectations is the political arena. Scientific claims are reproduced and the technology is framed in a certain way.

Favourable framing is probably necessary for performativity, in this case, for example, initiating the building of demonstration plants and formulating policies. Therefore, I will discuss how science is represented in the political context, illustrating the discussion with a few examples from the EU. Politicians are commonly viewed as having a major responsibility to promote and facilitate CCS.

CCS in policy documents

According to Markusson *et al.* (2011), the dominant view of CCS in the context of policies and regulatory issues is that CCS is at least reasonably robust and that the technology has garnered widespread support in international climate and energy policy discussions. This is also the case at the EU level. According to the proposal for the EU directive on geological storage, scenarios and modelling were crucial for the decision-making: 'The main areas where scientific expertise were required were: Energy modelling to predict the outcome of various options for promoting CCS deployment' (EC 2008a: 4).[8]

The draft directive's main statements regarding the various potential futures of CCS were based primarily on Capros *et al.*'s (2007) PRIMES modelling (EC 2008a: 4). Capros *et al.* (2007) present 16 scenarios up to 2030 based on a variety of assumptions and policies. The underlying data used in the study are not presented, so the transparency is very low. However, the data are discussed in another EU-funded report, the ACCSEPT study (2007: 139). In the so-called 'soaring-price scenario', the coal prices in 2030 is EUR 130 per toe – a price level actually already reached in 2008, only one year after the study was published. Volatile, dynamic or even more radical shifts in price are difficult to address in scenarios. In a review of how potential radical or permanent changes, or discontinuities, are managed in the futures literature, van Notten *et al.* (2005) revealed that discontinuities were omitted in 50 per cent of the analysed texts. Discontinuities were said to have been omitted because it was methodologically complex, or even inappropriate, to include them in the modelling and because discontinuities were associated with uncertainty and were valued negatively for that reason.

To continue the discussion of Capros *et al.*'s (2007) study, when considering that low coal prices make CCS and coal economically attractive, compared, for example, with renewables, the potential of CCS might have been overestimated. Furthermore, no externalities except unemployment are mentioned in the report, which in this case is a positive externality: CCS reduces unemployment (compared to the scenarios without CCS) due to lower electricity prices and a surviving mining industry. In scenario construction, sensitivity analysis is generally conducted, and this is also the case in Capros *et al.* (2007). On the other hand, Capros *et al.* (2007) elaborate on only two factors and they have no substantial impact on the results: (1) the higher cost of CO_2 storage in new member countries; and (2) no nuclear phase-out in Belgium and Germany. These factors seem very arbitrarily chosen, especially given that several recognized uncertainties influencing future costs are omitted or only slightly altered, i.e. learning rate, capital costs and life-cycle costs (see also Bryngelsson and Hansson

2009). The report concludes that not allowing CCS development is likely to imply significant higher costs. The European Commission seemed to consider the report robust enough, as it made the following statement at the beginning of the directive proposal: 'We cannot reduce the EU or world CO_2 emissions by 2050 if we do not also use the possibility to capture CO_2 from industrial installations and store it in geological formations' (EC 2008a: 2).

The European Commission also emphasized the need to address public acceptance. A public consultation – 'Capturing and storing CO_2 underground: Should we be concerned?' – was initiated via the Internet in 2007. In the consultation, information on CCS was initially presented to respondents, referring to the IPCC special report. It was even said that CCS is unavoidable, which is not actually stated in the special report (IPCC 2005),[9] and may significantly reduce stabilization costs. The stated advantages of CCS were balanced by mentioning a few disadvantages, including increased fuel consumption and risk of leakage, though these were presented as manageable. The risks of not implementing CCS were also described, but in a very unpleasant scenario. A seven-year delay in implementing CCS could cause an additional 90 Gt of CO_2 to be released into the atmosphere by 2050 – a worst-case situation that must be resolved by mandatory implementation of the technology (EC 2008a: 4–7). Furthermore, CCS is promoted not only to facilitate achieving EU stabilization targets but, more significantly, to enhance the development of developing countries.

Another interconnection between scenarios and politics can be exemplified by the previously mentioned paper by Stangeland (2007) with its emphasis on the European Technology Platform for Zero Emission Fossil Fuel Power Plants (ZEP 2008a)[10] and its proposal for a EU CCS demonstration programme. Even in the introduction's first section, explicit reference is made to Stangeland's conclusions regarding the future potential of CCS, and no mention is made of that study's (i.e. ZEP 2008a) major methodological limitations anywhere in the paper. It is simply cited as a fact that CCS can reduce EU's emissions by 400 Mt of CO_2 per year by 2030 (ZEP 2008a: 3). The executive summary, among other matters, concludes that the EU's emissions reduction targets cannot be reached by 2050 without CCS, that an 'EU-flagship programme' is needed to 'kick-start' 12 demonstration plants, and that EU financial support is needed (ZEP 2008b). Because the support expressed in scientific contexts regarding the future of CCS exerts such a large influence, or at least is emphasized, in central political documents and as information or basic assumptions in communication efforts, deeper reflection on expectations and the potential performativity of science is needed. In the final section, I will address some of these issues.

Concluding discussion

In this section, I will discuss the characteristics of the texts analysed and how they might construct the conditions for high or even potentially inflated expectations regarding the future of CCS. As a concluding remark, I will highlight a few lessons learned.

To begin with, I will examine the scientific texts. A paradox is evident in scenario science and in the texts discussed. On the one hand, there is a tendency to create realistic descriptions of the future to guide political and industrial decisions; on the other, there is reflexivity regarding the limitations of the chosen methodology and often explicit statements that the results probably indicate too large a future role for CCS – in other words, over-optimism.

A few researchers also claim that recent practical experience of CCS implementation indicates that the technology's pace of development was initially overestimated. This notion is compatible with Tichy's (2004) finding that experts tend to express over-optimism regarding the future of any technology they are involved in developing. This raises a number of problems. Letting the methodological restrictions inform the results, or presenting them adjacent to the results, would probably reduce the legitimacy of the reports as interpretations of the future. On the other hand, entirely excluding the methodological critique would almost certainly undermine the studies' trustworthiness. Hence, by presenting a carefully chosen balance between methodological critique and restrictions, the authors can more easily justify their sometimes far-reaching conclusions.

One pattern, in line with the sociology of expectations perspective, is that CCS is often presented as a necessity if we are to achieve ambitious climate stabilization targets – an argument strengthened by referring to the unavoidable growth of fossil fuel consumption and its interrelatedness to economic growth. Since this trend is established and CCS is the main technological option for managing CO_2 emissions, CCS is constructed as essential and the methodological restrictions and simplifications are justified. Managing the remaining real problems with CCS, which are almost entirely framed as non-technical, is consequently described as a collective societal responsibility. That is why politicians and the public must be provided with scientifically based information regarding CCS. But one problem remains: this scientific information rests on unstable grounds. This may be exemplified by estimates of the ever-changing capital costs, by cost reduction assumptions based mainly on analogies to successfully implemented technologies, and by the explicit exclusion of modelling factors that would probably have a significant negative impact on the results if included. Furthermore, a dominant assumption is that all major problems will be solved given time; however, the history of energy politics and technology in general and the limited amount of practical CCS experience to date suggest that this is unlikely to be the case.

Despite the methodological reservations, historical experience and present implementation problems, it is the optimistic conclusions that are primarily reproduced in the contexts analysed here, and often without significant reservations. I cannot say anything about the researchers' real and unexpressed expectations or intentions – that would require a complementary methodology. Previous research into the mass media portrayal of CCS finds that the media reflect in a balanced way how CCS is perceived by stakeholders (van Alphen *et al.* 2007); consequently, most mass media coverage presents a positive or neutral framing of the technology

(Mander and Gough 2006). However, this framing seems to be gradually shifting towards a more CCS-sceptical stance (Shackley and Evar, Chapter 9, this book). Nevertheless, with reference to the science, it is often claimed that climate change cannot be managed without CCS, so the importance of involving the major obstacle to CCS implementation – the public – is emphasized.

A number of so-called acceptance studies have been conducted to investigate how communication could be organized to involve the public in CCS-related decisions (see Bradbury, Chapter 4, this book) or how simply to inform the public of the merits and performance of CCS. In these contexts, CCS is often presented as a technology with huge potential and that if it were discarded as an option, major problems such as increasing emissions or reduced welfare would result. Interpretative flexibility regarding CCS seems to be high and context dependent: CCS can be presented as contributing to a variety of societal goals. Statements regarding the expectations and uncertainties concerning CCS also vary depending on the intended audience, and the reservations articulated in scientific contexts are underrepresented in acceptance studies. Not reporting these reservations to the broader public could well give rise to expectations that may turn out to be too high in retrospect.

On the other hand, high initial expectations are often necessary in order to gain financial, political and societal support for a nascent technology or idea and to define research priorities, inspire researchers, build coalitions with common goals and create confidence in decision-making (McDowall and Eames 2006). However, the potential problems of high expectations have been emphasized in previous research. They can produce an inferior basis for decision-making and unpreparedness for upcoming problems, ultimately leading to disappointment and reduced support. Another problematic outcome of the performativity of high expectations regarding the future of CCS could be, for example, that the potential of other mitigation options is downplayed. Hence, a situation of moral hazard might emerge: CCS may be considered insurance for society against extreme climate change and as an excuse not to make more radical changes to the energy system. This is a concern emphasized by many ENGOs (Anderson and Chiavari 2009).

What lessons can be learned from these discussions? Society faces a situation that calls for rapid and far-reaching decisions and actions regarding several acute environmental problems and technological options. Not choosing or taking a stand is not a constructive alternative. Realistic scientific knowledge regarding the future of CCS seems unattainable, and it is not in that direction that I am seeking answers. Costs are constantly changing and new and unexpected problems are emerging; capturing these changes by more frequently updating the scenario's input data or increasing the complexity of the models by including more variables might erode the models' trustworthiness and robustness by increasing their already extreme complexity and consequently low transparency and intelligibility. One cannot escape the fact that the scenarios have a short 'shelf life' and constitute questionable material with which to inform decision-making because of their methodological limitations. It is therefore better, as Brown and Michael (2003) point out, to construct acceptable, tolerable and sound expectations. But the question remains, how do we achieve that?

- Methodological reflections and uncertainties as well as comments regarding their influence on the results should be made more central in the scenario literature, and should be distributed to a wider range of societal stakeholders. It is often specifically these aspects that are of greatest concern to the public (e.g. leakage rates, risks, regulations, financing, liability and system boundaries in the models). It is important to communicate to the public that scientific uncertainty is not a problem that has to be solved, but an inevitable condition in science (see also Evar and Shackley, Chapter 10, this book).
- Present the conclusions with more humility and reflexivity; for example, not assuming that all problems are manageable or that no major uncertainties exist, especially not based on the justification that if they do exist they are mainly non-technical.
- It is problematic to divide problems into technical, political and societal categories, as technology is created and understood in a social context. As Markusson *et al.* (2011) note, by separating uncertainties from each other and not considering their inter-linkages and interactions, the opportunity to achieve an integrated analysis might be missed. Managing one uncertainty might well influence another.
- If the uncertainties are communicated, another problem arises. A general phenomenon in the mass media is that uncertainty, as framed in the scientific context, is often mistranslated when communicated to the public and politicians. Qualitative descriptions related to uncertainty, such as 'almost certain', 'likely', 'improbable' and so on, mean different things to different people – especially to people with non-scientific professions or perspectives. Hence, the tools for scientific communication with laypeople need to be enhanced (see also Bradbury, Chapter 4, this book)
- Create forums for formulating acceptable and robust expectations as well as opportunities to learn from historical cases (Brown and Michael 2003). As Borup *et al.* (2006) claim, past experience might not be as typical as is assumed. One option might be to practice robust decision-making (RDM) – a method claimed to be effective when decisions must be made under conditions of 'deep uncertainty' (i.e. conditions in which knowledge is lacking or parties do not agree) and the system is complex. RDM rests, among other things, on multiple views of the future and allowing stakeholders to understand the key assumptions underlying these alternative views (cf. Lempert and Collins 2007). Hence, scenarios and judgements regarding future developments are regarded as material for discussion rather than as assumptions to inform decision-making – thus facilitating the 'decolonization' of the future. As Adger *et al.* (2009) declare, inability to gain precise knowledge of the future does not justify passivity or not making decisions; instead, they propose that RDM circumvents the need for precise knowledge.

Notes

1 In this context, performativity refers to the fact that science plays a crucial role not only in describing a given technology, but that the framing created also affects the actual development (e.g. via policies, public acceptance and funding) of the technology (cf. Callon 1998).

2 The costs of the technology are reduced by *x* per cent every time the total installed capacity doubles.

3 In addition, external and social costs are explicitly excluded from the analysis in SRCCS (IPCC 2005).

4 The report from McKinsey and Company (2008) was presented by its authors to European Energy Commissioner Andris Piebalgs. The report presented estimates of when CCS might be commercially viable and how effective demonstration and implementation could be achieved. Piebalgs stated: 'This report is a valuable contribution to the discussions currently ongoing to find the best way to push this technology forward' (EC 2008b).

5 The cost increases amounted to 50 per cent for coal production, 80 per cent for power plant capital investment and 35 per cent for steel mill products, while the cost of IGCC plants tripled between 2004 and 2008 (McKinsey and Company 2008).

6 These cancellations or postponements include, according to Hawkins *et al.* (2009), large high-profile projects such as Statoil's Halten project in Norway, Saskpower's 300 MW oxyfuel plant in Canada, NRG's Huntley IGCC in New York State, Hydrogen Energy's Carson and Kwinana projects, and the restructuring of the FutureGen project in the US by the Department of Energy (DOE). The list of such projects could be much longer.

7 One exception regarding the cause of the failures is mentioned: the Kwinana project was said to have been cancelled because the storage reservoir was deemed inadequate.

8 The Final Directive on Geological Storage of Carbon Dioxide (EC 2009) does not make any specific reference to scientific reports; reference is only made to other policy or legal documents. In the spring of 2011, four guidance documents were published to support the implementation of the Directive across EU Member States. These documents do not contain any comments on the future potential, costs and role of CCS in energy systems. Instead, they are very specific or technically detailed and deal with issues such as risk management, characterization of storage management, responsibility and financial security (EC 2011a). Because of the characteristics and low transparency of these documents, they are not easy to analyse from a sociology of expectations perspective (or at least using textual analysis as the main methodology). Therefore, I chose documents containing more explicit and traceable references and discussing CCS in its wider context.

9 The SRCCS (IPCC 2005) states that it is more expensive to mitigate climate change if CCS is not implemented.

10 ZEP is a coalition of stakeholders who explicitly promote CCS as key technology for mitigating climate change. The report aims to provide recommendations regarding how the EU's CCS demonstration programme should be designed and how CCS deployment can be accelerated (ZEP 2008a).

6

BEYOND 'FOR OR AGAINST'

Environmental NGO-evaluations of CCS as a climate change solution

Olaf Corry and Hauke Riesch

Introduction

The environmental NGO (ENGO) community has previously had a considerable impact on public policy processes, not least through its ability to influence agenda-setting and problem-definition in policy processes (Togerson 1997; van der Heijden 2010). However, it represents a particularly critical constituency for technologies such as large-scale CCS for which there is little other rationale than an environmental one, namely that of mitigating climate change (e.g. Shell 2008: 3). At the same time CCS is a complex case for environmentalists who may be drawn to it by a sense of urgency related to possible climatic tipping points, yet often harbour deep suspicions of fossil fuel industry actors, 'end-of-pipe' solutions and express scepticism regarding whether and how fast CCS will be rolled out and to what effect (Corry and Reiner 2011a).

Rather than challenging or confirming ENGO views of CCS as accurate or misguided, this chapter analyses the discursive framing of the varied positions ENGOs have taken on carbon dioxide capture and storage technology (CCS). Despite agreeing on many of the 'facts' about CCS, they differ in terms of whether they oppose or support it partly because they differ more fundamentally in how they choose to view the complex of problems and solutions, e.g. in terms of 'the problem' that CCS should solve, what exactly CCS is (how is it portrayed as a technology or socio-economic arrangement) and how they justify their stances. Although research indicates that framings affect opinions on emergent technologies (Druckman and Bolsen 2011), existing research has most often deployed a 'knowledge-gap' approach to explaining attitudes to CCS that gauges whether or how knowledge of CCS affects attitudes (e.g. Ziogou *et al.* 2010; Wallquist *et al.* 2011; de Best-Waldhober *et al.* 2009; Malone *et al.* 2010). Reception analysis sheds light on one dimension of the politics of CCS in terms of how CCS is seen as a solution (e.g. whether it is seen as safe or not) but tends to bracket other framing issues such as how the problem that CCS supposedly tackles is understood, what the identities of the actors are, who makes the claims and even what CCS precisely 'is' in the policy arena. Metaphoric representations of CCS are rarely recognised as pivotal. The idea of 'acceptance' smuggles in an assumption that CCS technology is a given entity, rather than a discourse – i.e. a social fact that can be construed in fundamentally different ways (see also Feldpausch-Parker *et al.* 2011). Literature on NGO responses to CCS is sparse (Stephens and Verma 2006; Anderson and Chiavari 2009) and studies of informal social movement responses are even more so (Corry and Reiner 2011a) while attention has concentrated more on public and local community evaluations of CCS instead (Reiner *et al.* 2006; Ashworth *et al.* 2007; de Best-Waldhober *et al.* 2009; Eurobarometer 2011). This is often summarised in terms of acceptance or opinion 'for' or 'against' CCS.

In contrast, the approach taken here aims to go beyond this dichotomy to examine social representation of CCS, the way climate change and energy is construed as a problem in the first place, and the strategies and metaphors ENGOs use to assert their positions. The object under scrutiny is not directly the (often sophisticated) substantive reasons NGOs have for the positions they adopt on CCS, nor a study of individual 'perceptions' of CCS as such. Rather, it focuses instead on the level of public discourse about CCS: what informal rules can be identified about what can be said in a valid way, how and by whom? CCS policy discourse is not the result of narrowly rational processes of finding optimal solutions to given problems but rather a symbolic structure framing negotiations over language and competitions between multiple political rationalities (Hajer 2002; Stone 1989). For discourse analysts 'large proportion of our most commonplace thoughts make use of an extensive, but unconscious, system of metaphorical concepts, that is, concepts from a typically concrete realm of thought that are used to comprehend another, completely different domain' (Lakoff 1995: 177). Not only metaphors and individual concepts such as 'greenwash' or even 'capture' or 'storage', but the framing of entire issues or the 'language game' within which a political discussion is conducted – all these symbolic structures set boundaries and conditions for how problems and solutions can be evaluated or even debated. In line with social movement research, social movements and NGOs are viewed as active agents (re-) creating such language games rather than passive carriers of ready-made information or ideologies (Benford and Snow 2000). In other words ENGOs are treated in this chapter not just as 'for' or 'against' CCS but as creators and purveyors of policy discourses that can be influential in deciding how a technology such as CCS is evaluated by them and ultimately by others.

Understanding NGO perceptions and arguments about CCS is important not just because NGOs influence policymaking and the public perception of environmental issues. A fruitful engagement between NGOs and other actors relies on mutual understanding and awareness of the discursive parameters that frame a debate and this chapter aims at theorising NGO representations of CCS in a way that facilitates that. Equally, the relationship between the environmental movement and science is much more complex than a matter of siding with science or against it and our chapter aims at showing how it is not the case that there is just one side one could take while accepting the science itself. It thus aims to contribute to the ongoing debate about how technology in society should be studied, using problem-definition literature and discourse analysis, while casting light on the debate about CCS in particular.

First we explain the importance of political discourses for how CCS is evaluated and draw on problem-definition literature as well as theories of discourse suggesting that the politics of CCS is dependent upon social constructions of the problem of climate change and of CCS itself. Second, the discursive positions of a number of ENGOs regarding CCS are analysed in terms of (a) how the problem is construed, (b) how CCS is viewed as a solution and (c) how claims about the role of CCS are justified. From this analysis two dimensions and four ideal typical positions are extracted that can help us understand how different concerns and conclusions can be reached without necessarily disagreeing on the facts or science of CCS. This, it is suggested in the conclusion, will have implications more widely for those interested in how representation as well as information impacts technology policy, climate change and energy policy.

Material for the study consisted of official policy statements on CCS where these existed from ENGOs operating in English. An ENGO was defined as any non-state and non-corporate organisation advancing values and policies through public discourse primarily concerning the environment.[1] For certain ENGOs that did not have such documents, interviews were conducted with employees or former employees. Some of these were done on condition of anonymity. The readings were used to generate (rather than to test) a map to understand the debate about CCS amongst ENGOs, presented below, which sets up ideal-typical positions for how CCS is evaluated (Figure 6.1). This aims to facilitate a more complex and nuanced understanding of the debate about CCS while still providing a structure to make sense of that complexity. This in turn may help actors understand and act while being engaged in the politics of CCS. Debates about other emergent technologies designed to transform societies appropriately in relation to complex problems may display a similar structure.

Problem-definition, solution-construction and strategies of legitimation

The first thing to note in that regard is that ENGOs adopt very different stances in relation to CCS even when they do not disagree on the basic facts about the technology. The Norwegian climate change NGO ZERO presents a spread of ENGO positions on CCS from the very positive to very critical (eight broadly in favour, six sceptical (ZERO 2011)). Bellona states that 'CCS has a large potential. In fact, CCS alone can reduce global CO_2 emissions by one-third by 2050' (Bellona 2011a). For Bellona, CCS is not just an option but indispensable: 'without CCS it is not possible to reduce greenhouse gas emissions sufficiently to combat global warming' (Bellona 2011a). In contrast, Greenpeace concludes that 'CCS is unproven, risky and expensive and investing in it threatens to undermine the range of clean energy solutions which are available right now' (Greenpeace International 2008). Furthermore, 'CCS is being used as an excuse by power companies and utilities to push ahead with plans to build new coal-fired power plants; branding then "capture-ready"' (Greenpeace International 2008: 6). In between these two positions, organisations such as the WWF at times regards CCS as a possible 'bridge' between current fossil fuel dependency and a renewable future but remains sceptical about whether conditions are in place that would allow this to happen: 'CCS could play an important role as a transitional technology in the context of an ambitious climate strategy in particular in relation to fast growing economies such as China's' (WWF Germany 2008). This is made conditional on 'it working' and 'the risks being manageable' (ibid.). WWF-UK recommends more testing and demonstrations to assess the viability of CCS and can only support it 'if the technology is demonstrated to be effective and the transport and storage can be shown to be safe and environmentally benign (...) and if a sufficiently robust regulatory regime (including independent monitoring and evaluation) is established' (Allott and Kaszewski 2008: 9). How can ENGOs reach such diverse conclusions about the same technology and the same problem?

Part of the answer is that they are referring to different problems (or at least different parts of the problem) and different technologies (or different takes on CCS as a technology), and different assumptions about how technology and society affect each other. In a rationalist conception of politics, problems exist independently of language and mooted solutions either solve those problems, or they do not (Stone 1989). For the problem-definition approach or discursive policy analysis 'the first maxim of problem definition is the recognition that problems do not exist "out there", are not objective entities in their own right, but are analytic constructs, or conceptual entities' (Dery 1984: 40). Thus, to successfully define a political problem and to frame policy options in a certain way is a crucial part of the policy process. Thus, the problem of climate change can be defined in various ways, e.g. in scientific terms, as an economic problem or as a political problem of collective action, for instance (Brunner 1991; Hulme 2009). In the debate about CCS, different 'problems' are in play such as CO_2 emissions, energy security, the price of electricity, funding for renewables, the mining of coal, oil and gas, inequality and centralisation of energy production and corporate power (see below).

A brief look at problem definition literature reveals multiple approaches to studying problem-definition in policy processes but also common basic assumptions that will facilitate our subsequent analysis of the politics of CCS in environmentalist circles. First, there is no single way of representing 'a problem', and by implication anything else, since they are always social constructions. They are social *constructions* because, as Gusfield (1981: 3) puts it 'human problems do not spring up, full-blown and announced, into the consciousness of bystanders'. Dery makes this explicit: 'The very notion of problem definition suggests a constructionist (rather than an objectivist) view; that is, problems do not exist "out there", they are not objective entities in their own right' (1984: xi). Thus, a stone is potentially a projectile, a part of a wall, a jewel, a goal-post in a game of football, etc. What a particular piece of matter 'is' in a social sense depends on the system of meaning it is put into. CO_2 is potentially a drinks additive, something we breathe or a driver of global warming that can be traded on international markets, depending on the system of meanings (discourse) it is put into. Identities of actors as well as objects and structures are necessarily always contingent and unstable constructs since there is no necessity of them being understood as they are. Often politics is very much a battle about the meanings of different objects, in this case 'CCS' and identities such as 'NGOs'. Problem-definition literature has taken various problems and shown how they have been changed. Gusfield takes the case of auto-safety and shows how it changed slowly from being solely a question of 'the nut behind the wheel' to the 'unsafe car' to the 'unsafe roads' shifting responsibility and possible solutions each time (Gusfield 1981: 50). Similarly, AIDS in Denmark went from being defined as a 'homosexual' illness coming to be seen as a general public-health threat, independently of a change in infection patterns (Albæk 1999). Brunner showed how global warming was being framed as a scientific, political or economic problem and how this excluded other perspectives (1991).

Second, problem definition is not only a question of the representation of reality, but also always the strategic representation of situations – strategic because it is part of a conscious effort in political struggles (Stone 1989). For Rochefort and Cobb, political conflict 'is inherently spontaneous and confusing, but activists and organised interests attempt to direct its course by strategic maneuvers based on problem definition' (1994: 5) and Weiss (1989) sees problem definition as a process in which political values are strategically advanced and repressed. Thus, Gusfield's auto-safety problem changed character partly due to the conscious efforts of the consumer movement and Ralph Nader who wanted manufacturers and planners to bear some responsibility. AIDS in Denmark went from minority issue to general threat largely due to the efforts of the strategic campaigning of the haemophiliacs' association (Albæk 1999).

Third, problem definition theory tends to reject conventional models of politics and the policy process. This involves a two-pronged attack on (a) the ideas of rationality in decision-making whereby actors seek out information and optimise preferences and (b) the illusion of the neatly ordered policy process moving from grievance over mobilisation to policymaking and then implementation. According to Rochefort and Cobb, problem definition 'is fundamental to public policy making, intertwined as it is with the political process throughout the activities of issue initiation, program design, and legislative enactment' representing the most fundamental source of 'non-rational' political phenomena (1994: 57). Stone emphasises these themes most explicitly by rejecting what she terms the market model of politics in favour of the *polis*, a political community. In the latter, political reasoning is not conducted according to instrumentally rational criteria in a production-line fashion. Rather, it is done in the language of metaphor and analogy, by defining versions of reality and building coalitions behind constructs, stories and vague concepts rather than stringent or clear definitions. Political actors in politics are not exclusively self-serving individuals with predefined interests but groups and individuals who are influenced by loyalty, identity, culture, psychological and material interdependencies. They operate on the basis of incomplete and socially mediated information structures – discourses – rather than by way of perfect information. Public definitions of problems are thereby not simply the sum of individual interests, but the result of a collective process of meaning-making. And problems are, to a large extent, what political struggles are about. In sum, 'the traditional concept of rationality does not really exist in public policy development and analysis' (Stone 1988: 5). It therefore makes little sense to have a mechanical theory of politics which excludes the construction of problems and solutions from the political process from the beginning. Thus, an analysis of problem-definition in relation to CCS implies a discourse analysis of CCS asking how CCS is constructed as a solution to a specific representation of the problem.

Problem definition writers look for strategies such as 'blaming' where responsibility is placed with others and 'claiming' where actors assert authority over an issue or legitimation through science (Gusfield 1981). Similarly, discourse analysts speak of *strategies of discursive legitimation*: 'Political discourses which aim for resistance and contestation can be grounded in different ways. Discursive strategies can, for instance, be grounded by a claim to science, like Marxism, or through a claim to subjectivity, like much of feminist thought' (Mottier 1995: 522). CCS can thus be justified or challenged in different ways, each drawing on different systems of knowledge such as climate science or historical experience from other attempts at environmental regulation.

This highlights the fact that problem definition is *social* constructionist insofar as the process of definition is a collective one. An individual cannot arbitrarily decide the meaning of CCS but collective processes of validation, definition and construction affect what the term CCS is taken to mean, what can legitimately be said about it and what effects this has on the world of technology and materiality. The level of reality we are interested in here is thus not private attitudes to CCS or psychological beliefs as such but the public discourse, or rules for what can and cannot be said, about CCS as a climate change solution.

In the following we look at discursive processes affecting: (a) problem-definition i.e. what is the problem and where did it come from; (b) solution-definition i.e. definition of CCS (what it is and what it involves); and (c) strategies of legitimation for the positions adopted (what kinds of knowledge and arguments are drawn on to justify positions).

Ideal typical ENGO positions on CCS

Problem-definition of CCS: 'root cause' versus 'major effect' reasoning

NGOs largely agree that climate change is the problem, but specific constructions of problem causation place responsibility in different ways, as do different constructions of what precisely it is that is problematic about the task of tackling climate change and where in the causal chain of events that lead to global warming intervention is necessary. Differences in constructions of causation are determined not by disagreements about whether global warming is man-made or caused mainly by burning fossil fuels, but by differences in how far backwards in the causal chain to go. The field can be understood in terms of a continuum running between those who choose 'root causes' and those who focus on the 'major effect' models of reasoning. Without denying that carbon emissions cause global warming, the former chooses what it sees as the original cause as the locus for intervention. This could be a materialistic society that has grown dependent on cheap energy and burning large amounts of fossil fuel. The 'major effect' perspective does not involve denying the societal context but in terms of policy options has a propensity to look at the direct causes rather than the 'constitutive cause' (see Lebow 2009) of climatic change, i.e. CO_2 emissions, focusing on how to prevent these at that point in the causal chain where they are released into the atmosphere. Thus, while there is little disagreement on 'the facts', the problem can be framed differently leading to different conclusions regarding CCS.

Generally speaking, CCS supporters tend to view the problem implicitly as one of CO_2 emissions and their effects on the climate, rather than fossil fuels themselves or a specific type of society or economic system. Thus for Bellona, 'dangerous climate change is already taking place putting the health of people and ecosystems in grave danger. To change this, we need to eliminate the emission of carbon dioxide (CO_2) as quickly as possible all across the world' (Bellona 2011b). Fossil fuels are not 'the problem' per se, and for Bellona CCS promises to allow us to carry on burning them (implying this is something we wish to do). For Bellona, '50 percent of the world's man-made emissions of CO_2 comes from burning fossil fuels in power plants or factories. The remaining reserves of coal, oil and gas are large enough to last for many decades to come' (Bellona 2011b). The 'cause' of the use of fossil fuels is generally not entered into in this discourse. This articulates climate change in a technical problem definition whereby fossil fuels are a permanent fixture whereas 'emissions' need to be prevented. Thus, for Green Alliance, '[g]iven the energy security benefits of coal, and the inbuilt inertia of the energy system, the continued use of coal seems unavoidable' (Green Alliance 2008: 4). Solutions to this problem of emissions we are told exist already and 'the problem' is generally one of implementation of existing knowledge and technology, such as CCS, albeit necessarily in conjunction with renewable, energy conservation and other technologies that ultimately reduce emissions. This group's main worry is whether CCS will be recognised and accepted as a solution.

In contrast, those sceptical of CCS tend to frame the problem in wider terms and the source of the problem is traced further back such that the locus of appropriate intervention is causes of emissions (fossil fuels) or causes of causes of emissions (e.g. growth, materialism or capitalism). Whereas those favouring CCS tend to see causation in terms of the major immediate cause (e.g. CO_2 emissions), critics fix more firmly to the fossil fuel economy and some go one step further asking who has an interest in fossil fuel, which vested interests will work to perpetuate or increase CO_2 emissions and how power differentials will affect choices and the effect of various technologies that may change this (e.g. Corporate Watch and Climate Camp) or even what societal values lead us to expand economically regardless of sustainability. Greenpeace puts very heavy emphasis on the role of renewable and energy conservation, but ultimately takes one step backwards on the causal chain concluding that 'the world can fight climate change but only if it reduces its dependence on fossil fuels, particularly coal' (Greenpeace International 2008: 8). Emissions are in this case just as much a symptom as a problem as such. Fossil fuels and 'our addiction to old-fashioned energy sources' is, for Greenpeace, 'helping to drive conflict, climate change and putting our environment, health and fragile economy at risk' (Greenpeace 2011). This kind of root-cause analysis naturally evaluates CCS within a wider set of problems (conflict, health, economy) than just rising concentrations of atmospheric CO_2, since root causes easily cause more than one thing.

Going further back in the causal chain, others point to perceived causes of fossil-fuel dependency and look at built-in incentives such as the compulsion towards continuous growth and the structure of control over decisions over energy policy. This leads to questions of ownership and for some, a 'ruling ideology' which they believe will maintain dependency on fossil fuels and hence, necessarily, climate change. Corporate Watch states this explicitly in their evaluation of CCS: 'focusing on technological solutions ignores how the problem of climate change is caused, why it continues to worsen and how much needs to be done to stop it' (Fauset 2008: 2). Climate Campers preparing for demonstrations at the proposed Kingsnorth coal-fired power station in the UK focused on the energy company, the government and economic growth: 'E.ON and the government believe that you can have endless fossil-fuelled economic growth in a world of finite resources (...) People from all over the UK are here to create a democratic, low-carbon society in which our long term future on this planet is prioritised over the short term profit margins of the fossil fuel industry' (Climate Camp 2008).

'Dependence' is of course generally considered a bad thing in that it typically renders something or somebody vulnerable, by definition. Rather than 'allowing' us to continue using fossil fuels, as Bellona and Green Alliance have it, for opponents, CCS keeps us 'dependent'. The association of CCS with old style coal-fired power stations is therefore often enhanced through the visual representation such as on the front page of the report from Greenpeace which depicts power stations emitting clouds of smoke (Greenpeace International 2008: 2–4). Likewise for Corporate Watch: 'Choosing CCS means choosing coal' (Fauset 2008: 38). The continued use in the document of words like 'smokescreen' to describe the technology as well as the consistent use of scare-quote signs around 'capture ready' suggests that next to technical arguments made by Greenpeace, there is a rhetorical relation drawn between CCS, distrusted energy companies and the polluting fuel, coal. The idea that energy companies can be trusted to act on their green sounding promises is treated with scepticism or incredulity.

Furthermore, in this more expansive problem-definition of what CCS is about, climate change is not seen in isolation but as part of a larger complex of problems involving scarce resources and increasing environmental stresses:

> Oil depletion will become a major issue with the coming decade; while competition for land and water, de-forestation and destruction of ecosystems, soil fertility depletion and collapse of fisheries are posing increasing problems for food supply and survival in many parts of the world.
>
> (Fauset 2008: 2)

The basic driver behind all these interrelated problems for certain green NGOs is 'growth of the whole global economy' which 'means consumption of an ever-increasing amount of goods, using an ever-increasing quantity of energy, mineral, agricultural and forest resources' (Fauset 2008: 3).

In this framework, CCS makes limited sense because it perpetuates dependency on fossil fuels and fails to tackle the root cause as well as other problems associated with fossil fuels such as mining and high concentrations of economic power. In turn, though seen as a calamity in itself, climate change is a problem also because it worsens existing inequalities, exacerbates the vulnerability of less-well-off populations and countries and underlines the need for social change without which current problems will not be solved or simply reappear in a new guise when tackled.

Solution-definition: a technology or a social arrangement?

Emphasising paradox and negotiated social meanings, Stone notes that something can be 'two different things' at once: 'the world is composed of continua, not categories' and '[g]iven a world of continua, there is infinite choice about how to classify' and 'ultimately, political reasoning is a process of creating, changing and defending boundaries' (1998: 307, 309). A large amount of common ground does exists between ENGOs on the basics of how and to what extent CCS exists, e.g. how many demonstration projects exist, what the typical carbon dioxide capture rates are: 85–95 per cent in pre- and post-combustion capture and up to 98 per cent in oxyfuel combustion capture (Greenpeace International 2008: 11), what the energy efficiency penalty of capturing and storing CO_2 is (around 20–30 per cent) and that storage is likely to be over 99 per cent effective over a period of 100 years (Greenpeace International 2008). Formal definitions of CCS are usually also very similar. Nominally it is usually agreed that CCS refers to the triple processes of capture, transportation and storage of CO_2 from large point sources such as power stations.

However, in the detail and discussion about roll-out times and feasibility, significant differences in 'solution-definition' become apparent, including, most significantly, the boundaries or composition of 'CCS'. CCS is a relatively new construct, consisting in part of 'things' that have been around for a while, including various carbon sequestration techniques, infrastructure capable of transporting liquid CO_2 via pipes or in vehicles, off-shore and on-shore storage sites, techniques for injecting liquid CO_2 into geological formations, and perhaps Enhanced Oil Recovery (EOR) technologies originally designed to improve oil recovery rates by injecting CO_2 into oil fields to aid extraction. Added to this, 'CCS' is today used to refer also to newer social artefacts such as the legal frameworks being created to govern liability for potential large scale and long-term storage of CO_2 and the connection between these known technologies and knowledge and systems geared to tackling the problem of climate change.

Beyond the boundary issue of where CCS stops and starts as a category, the 'reality' of CCS is also disputed. For supporters CCS is generally considered 'a technology' that exists positively already, albeit not on a large scale: 'CO_2 Capture and Storage (CCS) is a technology that can reduce emissions from power plants and factories to almost zero' (Bellona 2011b). For Bellona, solving climate change is not strictly a purely technical issue: 'If the right regulatory framework and economic incentives are established, a large number of full-scale CCS projects could be commissioned within the next decade' (Bellona 2008: 81). For the pro-CCS ENGO The Clean Air Taskforce, 'Coal as used today is a uniquely dirty fuel and its environmental impacts are devastating, but the technology to make coal vastly cleaner is proven and largely available' (CATF 2011). The implementation of CCS in this case also involves regulatory frameworks and non-technical conditions but the core technology is not questioned. The politics of CCS becomes a question of 'rolling out' the pre-existing technology, implying perhaps a carpet, ready constituted in a certain length, in bulk – even if it needs the correct conditions or 'fittings', to extend that metaphor.

For those who rather view CCS as a social arrangement, in contrast, this idea of rolling out makes less sense since the very process of making it economically viable, sufficiently regulated and reliable, is included conceptually in what they think constitutes 'CCS'. If CCS is not a discrete technology, it can be developed, but not 'rolled out'. The difference is perhaps subtle but has consequences. One leading European environmental NGO figure (interview – anonymous, 2011) regards CCS as a hypothetical scenario, rather than an actually existing technology: 'the CCS debate is still very much in a place where it is an abstract proposition – a symbolic discussion – because implementation on a proper scale remains doubtful. If there were plenty coming on stream, then it might be a different discussion'. NOAH (Friends of the Earth Denmark), which opposes CCS, points out that 'CCS is not a single technology but a complex of technologies designed to capture CO_2 in power plants or heavy industry for subsequent underground storage. CCS is currently still at the demonstration stage' (NOAH 2011). Thus, CCS is in a way not constructed as a reality or as a discrete technology (see Markusson and Haszeldine 2010).

The boundaries of CCS as a technology are also disputed. For WWF 'CCS readiness' (the idea that power plants can be made ready to have capture and storage infrastructure added to it) is explicitly 'more than a technological issue' depending upon a host of other conditions being in place. Capture, transport and storage may be technically doable, but for it to happen the economic, business and regulatory conditions also need to be in place including modifications to power plants, 'detailed plans' showing 'how CO_2 will be transported', 'outline approval' of storage sites by regulators, 'credible business plans' to ensure retrofitting of CCS will be carried out and 'stringent regulatory criteria to enforce early conversion to full capture, transport and storage' are all necessary for 'CCS' to become a reality (Allott and Kaszewski 2008: 10). For sceptics, CCS as a technology would include these regulatory frameworks and infrastructures through which it works (or fails to work). Similarly, it can be asked whether 'CCS' includes EOR, for example. After noting that the CCS approach 'fails to recognise the wider systemic problem of over consumption or deal with the other impacts of that over consumption' Corporate Watch, a group which has common ground with Climate Camp, notes that 'CCS is also closely associated with enhanced oil recovery (EOR)' (Fauset 2008: 35). Thus, although everybody recognises that CCS needs both physical capture, transport and storage technology as well as regulation, associated infrastructure, specific economic dynamics and public acceptance, the dispute about whether CCS is a reality (e.g. when supporters argue that we already are doing each bit of the process of CCS, just not on a large scale and as an integrated system) or just a potential scenario is grounded in differences in which actors choose to draw the line between CCS and other structures and effect. Those who let the term denote the technology in the narrower sense feel justified in claiming that CCS is a reality that can reduce GHG emissions, whereas those who emphasise the wider set of factors necessary for CCS to work feel justified in saying that CCS is 'an abstract proposition' that does not necessarily end up in reduced emissions. In this light, assessing CCS is not simply a question of technical feasibility. For all environmentalists, the problem with CCS involves technical challenges as well as how to ensure that CCS is implemented without locking society into another generation of fossil fuels that ultimately ends up being 'unabated' if actors are able to later resist its implementation. But there is a difference in emphasis that reflects different discourses. For WWF '[t]here are real risks that a policy intended to accelerate development and learning of CCS technology is being used to legitimise much larger, investments in unabated (if 'capture ready') coal capacity' (Allott and Kaszewski 2008: 7). Once power stations are built, irrespective of whether space has been left for CCS equipment, 'the generators would be in a very strong negotiating position to press either for continued operation [despite not fulfilling promises to fit CCS] (in the name of keeping the lights on) or for the government to pick up the bill for the retrofit programme' (ibid.: 11). WWF draws an explicit parallel to the recent history of flue gas desulphurisation (FGD) in which the 'power sector faced costly, large-scale retrofits of environmental clean-up equipment' (ibid.) and resisted very effectively. Thus, the emphasis is on the problem of regulating the power sector and getting the institutional setup to overcome vested interests, rather than a question of the technological feasibility per se, e.g. in relation to storage. This is difficult and thus, reconciling 'coal and CO_2' is a question of 'squaring the circle' as they put it in large letters against a backdrop of a melting iceberg (as we know, squaring a circle is a tricky thing to do). The WWF briefing document on CCS itself is ominously entitled 'Evading Capture' alluding to the possibility of UK power sector actors avoiding CCS, the subtitle reading: 'Is the UK power sector ready for carbon capture and storage?' (ibid.: 1). Implementation is the main challenge.

In contrast, for some of Greenpeace's material, for example, the technology itself is directly questioned as 'unproven', storage is 'risky' posing 'a threat to health, ecosystems and the climate' and CCS 'will not be ready in time to save the climate': 'Concerns about the feasibility, costs, safety, and liability of CCS make it a dangerous gamble' (Greenpeace International 2008: 5). Against the backdrop of a picture of a chimney bellowing smoke (focusing on direct causes), Greenpeace promises to 'lift the smokescreen' (2008: 17) and sets out five reasons why 'CCS should not be accepted as either justification for building new coal-fired power plants or for continuing our dependence on coal in the longer term' (ibid.). Other groups are similarly concerned about the risks of CCS. While they admit that 'for some environmentalists, CCS is a lesser evil than either nuclear power or devastating climate change' Corporate Watch remain unconvinced asking 'could it work in time, and what are its side effects?' (Fauset 2008: 35).

Thus, the potential solution 'CCS' is clearly not just CCS – just as the problem of climate change is not unambiguous. If a two-by-two matrix is constructed with the two dimensions of the climate change problem (root cause to major effect approaches) on one axis and the two approaches to 'CCS' (technical versus societal notion of technology) on the other, four ideal typical positions are possible (Figure 6.1). First, those who frame climate change mainly in terms of its direct cause (e.g. emissions) and emphasise CCS as a discrete technology are represented by the first quadrant, and their main evaluation of CCS centres on the question of 'will it be accepted?' e.g. Bellona admits 'there are large barriers to achieving a fast and global deployment of CCS, but with a joint effort from industry, research institutes, politicians and NGOs, those barriers can be overcome' (Bellona 2011a). For this position, CCS is a smart fix that can work if it is funded and supported.

Those who still see CCS as primarily a technological solution (or as a group of technologies), but suggest the problem should be understood in broader terms such as in terms of use of fossil fuels (e.g. Greenpeace), would inhabit the upper right area. Their main evaluative question would logically tend to concern efficacy: 'will it work?' as intended since, for them, CCS is 'merely' a technology facing a deeper systemic problem of over-reliance on fossil fuels or high energy demand. This leads to 'end-of-pipe' metaphors indicating that even if the technology works, it would not solve the problem. Alternatively, this position allows CCS to be viewed through the metaphor of a 'bridge' allowing society to cross into a future renewable world even though it does not tackle the root cause immediately.

Third, those who focus on CCS as a broad social arrangement rather than a discrete technology but primarily worry about abating emissions rather than effecting 'root cause' radical socio-economic change (parts of WWF and FoE – lower left quadrant) see a potential for CCS but ask 'will it be implemented?', recognising that CCS is more than a piece of equipment, requiring a whole new legal and physical infrastructure if it is to be effective in abating emissions. For them the metaphors tend to be about 'capturing' or cornering the actors who might otherwise 'evade capture' – CCS is a regulatory regime disciplining coal.

Finally, those who both take a broad view of CCS as something with institutional and societal repercussions but also view climate change politics primarily in terms of tackling 'root causes' can be found in the bottom right area. Corporate Watch or Climate Camp are the clearest representatives of this discourse on CCS and their main concern regarding CCS, which they emphasise would have conservative societal impacts, would be its transformative potential along the lines of 'will CCS change our fundamental course'? i.e. would it – even if it technically 'worked' and could be implemented – exacerbate the root causes of climate change, namely the system that has created the problem in the first place (industrialism in the case of environmentalists, capitalism for the socialists, growth for ecologists). For them, the leading metaphor is 'greenwash', a 'smokescreen' obscuring a deception or CCS as a 'palliative' measure, giving the ultimately doomed coal industry a new lease of life.

In Figure 6.1 these different positions and the key concerns and metaphors associated with them are represented along two axes, according to whether they emphasise 'CCS the technology' or 'CCS the social arrangement' and whether they posit the locus of intervention at emissions or further back in the causal chain. Individual ENGOs may occupy more of these ideal-typical positions at different times and in different countries but tend to develop and rely upon certain metaphors in their argumentation. In general, the further from the top left position in Figure 6.1 one moves, the more likely the organisation is to question the viability of CCS as a climate solution.

Delimitation of 'CCS' (problem-definition)	**Locus of intervention against global warming (solution-definition)** Direct causes → *Emissions* *Growth/capitalism*	Root causes *Fossil fuel dependency*
CCS is a technology ↓	*Will it be accepted and funded?*	*Will it work?*
	CCS as a smart fix	'End of pipe' solution
	CCS as a 'bridge'	
CCS is a socio-economic arrangement	CCS as a regulatory regime *Will it be implemented safely, monitored properly?*	CCS as a greenwash *Transformative potential?*

FIGURE 6.1: Dimensions of the environmental NGO debate over CCS as a climate technology.

Key concerns about CCS are in italics and key metaphors are not italicized.

Strategies of legitimisation

Different framings of CCS appeal to various sources of authority and legitimisations to justify their different stances. At the same time all suggest uncertainty as a rationale for their adopted position in relation to CCS and refer to scientific authority, often IPCC judgements, for their positions and use the urgency of responding to climate change as a problem to justify their support or lack of support for CCS. However, this urgency is framed differently expressing uncertainty about different questions.

Sometimes, such as in material produced by NOAH and Greenpeace the uncertainty is about the effectiveness of CCS. Climate change must be tackled so urgently that CCS is deemed useless, even if it did technically work. 'Climate scientists warn global emissions must peak by 2015, just seven years away. CCS is unable to deliver the necessary greenhouse gas emission reductions to meet this goal' (Greenpeace International 2008: 18). CCS, although potentially a discrete technology, is considered to be far from being proven on a large scale and will 'arrive on the battlefield far too late to help the world avoid dangerous climate change' (ibid.: 17). For NOAH 'CCS is a technology that will not be able to deliver the reductions that are needed – and certainly not in the short term' (2009: 1). Greenpeace also relies on scientific or systematic evidence in refuting that CCS will be realised: 'A survey of 1000 "climate decision-makers and influencers" around the world reveals substantial doubt in the ability of CCS to deliver' (Greenpeace International 2008: 5).

At other times uncertainty about acceptance is the worry: will CCS be realised (in time)? (top left, Figure 6.1). In these cases, the urgency of the climate problem works in favour of CCS, though on a slightly longer time frame: 'it could take the edge off coal for the next 40 years' (Juniper 2011). Bellona takes a similar view, quoting the IPCC to argue that greenhouse gas emissions must be reduced by 50–80 per cent by 2050 to avoid dramatic consequences of global warming and citing a paper in the *International Journal of Greenhouse Gas Emissions* claiming that CCS could 'reduce global CO_2 emissions by one third by 2050' (Bellona 2007). Though not unconditionally in favour, Juniper asks 'what is the greater tragedy in this? We are looking at four degrees average warming with mass extinction of species and disruption of agricultural production, and we have to look at any practical way of preventing this' (Juniper 2011). This also places CCS within a wider set of problems than just global warming, including resource depletion, rising energy demand and the problems of alternatives such as nuclear power. The risk for the climate of leaving CCS unproven also figures in the debate. For example, a spokesperson for WWF Germany comments that 'it is not fruitful to vilify the technology [CCS] without testing it out and thereby carelessly wasting a chance to protect the climate' (WWF Germany 2008).

For others again, the uncertainty of the *net effect* of moving towards CCS is the main worry (bottom right, Figure 6.1). The main uncertainty pointed to is not so much to do with the technology itself, e.g. leakage (although that also figures), but whether CCS will in fact reduce CO_2 emissions because of the way it affects energy policy more broadly (diverting attention from renewables) and interacts with society in a wider way (prolonging fossil fuel use) rather than just whether storage sites will leak, for example. One interviewee stated that his organisation had let itself be 'guided by the science' on storage (which was deemed relatively safe) but that the business model and the way CCS encouraged no new economic structures or innovations was his main worry (interview – anonymous 2011). WWF, though noted as being one of the big environmental NGOs that is more welcoming of CCS, is wary of the technology for this reason – something they make plain in their otherwise cautious endorsement:

> [M]any observers fear that 'capture readiness' may be little more than a fig leaf that would open the door to a new generation of polluting coal stations while giving no assurance as to when – if ever – CCS will be fitted.
>
> (WWF Germany 2008: 3)

The German WWF website refers to CCS as potentially a 'Trojan horse' operated by energy companies: 'However the danger exists that energy companies would try to misuse CCS as a Trojan horse to achieve new coal fired power stations' (WWF Germany 2008, our translation[2]). Friends of the Earth (FoE), has differing opinions on CCS. While the UK branch is, like the WWF, cautiously approving, other national branches, like Denmark's 'NOAH' and Germany's 'BUND' are much more pessimistic about the technology. The Danish website, for example, features a CCS information page which is very critical of CCS, again featuring the prominent visual image of a smoking coal-fired power plant. Noah's overall position is that there are too many problems with CCS for it to be a solution, and that it is therefore 'contrary to the EU's and Denmark's policies for the environment' (NOAH 2009). Next to the other problems identified by the other opponents of CCS (risk of leakage, cost etc.), they identify that in particular

> The demand that new coal fired power plants have to be 'CCS-ready' will just open a huge door for even more coal fired plants. These plants may operate for many years without CCS while the technology is developed, storage sites are identified and financing is secured.
>
> (NOAH 2009: 1).

Similar to this is the uncertainty expressed about implementation (bottom right, Figure 6.1). The German branch of FoE, BUND, is wary of CCS, because it would effectively give a green light for the building of new coal plants without guaranteeing that there would necessarily be any added CCS.

> Most fossil fuelled power plants would be built within the next 15 to 20 years as a replacement for old power plants. During this time there would be no industrial-scale capture of CO_2 at power plants. These conventional new

> projects would however cement in a large amount of CO_2 emissions for decades – a significant reduction in CO_2 emissions in the energy sector would not be possible for a long time.
>
> (BUND 2006, our translation)

Though the UK branch is more welcoming of CCS, similar concerns are identified by their Welsh press office: 'The promise of fitting CCS in the future cannot be an excuse to build new coal burning power stations today' (FoE Cymru 2010) and therefore, among other reasons, '[c]oal use must be dramatically reduced everywhere if we are to stand a chance of overcoming the immense threat of climate change' (ibid.).

This risk of CCS failing to reduce CO_2 emissions and perpetuating the burning of coal make even relatively positive green NGOs nervous about endorsing CCS too wholeheartedly. Even Green Alliance emphasises that 'capture ready' is not sufficient and that although research and development must go ahead, it must not lead to an increase in fossil fuel use or 'unabated coal' stations.

Several NGOs point to CCS as a possible 'bridge' to a renewable energy future but emphasise that to be acceptable in these terms, CCS has to go hand in hand with a concerted effort to build the other shore of the river – the renewable future. Thus for WWF Germany 'CCS could play an important role as a transitional technology within the framework of an ambitious climate protection policy' (WWF Germany 2008, our translation). WWF takes this stance too as did an anonymous green NGO figure interviewed. Tony Juniper, formerly director of FoE, saw it in these terms as a necessary evil to get past a particularly dangerous juncture involving a climate and an energy 'crunch' but does not deny that a long term solution must be found, nor that CCS has other drawbacks associated with recovering and burning fossil fuels.

Another possible advantage associated with CCS (demonstration projects) is the potential for job creation and the possible export of technology and know how. The Green Alliance, with its close ties to industry and government (perhaps not unequivocally an 'NGO'), makes the clearest statement on this: CCS 'could provide the means to develop supply chains and infrastructure hubs; and provide the means to create the new green collar jobs our political leaders so frequently highlight' (Green Alliance 2008: 5). For Climate Camp, however, this is part of the problem: that CCS is tied to large corporate interests who ultimately have no interest in climate change. Mostly, however, despite also being linked to local pollution issues and wider problems associated with burning fossil fuels, such as mining, CCS is seen overwhelmingly in terms of its effect (for or against) solutions to climate change.

In sum, positions on CCS are legitimised almost entirely through references to climate change and rely heavily on scientific sources of authority and different types of uncertainty. The main risks considered are not related to natural science, but rather to business models; and political dynamics to do with implementation, regulation and economic incentives and they thus represent 'deeper' uncertainties which are not easily addressed by closing a knowledge gap through more technical information about the safety/risks of CCS. Perhaps with the exception of Greenpeace, a surprisingly small amount of time is spent on the technical risks of actually capturing, transporting and storing CO_2, while the 'Trojan horse' argument of CCS leading to more fossil fuel emissions is widespread. As a specifically climate-related technology, CCS is vulnerable to arguments claiming it has a negative impact on climate change. Actors who concentrate on these 'societal risks' of CCS backfiring rather than more technical or quantifiable risks such as those associated with storage of CO_2 are therefore able to take a precautionary approach to CCS even as they rely on scientific discourse and avoid a direct challenge to geologists' claims that CCS is 'safe'. Paradoxically, much communication material promoting CCS focuses on the engineering risks rather than the societal ones (Corry and Reiner 2011b).

Conclusion

ENGOs are positioned to have a significant impact on public policy towards CCS yet analysis of their positions has been overshadowed in the literature by public opinion and local resident acceptance studies. Also, acceptance of CCS has been discussed largely in terms of the level and quality of information supplied about the technology to (generally sceptical) public opinion. This information-gap debate has often assumed a rationalist model of politics whereby solutions are assessed in terms of essentially neutral information concerning criteria related to whether they work or help achieve goals to predefined political problems. In contrast, this chapter has sought to show that different political rationalities are involved in the policy debate about climate change, including CCS, that goes beyond stringent assessments of cost or risk and concern the discursive construction of the problems, solutions and justifications for a greater variety of positions that simply 'for or against' CCS. What is actually meant by 'CCS' differs and the nature of the problem and its causation is interpreted quite differently, even within the ENGO community.

Some general patterns also emerged. Those who chose the point of emissions of greenhouse gases into the atmosphere (the top of the smokestack, so to speak) as the relevant point for political intervention tended to be keener on CCS than those who saw fossil fuels or more fundamental socio-economic structures as the real cause of climate change. The former worried mostly about public acceptance when CCS was seen as a smart fix or the effective implementation of it when CCS was basically seen as regulatory regime. Those who chose to see causation in terms of the underlying drivers of the burning of (unabated) fossil fuel were generally less supportive of CCS, doubting that it had a net mitigating effect either because it was 'end of pipe' which is ineffective or 'greenwash' which is directly counterproductive. The varying stances towards CCS are thus underlined by varying interpretations of the problems and solution, and these lead to different conceptualisations of its general suitability.

The metaphor of CCS as a 'bridge' to a different energy future is interesting here because it effectively combines the 'smart fix' idea with the idea of societal transformation, sidestepping the 'end of pipe' objection (the problem will be solved eventually, though not directly by CCS). The main challenge for CCS in terms of this metaphor is whether CCS-the-regulatory-regime will be resisted by the public, industry or politicians, rendering CCS ultimately a 'greenwash' that hinders root cause-transformation while facilitating more fossil fuel consumption.

The politics of CCS – and by extension other emerging technologies – is thus about more than technology and narrowly technological risks, e.g. of leakage. It is also about analyses of economic systems, social change, claims about causality, conceptualisations of uncertainties and strategies of legitimation and how these result in different interpretations of the motives of companies, governments and other actors. CCS communications efforts concentrate heavily on explaining technological processes and principles but generally skirt around deeper questions e.g. of implementation and the dilemma of root causes versus direct causes that runs through the heart of the debate (Corry and Reiner 2011b). This means that the technological debate is as yet poorly aligned with the political debate about CCS going on amongst ENGOs – crucial stakeholders for the future of CCS.

Notes

1 It can be difficult to draw the line between state and non-state and non-corporate since funding and personnel often cross the boundary lines. However, along with institutional independence, we let organisational self-description be decisive, and hence allow e.g. Green Alliance to be classed as an ENGO, despite close links to business and state institutions.

2 Allerdings bestehe die Gefahr, dass die Energiekonzerne versuchen, CCS als trojanisches Pferd zu missbrauchen, um neue Kohlekraftwerke durchzusetzen.

PART II
Governance

7

INTRODUCTION TO PART II

Governance

Simon Shackley

Topic/scope

The term governance is used to denote an extension in decision-making beyond the narrow confines of 'government' (political appointees plus civil servants and their technocratic agencies, advisory bodies and 'quangos') to embrace also industry, non-government organisations (NGOs), professional groups (e.g. scientific and technical societies) and the wider lay public, or representatives of this wider public. Use of the term governance is also meant to capture a transition from a 'command and control' state to a more collaborative, less powerful ('hollowed-out') one.

Governance is clearly a very large topic and highly heterogeneous (e.g. distinct systems of governance exist across nation states and there are, furthermore, international systems of governance which (in some accounts at least) are semi-autonomous from national interests). A classic example is the European institutions, which form a super-national level of governance that interacts in complex and variable ways with the 27 Member States of the EU. There are also complex relationships between the nation state and the sub-national regions or states, depending upon the country. Clearly, federal systems have more decentralised decision-making in which the states have more powers and where negotiation between the federal and state level becomes an important feature of policy making. Classic examples are the USA and Germany.

The CCS law being negotiated in Germany during 2010–2011 is a good case in point since much of the political debate has revolved around the decision-making powers of the *Bundesländer* (or states) and the federal level. Several *Länder* maintained that they should have veto powers over approval of CCS projects within their territory and this view won out in the law proposed in 2011. This distribution of power has major consequences for CCS in Germany since the two *Länder* where there are more abundant opportunities for geological storage (Schleswig-Holstein and Niedersachsen (Lower Saxony)) appear to be adopting a more negative perspective on CCS projects. Without their participation, it is not clear whether CCS in Germany utilising reservoirs below the land surface could really take off. In September 2011, the *Bundesrat*, the upper house of parliament in Germany which is made up of representatives of the *Länder*, rejected the proposed law. As a further example, the United States Environmental Protection Agency (EPA) has also been formulating a position on storage permits and liability, separately from negotiations taking place on State level throughout the USA.

Governance is also subject to changes over time due to a whole set of socio-economic, political and environmental processes. These changes vary from the reasonably straightforward and semi-predictable, such as the fact of an aging population in certain countries due to demographics, to the completely unpredictable and contingent. Highly contingent processes of change include elections, financial crises, wars, insurrections, assassinations and resignations. Somewhat less contingent, but still far from predictable, changes include the vast majority of events and processes which clutter the everyday life of policy and 'politicking'.

This part covers only some selected aspects of governance at the international scale (Stephens and Liu in Chapter 8, Shackley and Evar in Chapter 9) and, in more detail, within a selection of countries (Evar and Shackley, Chapter 10). The main countries covered are the USA, the UK and Norway, with briefer mentions of the Netherlands, Australia, Germany and a few other countries. The scope is therefore selective, but does include in-depth empirical analysis of the policy and political processes and dynamics at the international level and in a few of the countries that have been important to the emergence of CCS to date.

Relevance to non-academics

We can perhaps distinguish between two types of relevance to non-academic audiences: instrumental and interpretative. Instrumental relevance refers to the achievement of goals and agendas defined by policy- and decision-makers. The more common tradition within the social sciences is the interpretative one, whereby social science theory and methods are used to address and better understand 'how' and 'why' questions: why particular decisions are taken; why certain policy actors adopt the positions that they do; how and why negotiations proceed in a particular way; and so forth. Ultimately, the aim of interpretative social sciences is to test and advance social theory.

The main question being addressed from the instrumental perspective is: 'how does the research contribute to effective achievement of goals and objectives?' (where those objectives are considered to be pre-determined). The usual perspective from which to advance an answer to the instrumental question is economics. Hence, we can use economics to investigate the most cost-effective (efficient) way of reducing a given quantity of carbon dioxide (equivalent) emissions, allowing the role of different low-carbon energy technologies (including CCS) to be explored (alongside renewables, nuclear, demand reduction and so on). Such work has been

undertaken by the likes of the IEA (international scale), EU Commission (EU-level) and, at the national scale (e.g. the UK Energy Research Centre modelling) (e.g. IEA 2011b; EC 2011b; Strachan 2011).

More specifically, the instrumental approach is also useful for devising the 'optimal' (i.e. cheapest or otherwise utility maximising) way for matching-up point sources of CO_2 with potential geological sinks, given a variety of pre-identified constraints (e.g. on pipeline route options) (e.g. Kemp and Kasim 2010). A commonly used tool in such analyses is linear optimisation, whereby an optimal solution between several competing objectives (usually cost minimisation given a predefined set of constraints) is sought. This is, for example, the basis of the MARKAL energy system modelling that has been applied to identification of an appropriate mix of low-carbon energy technologies given differential costs and cost-learning curves. Instrumental economic approaches have been covered well elsewhere (e.g. Finkenrath 2011; ZEP 2011b) and will not be considered in detail any further here. While economics is the most common amongst instrumental approaches, there are several traditions within policy sciences, which grapple with the question of how decisions should be made such that better outcomes result. We will explore one of the main approaches here in the subsequent theoretical section.

Turning to the interpretative social sciences, these are usually rooted in particular theoretical traditions, to be discussed below. Policymakers and other potential 'users' of social sciences frequently find it hard to engage with less instrumental approaches because it is not always clear how to 'use' the insights that are presented in the same way that, say, economic understanding can be 'applied'. Yet, to entertain such a literal understanding of 'usefulness' would be to impoverish the participants of policymaking and to reduce policy to a simplistic and mechanistic exercise of evaluating and selecting options. This is what could be termed the 'McKinsey fallacy', after the 'marginal abatement cost curve' (MACC) that the consultancy firm has made famous. The fallacy is the notion that we can realistically and sensibly plot all carbon abatement technologies on a graph, indicating the level of carbon abatement against cost per tonne abatement for a given option. The implication that this is not as straightforward as it at first appears is founded on a recognition of uncertainty, indeterminacy and interdependency. The assumption of a MACC is that the information being plotted is sufficiently certain to permit ranking, but if the modelling uncertainty is larger than the cost difference between technologies/options, then the ranking is not robust. In addition, a MACC only models monetary resource requirements, or assumes that most limitations are reducible to a monetary value. This is often not the case because policy inertia, successful innovation, political manoeuvring and public acceptance cannot be priced into such models. A yet further problem is that options are frequently interdependent, not independent – which is really the issue of lock-in that is discussed in Stephens and Liu (Chapter 8), introduction to Part III on innovation (Chapter 11) and elsewhere in the literature (e.g. Shackley and Thompson 2012; Markusson and Haszeldine 2010). Box 7.1 makes the case that utilitarian social sciences cannot provide 'all the answers', at least not without causing policy makers problems.

BOX 7.1 THE LIMITATIONS OF UTILITARIANISM EXPANDED

Rather than providing the policymaker with a clear 'answer', the simplistic MACC comparison could actually be misleading and result (in retrospect) in the wrong technologies/options being prioritised. What is more, not all climate change mitigation options are known at a given point in time and some cut-off has to be made regarding what is 'in' and what is 'out'. Yet, what turns out to be the most important option at a later stage may not even end up being included in the MACC curve, because only 'accepted' options are presented.

Much energy system modelling is subject to problems similar to the McKinsey fallacy. Linear optimisation methods such as MARKAL are frequently prone to sometimes dramatic 'switching-behaviour'. This is a function of the relative prices of different energy technologies (or of technology-learning effects) that are inputs to the models. Even a small cost margin between options can have a large effect on the apparent preference for one technology over another. Such switching is far more likely to be an artefact of the model than a robust representation of what will happen in reality. The source of this fallacy is the concept that tools designed for a reasonably straightforward, predictable system (say a jet engine) can be applied with confidence or robustness to a system, such as the energy one (a tightly coupled amalgam of technologies, institutions, social practices, regulations, user expectations etc.), which is highly complex.

In reality, policymakers are agents within a complex system, where information is treated as subject to multiple levels of uncertainty and contingency and is generally highly incomplete. In practice the numbers provided by researchers will likely not be 'believed' and acted upon. Therefore, the instrumental version of knowledge for policy is frequently over-simplistic, though it may be maintained and supported by policymakers due to the manifest and latent functions of apparently quantitative foundations of policy decisions (Ezrahi 1990; Porter 1995).

Linear cause-effect is not the modus operandi of operations in complex systems. Instead, effective policymakers use reflexivity in practice, whether or not they choose to theorise or even acknowledge it. Reflexivity is a hotly contested term (e.g. Beck *et al.* 1994) but is here (more conservatively) defined as disengaging from the immediate requirements of practice and action and applying a critical stance towards one's actions and underpinning assumptions: i.e. it is the 'second-order' learning that enables us to critically formulate and evaluate strategy and tactics and, at a deeper level, wider cultural, political and technological trends and tensions.

A legitimate role of social science enquiry is precisely to ask questions about how and by whom knowledge for policy is produced and used, and with what effects. The relations between knowledge-producers and policymakers reveal much about the policymaking process itself (e.g. its underpinning practices and assumptions). Another rich line of enquiry is to interrogate the underpinning dynamics of policymaking. How is a particular policy frame produced, by whom, why and with what effects? What is the relationship between scientific discourse and development of policy positions?

How can policymakers make use of interpretative social science?

It is often not possible for policymakers to 'use' the insights that derive from interpretative social science research in a linear fashion to make 'better decisions'. It is also unlikely that seasoned policymakers will learn much about the 'ins and outs' of political deal-making from this part of the book, though it may at least provide useful background for the less-experienced policymaker.

One common strategy on the part of policymakers, for example, is to apply a high level of scepticism towards information and perspectives provided by external scientists, industry, enthusiasts, environmental groups, consultants and so on. Only those ideas which gain purchase with a sufficiently wide and influential group of stakeholders, and which can also be backed-up by a reasonable evidence basis and narrative, will survive. This is akin to the evolutionary theory of variation and selection, whereby 'fit' between the policy idea and the receiving, selection environment of the policy domain will determine whether the idea carries forward or is abandoned. A case in point is the shift that has been underway in the past several years from a strong focus on 'market-based mechanisms' as the answer to environmental (and other) problems to more state-interventionist approaches. This political shift is changing the receiving context for policy ideas, mechanisms and measures. To summarise, the real value of this part of the book lies in its ability to promote and stimulate reflexivity in the policy process, however un-measurable and intangible this may be in comparison with guidance figures produced by economic modelling.

Theoretical perspectives on governance which focus on understanding past and present political processes

As yet, relatively little has been written by academics on the governance of CCS, even though policy and governance has been the main preoccupation of many CCS industry and government stakeholders. An exception is the edited collection *Caching the Carbon* (Meadowcroft and Langhelle 2009a) which draws upon the political and policy sciences. The main theoretical approaches adopted in that book are the 'interests model', with some use also of coalitions models. (For a glossary see Appendix 7.1 to this chapter).

The notion of decision-making arising from competing political interests is still widely followed by many expert social scientists and commentators (journalists, participants in policymaking or evaluating, etc.). From this perspective, those with power, who are able and skilled at exerting it, will go head-to-head in a political contestation over how policy is formulated. The result of such manoeuvrings will be influenced by the capability of actors to enrol others to their cause as well as by external events beyond the control of the political actors. The 'interests model' is the one most frequently deployed in understanding the politics and policy surrounding CCS. Meadowcroft and Langhelle (2009a) apply an understanding of economic and/or political interests in interpreting different national debates on CCS. The fact that some countries have a large indigenous fossil fuel resource, for example, is assumed to be an important factor in shaping the politics of CCS within those nations. For nations with large fossil fuel reserves, such as Australia, Canada, the USA and South Africa, CCS becomes more important from an energy security perspective and/or as a defensive posture vis-à-vis potential for deep carbon cuts in the future in domestic or export markets.

Other interests include concerns about energy security, opportunities vis-à-vis technological and skills capabilities, the strength of the climate change lobby and of the renewable energy lobby, etc. A further example from Meadowcroft and Langhelle (2009a) is the suggestion that those countries with a stronger renewables sector, such as Germany and the UK, have a more powerful renewables lobby which is able to counter arguments from the CCS lobby. Hence, it is proposed that an explicitly negative CCS perspective will be stronger where renewables represent a larger percentage of the energy sector.

In the field of international relations there are two versions of the interests model: the 'realist' account, which maintains that countries will define, and be primarily motivated by, their self-interest (e.g. for wealth, power, influence etc.); and the 'liberal-institutional' account, which is concerned with the demand for international organisations and regime (defined as a 'set of implicit and explicit principles, norms, rules, decision-making procedures around which actor expectations converge in a given issue area' (Krasner (1983), quoted in de Coninck and Bäckstrand 2011: 370) and regime complexes. Regime and regime complexes emerge because international institutions can resolve certain problems more effectively, efficiently and/or legitimately than nation states acting in isolation. This book, though, does not cover CCS from the international relations perspective and the reader is referred to de Coninck and Bäckstrand (2011) for a useful timely review.

Obtaining robust evidence in support of such interest models is frequently difficult. While large fossil fuel reserves might help to explain the enthusiastic support for CCS in some countries (Canada, Australia, USA), it cannot be used to explain support for CCS in countries with relatively small and dwindling fossil fuel supplies, such as the UK, Germany and France (though these countries are characterised by Meadowcroft and Langhelle (2009a) as engaging with CCS somewhat later than the aforementioned). It also does not explain why some countries with large fossil fuel supplies, such as Russia, Ukraine and Poland are not more enthusiastic about CCS. Even if they are not UNFCCC Annex 1 countries, their fossil fuel export markets could be affected by tough carbon cuts in key export markets. In a comparison of the opinions of stakeholders within different EU countries, the existence of a fossil fuel sector did help to sway opinion on CCS more favourably, though the differences were not large (Shackley *et al.* 2007). That survey also found a statistically significant correlation that 'respondents from countries with less penetration of renewable energy tended to perceive a greater need for CCS in their own country, and vice versa' (Shackley *et al.* 2007: 5106), supporting an interests model that the presence of a stronger renewables lobby results in relatively less support for CCS. Yet, the size of the renewables sector is still relatively small, even in Germany, while Bellona, the most pro-CCS NGO in Europe, is from Norway, a country that relies almost entirely on renewable energy for its domestic electricity needs. It appears that it is entirely feasible for key stakeholders to support both CCS and renewables as key technologies for the transition to low-carbon economy, as evidenced in Scotland, Norway and Denmark for example.

Other correlations identified by Shackley *et al.* (2007) include greater support for CCS in richer countries, from countries with an oil and gas sector and easy access to the North Sea. There was, however, no clear correlation between support for CCS and the difficulty of meeting Kyoto targets (which might be regarded as a further 'interest' to aim to satisfy, especially given the expectation at the time of the survey (2006) that more challenging post-Kyoto targets would be set) (Shackley *et al.* 2007).

The critique of the 'interests model' has been mounted by social scientists who have argued that interests cannot be taken as a pre-given but are actively produced through social processes and negotiation – hence are potentially more flexible and adaptable than allowed for by the interests model. Furthermore, pre-existing interests cannot be assumed according to this school of thought; rather, researchers have to demonstrate and provide evidence for their existence. At the very least, interests models need to be interrogated carefully to ensure that there is not simply a selection of evidence that supports a particular interpretation of what the interests are, while ignoring equally legitimate but different interpretations of those interests. In reality, policymaking and politics are confronted with different understandings of interests and, while a clear 'national interest' may sometimes emerge as the 'winner' for a time, it is never set in stone and can always be (and usually is) subject to renegotiation at some stage.

Alternatives to interest-based accounts tend to follow some variation of (social) constructivism which is the position that society and social reality emerge from the bottom up through endless processes of interaction and negotiation. It is only in and through these micro-encounters that 'social reality' has any meaning. To the extent that pre-existing 'structures' and 'frameworks' have any meaning, it is through how understanding and beliefs about such structures on the part of social actor enter into the micro-social processes of interaction which produce reality (a very different thing from claiming that 'structures' have no relevance). From this perspective, explanation cannot rely upon the existence of assumed powers, interests, values or other abstract meso- or macro-level concepts unless it can be demonstrated that these are being substantiated during social interaction. This imposes a fairly formidable methodological constraint on constructivist accounts since collection of empirical data of sufficient quality and volume is very resource-intensive and frequently impossible. In reality, constructivist-influenced accounts are frequently a hybrid including elements of an interests-based explanation.

One of the earliest constructivist-influenced accounts was Peter Haas's concept of an 'epistemic community' (Haas 1990, 1992) to describe a broad coalition of actors including scientists, government and other public sector officials, and politicians who come to share a common interpretation of 'the problem' and of the broad policy and political requirements in response. As the name implies, epistemic communities are knowledge-based and, according to Haas (1992), the participants have four features in common, namely a shared set of: (1) normative and principled beliefs; (2) causal beliefs; (3) validity standards; and (4) a common policy enterprise. It has been argued that the Intergovernmental Panel on Climate Change (IPCC) represents an epistemic community, since it is a transnational network of scientists and associated knowledge-brokers which claims political authority and authenticity through its recourse to the best global science and is motivated by shared causal and normative beliefs and a common policy enterprise (i.e. to alleviate anthropogenic climate change) (Bulkeley and Newell 2010; Shackley 1996).

In Chapter 8 of this book, Stephens and Liu advance an account of a CCS epistemic community and provide quantitative and qualitative information which builds their case. Other contributions invoke the importance of the IPCC's Special Report on Carbon Dioxide Capture and Storage (2005) as 'the' authoritative voice of science that changed perceptions of CCS, raising its profile and getting the topic taken more seriously (de Coninck and Bäckstrand 2011; Meadowcroft and Langhelle 2009a). According to one set of authors, the IPCC's 2005 report:

> represented the culmination of a discursive *rapprochement* between international energy and climate policy circles, with the fossil energy sector coming to acknowledge more explicitly the dangers of climate change while the climate community accepted CCS as a significant abatement option (Hellin 2008).
>
> (Meadowcroft and Langhelle 2009a: 267–268)

The CCS field has generally been characterised by a lack of peer-reviewed journal articles in the conventional engineering and scientific journals (Page *et al.* 2009), in part perhaps because of the commercial-in-confidence character of some of the underpinning data, analytical and modelling tools. The IPCC Special Report helped to bring academic scientific respectability to CCS. It can, however, be questioned whether techno-scientific knowledge per se has been the *key* driving force behind emergence of the CCS epistemic community. For example, it is questionable whether the energy sector, and its peak representative organisations such as the International Energy Agency (IEA), came to adopt a more positive position on CCS primarily as a consequence of the IPCC's special report rather than as a result of multiple, intersecting political, economic, policy and commercial considerations.

Another influential theoretical approach to emerge out of the dissatisfaction with traditional interest models has been the concept of 'advocacy coalitions' (Sabatier, Jenkins-Smith etc.). Pollak *et al.* (2011) describe the advocacy coalitions framework (ACF) as follows:

> The ACF defines the process of policy change as a function of the interaction between powerful coalitions composed of actors from different institutions within a certain geographic area who are interested in a specific policy arena and share key beliefs (Sabatier and Weible 2007). It holds that the most useful unit of analysis to understand the process of policy change is the policy subsystem comprised of the coalitions interested in a specific policy area (Sabatier 1988). Policy subsystems are of interest because they represent the locus of policy conflict, policy learning, and policy change. Rather than assuming that actors' orientations and preferences are primarily a function of the institution with which they are associated (Scharpf 1997), ACF theory postulates that actors form coalitions based on their policy beliefs and that the success of these coalitions in getting components of their belief systems translated into actual policy is a function of resource availability, policy-oriented learning and external shocks (Sabatier and Weible 2007).
>
> (Pollak *et al.* 2011: 314)

The ACF proposes the existence of three levels of beliefs: 'deep core beliefs' which are non-negotiable and critical to the existence of the coalition; 'policy core beliefs', which are applications of the deep core beliefs to some policy subsystem; and 'secondary beliefs' which are less normative, more specific to narrower issues and subject to change from external events and/or learning within the policy subsystem. It follows that these subsidiary elements can be fairly readily changed and discarded if need be, but the core beliefs need to be maintained or else the entire programme is threatened. (In making such distinctions between belief-levels, the 'advocacy coalitions' framework draws heavily upon social studies of science, in particular Lakatos and Kuhn) (for more information on whom, see Chalmers 1999.)

A good application of the ACF in the CCS field is the analysis by (Pollak *et al.* 2011) of two distinct coalitions in the US. The context is that CO_2 has been used for 35 years in the USA for Enhanced Oil Recovery (EOR) and that this practice is increasing as oil becomes ever more valuable. The 'energy coalition' has focused on preserving the status quo surrounding CO_2-EOR, meaning that the requirements for monitoring and verification and for well construction and completion are much less onerous that is the case for geological CO_2 storage where the main motivation is to avoid anthropogenic climate change – as advocated by the alternative 'climate coalition'. The motivation of the energy coalition has been defensive – as the climate change issue grew in importance in the US, the coalition wished to ensure that CCS would not come to disrupt existing CO_2-EOR operations and that operators might even benefit from carbon credits but without the onerous requirements implied by the climate coalition (e.g. as represented in the EU CCS Directive). Pollak *et al.* (2011) argue that the energy coalition has, to date, broadly won out over the climate coalition in the arguments over the perceived role of CCS, i.e. as more of an adjunct to the existing oil industry than as the start of a new CO_2 storage activity driven by the anthropogenic climate change motivation to avoid introducing CO_2 to the atmosphere.

It is useful to ask whether the traditional interests-model might have equally well accounted for the difference between these two coalitions. Clearly the CO_2-EOR community has a fairly well-defined interest in trying to maintain 'business as usual' as far as possible, while those who define their interest as 'combating anthropogenic climate change' would be pushing for something more ambitious. To resolve this question, it would be necessary to turn to the detailed methods and evidence that was deployed in a given account and to ask the question: to what extent have coalitions been demonstrated from empirical evidence according to norms in the field?

Hajer (1995) took the ideas of advocacy coalitions in a somewhat different direction in his notion of a 'discourse coalition'. Notable in Hajer's account is the role of scientifically based 'storylines' as the glue that holds coalitions together. As with advocacy coalitions, an important feature of discourse coalitions is that the core storyline is not overly complex or detailed. The detail behind positions comes in the surrounding auxiliary beliefs and assumptions. It would not be possible for broad coalitions to emerge or survive were the storylines to become too complex and detailed as it is very unlikely that a wide array of different policy actors could ever agree on the specifics.

In Hajer's (1995) seminal study, the role of sulphur dioxide and nitrous oxides in causing acidification of ecosystems was the key storyline which held together a wide array of policy stakeholders. In the case of CCS, we can identify a number of interlinked elements to the storyline. First, that anthropogenic climate change is a real problem; second, and consequently, deep cuts in greenhouse gas emissions, in particular CO_2, are required. The third element is the belief that CCS can make a significant and broadly positive contribution to the required deep cuts in carbon emissions, both for existing plant and future fossil fuel combustion. This implies a belief that CCS can be undertaken cost effectively and without incurring unacceptable health, safety or environmental risks (according to the coalition's own understanding of what is an acceptable risk). For a discourse coalition to operate effectively, it is not necessary or indeed possible for the precise interpretation of terms such as 'unacceptable risks' to be formulated and agreed. Some policy actors will consider that some environmental risks incurred by CCS are acceptable because of the 'greater net benefit' of CO_2 storage compared to emission to atmosphere. Other policy actors will adopt a more precautionary approach whereby only a very low level of leakage of stored CO_2 would be regarded as acceptable.

Provided that these different actors do not delve into the detail of exactly what level of precaution should be adopted in considering CO_2 leakage (hence do not aim to define the detailed meaning of 'unacceptable risks') they can still maintain the simple belief that CCS development should not result in any unacceptable risks.

As Meadowcroft and Langhelle (2009a) note, there is also a negative storyline regarding CCS adopted by some environmental groups and possibly also by advocates of other low-carbon technologies. Examples are the 2008 report *False Hope* by Greenpeace International which criticised CCS in terms of being: (a) an untested technology; (b) likely to be very costly; (c) likely to give rise to a set of unknown and unacceptable risks and hazards; and (d) detracting policy and funding attention from other low-carbon technologies which require support, especially renewables (the so-called 'moral hazards' argument). Surprisingly, not dissimilar arguments were deployed in two articles that were published by *The Economist* in 2009 (*Economist* 2009a, 2009b). The main point being made in these *The Economist* articles was that the technology remains unproven at scale on coal power plants and that there has been a lot of optimism on costs which cannot presently be validated and hence justified. A few academics have presented more scholarly critiques using similar arguments (Tsouris *et al.* 2010), but also questioning whether the analysis of the size of the energy penalty in existing engineering R&D has been overly-optimistic (Page *et al.* 2009). Part of the motivation behind the *Economist* articles appears to be suspicion of the need for expensive direct subsidies to the coal sector in order to pay for CCS and a strong preference for market-driven solutions using a carbon price. This is not to suggest that there is, as yet, a discourse coalition that is explicitly opposed to CCS, since there are clearly large differences between environmental NGOs and the interests represented by right-of-centre editorial writers for political magazines.

The theoretical approaches briefly reviewed here should encourage reflexivity on the part of policymakers by – hopefully – stimulating interest in understanding how and why certain outcomes are reached and why CCS is perceived in a particular fashion by different policy actors. We now turn to a different, perhaps more instrumental, theoretical tradition in the social sciences that asks how decision-making on new technologies can be improved.

Theoretical perspectives on governance which focus on understanding how to improve political processes

In a book entitled *The Social Control of Technology*, written three decades ago, David Collingridge (1982) posed a key challenge in knowing how to govern new technologies. The 'control dilemma' notes that before a new technology is reasonably well developed, its impacts are not known with sufficient precision for any negative impacts to be properly understood, assessed and regulated. Yet, without such regulation early on in its development, a new technology will (if successful) be deployed rapidly and become difficult to regulate once its impacts are well understood. Collingridge argues that the temptation to employ a predictive approach to *anticipate* the impacts of a new technology should be resisted because such impacts are largely unpredictable. Just like attempting to predict the oil price or economic growth is liable to be a fruitless enterprise, so are attempts to predict the socio-economic, and even environmental, consequences of new technologies. Collingridge proposes what he terms a 'fallibilist' approach instead – this involves a process of learning-by-doing and trial-and-error. Small-scale experiments with new technologies should be undertaken and an effort made to learn as much as possible from such interventions. This learning can then be put to good effect in designing the next stage of technology deployment. Clearly, not all technologies would lend themselves to this type of intervention. Even small-scale experimentation with nuclear technologies could have very serious impacts on human health and safety. For this reason, Collingridge proposes that technology selection should take account of 'error-costs' – that is the costs if a wrong decision is made regarding a technology. Finally, he presents specific indicators by which to measure learning potential and error-cost.

Collingridge's work aims to shift the debate on decision-making away from technology assessment in the conventional sense (i.e. asking 'what are the impacts of this new technology') to technology governance (i.e. asking 'how do we control development of this new technology and how do we acquire the necessary information and understanding by which to do so?'). This places the issue squarely in the 'governance' camp. The two approaches – anticipative and fallibilist – map closely to two of the key traditions within public policy analysis: the 'synoptic rationalist' versus the 'disjointed incrementalist' approach (Ham and Hill 1984). From observing attempts to plan socio-economic development, some policy analysts realised early on that prediction is impossible. Even (or especially?) where elaborate new tools such as computer models and large data sets were employed, enormous errors cropped-up in predictive forecasting (Shackley 1998).

Over 50 years ago, Charles Lindblom (1959, 1979) led the critique on rationalism in his seminal article 'The Politics of Muddling Through' in which he advocates an incrementalist approach to decision-making. Incrementalism implies reduced error-costs as smaller steps are taken, less is risked prior to some intermediate outcome being achieved, at which juncture the wisdom of the change or investment can be evaluated before more change or investment is undertaken. This allows political consensus to be more readily forged between disparate actors. More radical change instituted all at once, by contrast, encounters more opposition and often fails or is scaled back. Radical changes also have a tendency to go wrong because they presuppose knowledge of impacts, which is impossible to obtain and therefore typically encounter unanticipated effects with adverse consequences for development. The introduction of new technology appears to be little different from other policymaking activities in these regards, except for the enhanced risk of technological lock-in. Two obvious criticisms of Collingridge and other incrementalists are the following:

1. Their ideas and prescriptions do not always fit comfortably with observed processes of social and political change and their underpinning dynamics (e.g. politicians frequently want to get behind 'big radical projects' for reasons of acquiring political capital and status that is less driven by asking the question, 'will it work and be a sensible investment of limited resources?)
2. Their ideas are too closely tied to maintenance of the status quo and preclude the more radical changes which some argue are necessary to overcome the climate change conundrum.

Theorists of change and complex systems have observed how system equilibria are sometimes disrupted by unpredictable change – socio-economic, political, technological, environmental, and frequently some interaction of all these (e.g. Diamond 2006). Such change may sometimes result in system collapse and/or the emergence of a new (temporarily) stable equilibrium. Thus movement from one stable configuration to another is driven by high instability. Such ideas have a close conceptual relationship to the ideas of 'punctuated equilibria' in evolutionary theory. Within ecological and complexity science, a noteworthy application of these ideas is found in the theory of 'Panarchy' (Gunderson and Holling 2002). This work posits that system change across ecological, economic, political and social structures generally can be understood as a cyclical process. The Panarchy model proposes that hierarchical, and apparently stable, structures collapse through a process of (relatively) sudden energy and materials release that leads to a period of wide resource exploitation as new niches form, and selectivity and system complexity again follows. Complexity might increase system resilience to *known* ranges of change and uncertainty, but conversely increase vulnerability where those changes lie outside of such assumed ranges.

Incrementalists might argue that their principal interest is not in *explaining* technological and socio-economic change, but in addressing the question of *how* to govern such change. Even if major change does occur through unexpected, non-linear, abrupt processes, this does not mean that policymakers are in a position to make their decisions following such precepts. By adopting incremental-type decisions, the policy space is, arguably, left sufficiently open for the influence of unexpected, non-linear processes to flourish – or possibly (to continue) to cause serious problems. The incrementalists' notion of 'good governance' is, consequently, somewhat limited. It is implicitly assumed that the best advice that can be given to the vast majority of policymakers in the vast majority of circumstances is for decisions to be taken incrementally to allow flexibility (though precisely what is meant by 'incrementally' could be open to different interpretations). Chapter 10 in this book, by Evar and Shackley, is a good example of the application of neo-incrementalist thinking to a real-world CCS challenge.

A somewhat more ambitious position is adopted by the advocates of 'clumsy decision-making' or 'clumsy governance' (Verweij and Thompson 2006). This idea is rooted in both the theory of socio-cultural variability (sometimes simply called cultural theory) (Thompson *et al.* 1990) and the theory of complex systems.

It maintains that there are irreconcilable culturally driven perspectives in policymaking on complex issues – hence a rationalist position of consensus cannot usually be produced and we should be suspicious where this appears to have emerged. Does this therefore consign governance to an unmanageable chaotic or capricious fate, prone to the worst vagaries of political or economic interests? Or maybe prone to cultural fads, whether that be individualism, hierarchy or some other flavour?

The theory of socio-cultural variability proposes that there are four culturally distinctive positions, biases or tendencies which continually interact with one another (individualist, hierarchist, egalitarian and fatalist). From such dynamic interplay emerges a more systemically beneficial result and decision-making: *Divided We Stand* goes the title of one influential text (Schwarz and Thompson 1990). The proponents of clumsy governance argue that by ensuring that all four cultural tendencies are involved in decision-making, a better set of decisions will tend to be the outcome. Exactly how clumsy governance should work remains to be properly worked out and remains contested (Verweij and Thompson 2006). Some argue that each of the cultural tendencies needs to fathom out its own perspective (i.e. internal dialogue first) and that there is then a 'battle of the perspectives' through a political process to decide a policy outcome (i.e. through good old-fashioned politics). In other accounts it has been suggested that a more explicit and deliberate interaction between the cultural perspectives through a structured political process is what is required (i.e. external dialogue between the cultural biases). This disparity in accounts of how 'to do' clumsiness means that there is currently no straightforward translation of the theory into practice in the specific case of CCS.

The contributions of the chapters in this Part

Stephens and Liu provide a comprehensive account of how the CCS community has emerged over the past two decades and make a strong case that it is appropriate to think of it as an 'epistemic community'. They describe in more detail some of the theories mentioned above and explain why epistemic communities matter to the future development of CCS. They marshal an impressive suite of empirical data in support of their case, including analysis of conference participation, uncovering of data on public and private funding of CCS research, development and demonstration (RD&D) and analysis of dedicated CCS organisations. Recognising that a CCS epistemic community exists allows some precaution to be entertained vis-à-vis acknowledged problems such as 'group-think'.

Shackley and Evar use one of the oldest theories in environmental policy analysis, the issue-attention cycle (1972), to make an ambitious argument regarding CCS. They argue that CCS has been 'on the up' from the 1990s to about 2008, but has afterwards been waning. Deliberately polemical, their argument is hard to prove or disprove definitely at the current time, though evidence of the 2008–2011 period does seem to lend it support. The analysis points to a change in the type of knowledge that is required from the CCS community from academic and disciplinary, to applied, multi- and trans-disciplinary and policy-relevant.

Evar and Shackley's chapter is a detailed case-study of a fascinating controversy over the potential health impacts arising from the use of amines as a solvent for capturing CO_2 in the specific case of the development of the Norwegian Mongstad (CCS) Test Centre. The potential hazards of the degraded products of amines has come as something of a 'side-swipe' to the CCS community but this debate has come to play an important role in determining the future of one of Norway's flagship CCS initiatives. The authors apply the ideas of error-cost, under- and over-critical scrutiny of science for policy and neo-incrementalism in an effort to put the health controversy into perspective.

Appendix 7.1 Glossary

Term	*What it means*
Clumsy governance	The most adequate systems of governance emerge from the recognition of distinctive cultural perspectives that are, ultimately, incommensurable, founded as they are upon different values, views of nature, and notions of what the 'good life' consists of.
Cultural theory	Also known as the theory of socio-cultural variability. Cultural theory maintains that there are (only) four resilient ways of organising (described by the grid-group heuristic) and associated with each mode of organising is a distinctive set of coherent beliefs, values and knowledges.
Error-costs	The economic, political or social cost(s) of incurring an error. It usually refers to the costs of making a wrong policy or commercial decision, e.g. with respect to implementing a new technology.
Fallibilism	The theory that decision-making should proceed through the process of learning-by-doing, also known as trial-and-error; this being because having sufficiently accurate information in advance of experience is held to be impossible.
Governance	The broad system by which political decisions are taken. Use of the term 'government decision-making' implies that it is elected parliamentarians together with civil servants who take decisions. Using the term governance implies that there are, in addition, other stakeholders who have a constitutive role in decision-making: i.e. they come to shape and influence decisions taken.
Incrementalism	The theory that the best decisions are always ones which entail the minimum, or at least a small, departure from the status quo. Incrementalism does not assume that no large changes are required, but does maintain that such change is best accomplished via a larger number of small steps rather than via a single step or small number of large steps.

Term	*What it means*
Interests model	The theory that discrete entities (individuals, organisations, firms, government departments etc.) have identifiable interests which motivate their actions and behaviours, that political outcomes can be explained by the interplay of different interest groups in society and that the interests of the more powerful groups will tend to prevail.
Reflexivity	A much contested term, here it is used to mean disengaging from the immediate requirements of practice and action and applying a critical, reflexive stance towards one's actions and underpinning assumptions (whether this be at the level of the individual, institution or even system).

8

THE EVOLVING INTERNATIONAL CCS COMMUNITY

Jennie C. Stephens and Yue Liu

Introduction

In the past decade, carbon dioxide capture and storage (CCS) has emerged as a controversial climate mitigation technology with important political salience. Visions of 'clean' coal-fired power plants that will not emit CO_2 into the atmosphere have provided powerful motivation for large public and private investments in CCS technology (Bäckstrand *et al.* 2011; Meadowcroft and Langhelle 2009a). The scale of CO_2 emission reductions deemed necessary for climate stabilization is so large that some CCS experts and scientists consider CCS a necessary future technology without which society may not be able to achieve acceptable emissions reductions to mitigate climate change (Pacala and Socolow 2004; Schrag 2009). On the other hand, some have questioned whether CCS could actually slow down a necessary transition toward renewable-based energy systems (Greenpeace International 2008; Shackley and Thompson 2012; Vergragt *et al.* 2011).

Recent growth in public and private investment in CCS technology has led to an increase in the number of individuals whose careers are focused on the advancement of this climate mitigation technology (Stephens *et al.* 2011). Communication among these individuals has created a diverse international network of professionals focused on the technology's advancement; an informal network that we refer to in this chapter as the international CCS community. This community is dynamic and includes scientific and technical experts, as well as representatives from business, government, academia and non-governmental organizations.

In addition to growth in the number of individuals who are now involved in this international CCS community, the community has also expanded its breadth of expertise, its geographic scope and capacity to influence policy. Despite this growth and expansion, the pace of technological advancement of CCS technology has been slower than some anticipated five or ten years ago, and the future of CCS advancement remains uncertain. A fully integrated commercial-scale power plant with CCS has not yet been built and demonstrated (de Coninck, Stephens *et al.* 2009), several CCS projects have recently been cancelled, and millions of dollars in investment commitment to CCS has not come through (GCCSI 2011b). Given the complicated politics surrounding who is supporting the development of CCS and why (Meadowcroft and Langhelle 2009a), understanding the evolution of the international CCS community provides valuable perspective on how networks of professionals focused on specific technologies change over time.

Recognizing the reinforcing, cyclical ways in which communities of technological experts may influence policymakers and investors to continue advancement of the technology that they are working on (Adler and Haas 1992; Ehrmann and Stinson 1999; Haas 1992; Nelkin 1975), this chapter explores how CCS advancement has changed over the past decade by analysing the development and expansion of the international CCS community. First, theoretical frameworks relevant to considering the evolution of a community of technological experts like the international CCS community will be presented briefly. Then, the methodology will be described. The changing composition of the international network of actors will then be reviewed as will the emerging mechanisms of communication within the community. This is followed by an examination of the changing CCS investment landscape, both public and private. A discussion about what these empirical insights suggest about the evolution of the CCS community concludes the chapter.

Theoretical perspectives relevant to technology communities

This review and assessment of the dynamic expansion of the international CCS community can be situated within theories of different types of communities that are focused on promoting specific technologies or policies. Such communities can evolve differently depending on their specific circumstances and the problem the community is trying to address. Among the theoretical frameworks developed to help understand different types of communities are 'thought collectives' (Fleck 1979), 'advocacy coalitions' (Ellison and Newmark 2010; Sabatier and Jenkins-Smith 1993), and 'epistemic communities' (Antoniades 2003; Haas 1992).

The notion of thought collectives was described by Fleck (1979) as communities primarily involved in the production of knowledge and facts. Given the societal urgency of the climate mitigation motivation for considering CCS, it could be argued that the international CCS community is more than a thought collective because it is clearly involved in more than just the production of knowledge and facts.

Another type of community, an advocacy coalition, has been described as a community that includes a variety of actors, including journalists, governments, policy analysts and other stakeholders with diverse social interests, who are advocating for the same goal (Sabatier and Jenkins-Smith 1993). The CCS community could be considered an advocacy coalition, but some opposition to this classification is likely as it could be argued that a single goal is not defined within this community; not all actors within the CCS community are promoting the technology.

Finally, the epistemic community has been described by Haas (1992: 3) as 'a network of professionals with recognized expertise in a particular domain and an authoritative claim to policy relevant knowledge within that domain or issue-area.' The extent to which the international CCS community can be considered an epistemic community has been explored previously (Stephens *et al.* 2011), and raises some interesting questions about the mechanisms for CCS advocacy within the community. Considering whether a community might be characterized as an epistemic community can lead to explorations about the capacity of the community's policy influence, its association with the broader public and the success of communicating its agendas widely.

When considering types of communities and characterizations of different communities, it is important to reflect on the heterogeneity of the international CCS community. For the purposes of this chapter, the international community includes all academics, policymakers, regulators, researchers, engineers and other professionals from both the public and private sectors who have a central focus in their work on CCS. Within this community, the 'Triple Helix' of government, industry and academia is represented in interactions and communications (Etzkowitz 2003). Within this community, a number of diversified sub-communities exist, including scientific and technical communities, a social research community, advocacy communities, an education community, as well as nationally and regionally focused sub-communities including CCS actors from Australia, Europe, the US etc. And a multitude of intermediate organizations facilitating communication among the different entities within the Triple Helix have emerged (Metcalfe 2010). Variation in scale and scope of these loosely defined sub-communities makes it difficult to assess each separately, however in this assessment of the broader international community consideration of this internal heterogeneity is critical, as different actors and organizations play different roles (Etzkowitz 2003).

The theoretical concept of innovation systems also provides helpful justification and context for analysing the dynamics of the evolving CCS community. Innovation systems have been described as networks of 'institutions in the public and private sectors whose activities and interactions initiate, import, modify, and diffuse new technologies' (Freeman 1987: 1). Theory of innovation systems highlights that technological advancement involves both individual and collective actions, and that technology innovation systems are composed of actors, institutions and their relationships (van Alphen *et al.* 2009). Within the innovation literature, the importance of creating more insight into the dynamics of innovation systems has been highlighted (Hekkert *et al.* 2007; Lundvall 1992). A framework that focuses on specific processes that are critical to the functioning of innovation systems has been proposed (Hekkert *et al.* 2007) and includes seven functions: (1) entrepreneurial activity; (2) knowledge development; (3) knowledge diffusion through networks; (4) guidance of the search; (5) market formation; (6) resource mobilization; and (7) the creation of legitimacy to counteract resistance to change. Understanding the changing composition and interactions within the community of individuals involved in the advancement of any specific technology provides insights relevant to several of these functions: both the most obvious ones relating to knowledge production (2 and 3), as well as policy advocacy (7) and the involvement of resourceful energy sector actors (6) in the formation of new markets (5). While several researchers have used the structure of these seven functions to analyse innovation systems in particular countries with a variety of specific technologies (Hillman *et al.* 2008; Suurs and Hekkert 2008; van Alphen, Noothout *et al.* 2010; van Alphen *et al.* 2009), when considering an international community of actors focused on a specific technology, the diversity and breadth of the local, regional and national contexts within which the members of the community are situated limits the direct applicability of these seven functions. Given these limits, characterizing the networks of actors and institutions and their relationships using other approaches may be helpful.

Within this context, this chapter empirically explores some key changes within the international CCS community throughout the past decade. The chapter contributes to the development of theory on various types of communities, draws on the Triple Helix concept from innovation theory as a theoretical model to explain the observed interactions between public, private and research communities, and considers functions of technology innovation systems. The chapter also provides insights with broad relevance to science, technology and society literature on the role of technological experts in advancing emerging technologies.

Methodology

This chapter explores how CCS advancement has evolved over the past decade by analysing the development and expansion of the international CCS community, as seen through the theoretical lenses of the epistemic community and innovation systems theory as described above. Specific characteristics of the CCS community to be assessed include the community's size as well as the national and sectoral representation of the individuals within the community. Mechanisms of communication within the community are explored, in addition to investment levels – both public and private – supporting the technology's advancement and the community's development.

Empirical evidence to provide a representation of these characteristics has been collected from multiple sources, including attendance records of the major international meeting of the international CCS community, the International Energy Agency (IEA) International Greenhouse Gas Control Technology (GHGT) conference series, as well as various publications and reports. The GHGT conference series has provided a critical community-developing forum with a conference held every two years hosted by various different countries with interest in CCS. By reviewing changes in the number of attendees at this conference as well as the national and sectoral representation of individuals attending this conference series, patterns of growth and expansion in the broader CCS community can be inferred. Changes in levels of public investment in CCS are assessed using several different sources including an energy R&D statistics database maintained by the International Energy Agency's Greenhouse Gas Programme (IEAGHG) and a 2011 Global CCS Institute report that reviews public financing of CCS over time (GCCSI 2011b). Mechanisms for communication within the international CCS community are analysed and documented by reviewing changes over time in different newsletters and communication outlets.

Changing composition of the international network of actors

Participation at the GHGT Conference Series

One method of characterizing the changing composition of the international CCS community is to assess the conference attendance records for the GHGT conference series. The GHGT conference has occurred every two years in different geographic locations throughout the world (see Table 8.1). This conference series was formed in 1997 following the merger of the International Conference on Carbon-Dioxide Removal (ICCDR) and the Greenhouse Gas: Mitigation Options conference. Table 8.1 demonstrates relatively consistent growth in the attendance records of the conference throughout the past decade.

While attendance at this international conference is limited to those with access to financial resources to cover the travel costs and the conference registration fee, the changing composition of attendees provides an imperfect proxy for changes in the CCS community overall. Given the unique network building opportunities associated with the conference, attendance records are likely to represent the numbers and sectoral proportions of internationally engaged actors within the CCS community (although we acknowledge that not all individuals who may be considered part of the international community attend this biannual meeting).

TABLE 8.1: Location and number of registered attendees at the International Greenhouse Gas Control Technology Conferences from 1992–2012]

Year	*Conference name*	*Hosting city*	*Number of attendees*
1992	ICCDR-1	Amsterdam, Netherlands	250
1994	ICCDR-2	Kyoto, Japan	400
1996	ICCDR-3	Cambridge, USA	250
1998	GHGT- 4	Interlaken, Switzerland	530
2000	GHGT-5	Cairns, Australia	350–400
2002	GHGT-6	Kyoto, Japan	517
2004	GHGT-7	Vancouver, Canada	648
2006	GHGT-8	Trondheim, Norway	930
2008	GHGT-9	Washington DC, USA	1,216
2010	GHGT-10	Amsterdam, Netherlands	1,528
2012	GHGT-11	Kyoto, Japan	TBD

Analysis of the sectoral representation of the attendees at these meetings demonstrates that there has been an increase in the percentage of delegates representing the private sector, while the percentage of government attendees has declined over time (Figure 8.1). Participation by non-governmental organizations (NGOs) has increased, perhaps reflecting more public engagement on this issue. In 2002, of the 517 delegates who attended the GHGT-6 conference, about 25 per cent (132) were academics, 42 per cent (219) were government affiliated, 30 per cent (157) were from the private sector, and less than 2 per cent (9) were NGO representatives (Figure 8.1A). In 2010, of the 1,528 delegates who attended GHGT-10, about 26 per cent (403) were academics, 26 per cent (396) were government affiliated, 44 per cent (676) were from the private sector, and 3.5 per cent (33) were NGO representatives. Comparing these two conferences, the most recent 2010 conference had a greater percentage of business and NGO representatives, about the same percentage of academics, and a lower percentage of government representatives (Figure 8.1).

The sectoral distribution of the past five GHGT conferences (2002 to 2010) represented in Figure 8.2, demonstrates a consistent growth in private sector and NGO involvement. The relative proportion of academic attendees has remained steady throughout this period at around 25–26 per cent for each conference. The relative number of government sector representatives dropped from 42 per cent in 2002 to 25.4 per cent in 2006 where it has stayed stable for the most recent conferences.

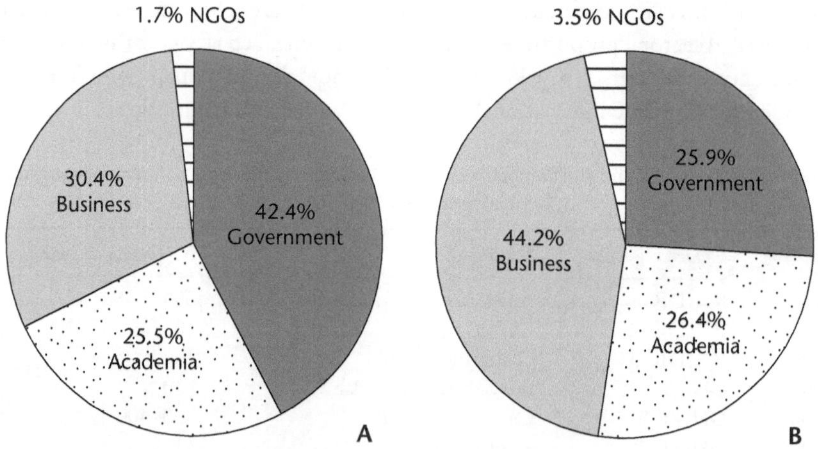

FIGURE 8.1: The sectoral distribution in percentages of: (A) registered attendees at the GHGT-6 conference in Kyoto, Japan in 2002, and (B) registered attendees at the GHGT-10 conference in Amsterdam, the Netherlands in 2010.

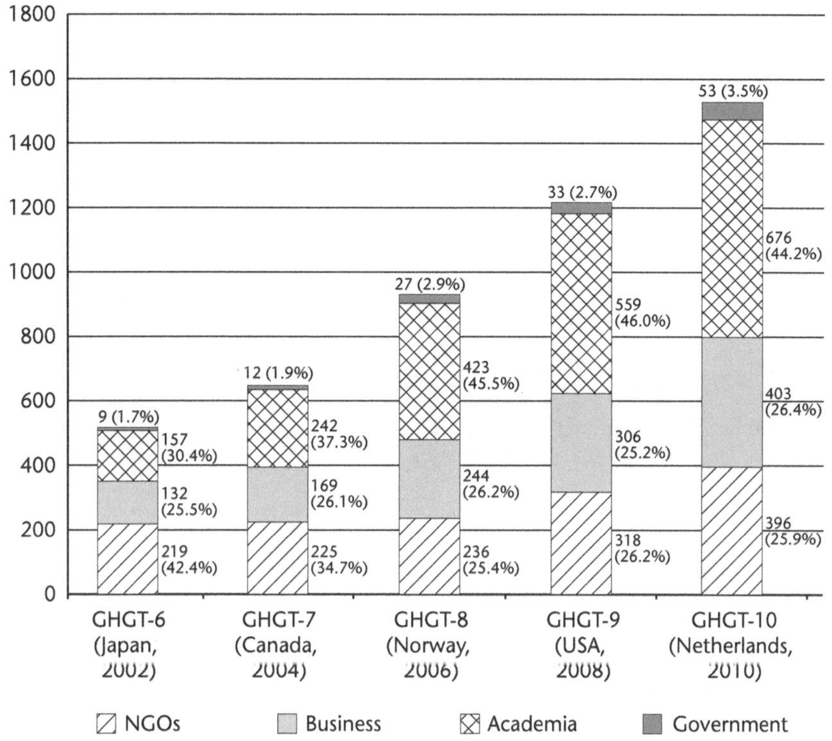

FIGURE 8.2: The sectoral distribution of registered delegates who have attended the GHGT-6, GHGT-7, GHGT-8, GHGT-9 and GHGT-10 conferences.

Changes in the national and regional representation of the attendees at the GHGT conference series are represented in Figure 8.3. The registered attendees for GHGT-6, GHGT-7, GHGT-8, GHGT-9 and GHGT-10 conferences represented 35, 36, 49, 40 and 55 different countries, respectively. For each conference, more than 80 per cent of the attendees were from Europe, North America and the Asia Pacific region (including both Japan and Australia). The location of the meeting, which has changed with every event, clearly influences the distribution of attendees from different regions of the world. For example, in 2002 when Japan hosted GHGT-6 most attendees were from the Asia Pacific region. There were 226 Japanese attendees for GHGT-6 which accounted for 43.7 per cent of the total number of conference attendees. GHGT-7 and GHGT-9 took place in Canada in 2004 and in the United States in 2008. These two conferences were dominated by attendees from the North American continent. GHGT-8 and GHGT-10 were both held in Europe, in Norway in 2006 and in the Netherlands in 2010. European attendees represented the largest regional cohort at these two conferences (Table 8.2). For all of the conferences from 2002 (GHT-6) to 2010 (GHGT-10) the countries with the highest numbers of attendees included the host country and the United States.

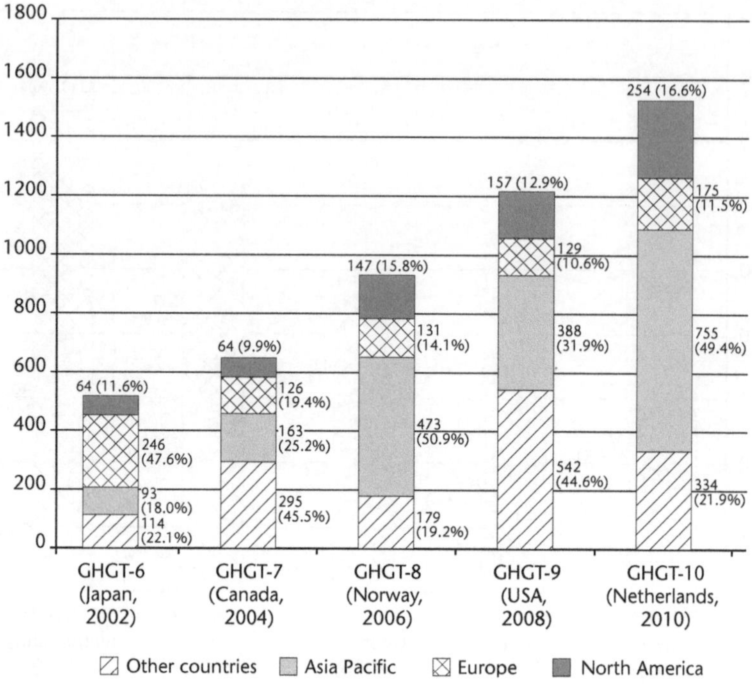

FIGURE 8.3: The regional affiliation of attendees at the five most recent GHGT conferences. The Asia-Pacific region includes Japan and Australia, North America includes the USA and Canada, the European region includes the United Kingdom, Netherlands, France, Norway, Sweden and Germany. Included in the 'other countries' category are representatives from countries in South and Central America and Africa.

Representation at these conferences of individuals from non-OECD (Organization for Economic Co-operation and Development) countries is consistently quite low. Just a few individuals from India and South Africa have attended all of these conferences, whereas an increase in representatives from China has occurred; in 2002 only one Chinese national attended GHGT-6, whereas at GHGT-10 in 2010 there were 21 Chinese representatives present.

Although NGOs have consistently represented a small percentage of GHGT conference attendees, the growing (absolute) number of delegates from the NGO sector may reflect increased public interest and concern about CCS. NGO representatives are essential members of the CCS community, and they have increasing importance in bridging communication gaps between CCS experts and the public. NGOs' positions and actions have strong influence on public opinion and affect policy networks and coalitions directly and indirectly (Gough and Shackley 2001; Hammond and Shackley 2010). Continued increase of NGO engagement within the CCS community may be likely given increasing public opposition to proposed and planned large-scale CCS projects throughout the world (Shackley and Evar, Chapter 9, this book).

TABLE 8.2: Number of attendees from the top-ten countries (nations with the largest representation in terms of number of registered attendees) for each conference (2002–2010)

GHGT-6 (Japan, 2002)	Japan	USA	Canada	UK	Australia	Netherlands	France	Norway	Sweden	Germany
Number of attendees	226	80	34	29	20	17	17	14	12	4
Percentage	43.7%	15.5%	6.6%	5.6%	3.9%	3.3%	3.3%	2.7%	2.3%	0.8%
GHGT-7 (Canada, 2004)	USA	Canada	Japan	UK	Norway	France	Netherlands	Australia	Sweden	Germany
Number of attendees	152	143	104	42	40	28	27	22	13	13
Percentage	23.5%	22.1%	16.0%	6.5%	6.2%	4.3%	4.2%	3.4%	2.0%	2.0%
GHGT-8 (Norway, 2006)	Norway	USA	Japan	UK	France	Netherlands	Germany	Australia	Canada	Sweden
Number of attendees	178	139	90	77	74	64	51	41	40	29
Percentage	19.1%	14.9%	9.7%	8.3%	8.0%	6.9%	5.5%	4.4%	4.3%	3.1%
GHGT-9 (USA, 2008)	USA	UK	Canada	Norway	Japan	France	Netherlands	Germany	Australia	Sweden
Number of attendees	454	92	88	86	82	75	60	57	47	18
Percentage	37.3%	7.6%	7.2%	7.1%	6.7%	6.2%	4.9%	4.7%	3.9%	1.5%
GHGT-10 (Netherlands, 2010)	USA	Netherlands	Norway	UK	France	Germany	Japan	Australia	Canada	Sweden
Number of attendees	270	193	155	137	125	119	104	71	64	26
Percentage	17.7%	12.6%	10.1%	9.0%	8.2%	7.8%	6.8%	4.6%	4.2%	1.7%

Changing mechanisms of communication

As the international CCS community has grown throughout the past decade, multiple different kinds of mechanisms for communication within the community have evolved. In addition to the GHGT conference series that has occurred every two years, multiple newsletters, websites and other publications have begun to facilitate communication among individuals in the community (Table 8.3). The different publications represent the diversity of sectoral representation within the international CCS community. Some are geared more toward academic researchers (the peer-reviewed *International Journal of Greenhouse Gas Control* launched in 2007), some target the private sector (i.e. *The Carbon Capture Journal*) and some anticipate more of a government, public-sector audience (i.e. *The Carbon Sequestration Newsletter*). The variety of platforms for communication that have emerged demonstrates diversity in communication within the CCS community. The different communication platforms described in Table 8.3 represent a diversity of multinational and inter-sectoral communication channels within the CCS community. These provide multiple mechanisms for international sharing of information and exchange of both technical and social learning about CCS.

TABLE 8.3: International CCS community communication platforms

Organization	*Communication platforms*
The IEA Greenhouse Gas R&D Programme established in 1991	• The Greenhouse Gas Control Technologies (GHGT) conference series was formed in 1997. Held every two years in different members' countries • Coordinates CCS international research networks in biofixation, CO_2 capture, monitoring, oxy-fuel combustion, risk assessment, wellbore integrity, modelling, high temperature solid looping cycles, and social research • Launched *Greenhouse News* (formerly *Greenhouse Issues*), a newsletter with updates in greenhouse gas abatement and mitigation • Peer-reviewed *International Journal on Greenhouse Gas Control* launched in 2007 • Annual summer school on Carbon Capture and Storage taken place since 2007
US Department of Energy	The *Carbon Sequestration Newsletter* launched in 2001. The United States DoE publishes a short newsletter monthly describing significant events related to carbon sequestration that have taken place over the past month

Organization	*Communication platforms*
The Zero Emissions Platform founded in 2005	The European Technology Platform for Zero Emission Fossil Fuel Power Plants (ZEP) is a coalition of stakeholders including the European utilities, petroleum companies, equipment suppliers, scientists, academics and environmental NGOs, united in their support for CO_2 Capture and Storage (CCS) as a key technology for combating climate change. ZEP serves as advisor to the European Commission on the research, demonstration and deployment of CCS
Carbon Capture Journal launched in 2007	A magazine featuring developments in industrial-scale carbon capture and geological storage technology, with news about the major projects and development with government policy
The Carbon Sequestration Leadership Forum (CSLF) established in 2003	Thirteen countries and the European Commission signed the CSLF charter as part of this ministerial summit (2003). An international climate change initiative focusing on the development of CCS technologies as a means to accomplishing long-term stabilization of GHG levels in the atmosphere
Global CCS Institute launched in 2009	Not-for-profit organization that connects individuals and organizations around the world to accelerate the deployment of CCS projects. It contains special body of community participation to support discussion between peer projects and between different stakeholders within the broader CCS community
The European CCS Demonstration Project Network launched in 2009	This network will play a central role in the demonstration of CCS technologies in the EU with its primary and initial aim of enhancing coordination among early players in European CCS demonstration projects
A global network established by IPAC-CO_2 in 2010.	A network initiated by the International Performance Assessment Centre for Geologic Storage of Carbon Dioxide (IPAC-CO_2) links organizations in eight countries which conduct research into the geological storage of CO_2

A related important issue to consider is the effectiveness of the community's communication outside of its own community, with policymakers and/or the general public. The increase in communication platforms (detailed in Table 8.3) represents enhanced internal communication, but also expanded capacity to influence policy. Many of the publications are publicly available and can be used in policy advocacy.

Another dimension is the capacity of the CCS community to communicate with the public, particularly considering the strong public opposition to specific CCS projects that has occurred in a handful of places, including Vattenfall's cancelled and postponed storage projects in Denmark and Germany (NyTeknik 2009) and the cancelled Barendrecht project in the Netherlands (Feenstra *et al.* 2010; van Noorden 2010). Recently, in January 2011, alleged leakage of CO_2 at Canada's Weyburn CCS project claimed by a couple whose land is in close proximity to the CO_2 injection, captured the international CCS community's attention. This alleged leakage raises concerns about whether or not CCS is safe. Currently, this leak claim is under investigation by the International Performance Assessment Centre for geologic storage of Carbon Dioxide (IPAC-CO_2), an environmental, non-government organization (Graham 2011), and the CCS community has included only minimal mention of this controversy in their publicly available communication platforms. Anecdotal information accumulated from discussing the Weyburn leak with various members of the CCS community suggests, however, that many individuals within the community have been playing close attention to the Weyburn case and informal, private communication among members of the community appears to have been quite extensive. The limited public communication by members of the CCS community on the Weyburn controversy demonstrates the highly selective and cautious nature of external communication.

Engagement of developing countries

A review of GHGT conference attendance data and the emerging communication platforms demonstrates that active participation in the international CCS community is dominated by individuals from countries whose governments have a strategic interest and investment in CCS; these are predominantly developed countries.

Among developing countries, China is among those that have demonstrated the most interest in CCS. In China, the national government has acknowledged CCS as a key technology that could potentially decouple the use of coal and other fossil fuels from atmospheric emissions of CO_2 (Wilson *et al.* 2011). Since 2005, China has collaborated with the United Kingdom and the Europe Union in the Near Zero Emissions Coal project (NZEC) to introduce CCS technology (NZEC 2007). The Chinese government, through its national energy companies, has been testing different elements of CCS technologies in the fields of coal-fired power plants, coal-to-liquids, enhanced oil recovery and enhanced coal bed methane production (Jaccard and Tu 2011). Also, in Southern Asia, Malaysia and Indonesia have developed strategic initiatives regarding CCS (Othman 2009), and there is growing interest in CCS in the Middle East.

Even in Africa, there is some engagement in CCS. One initiative, CCS-Africa, was developed in 2007 by researchers from the Energy Research Centre for the Netherlands (ECN), with funding from Shell, Statoil, the Norwegian Ministry of Foreign Affairs and the United Kingdom Department for Business, Enterprise and

Regulatory Reform. Through regional workshops in Dakar (Senegal) and Gaborone (Botswana), this CCS-Africa project initiated discussion on the level of knowledge and access to information on CCS in African countries (Bakker *et al.* 2007). Within South America, Brazil currently has a number of operating and planned CCS demonstration projects with a significant research and development programme (Román 2011). Two demonstration projects, the Petrobras Miranga Project and the CEPAC Carbometano Porto Batista Project, integrate different types of carbon storage. The Petrobras Miranga Project has three different storage scenarios: enhanced oil recovery (EOR), depleted gas reservoirs and saline aquifers. The CEPAC Carbometano Porto Batista Project is being developed to demonstrate storage associated with enhanced coal bed methane production (Beck *et al.* 2011).

The changing composition of the CCS community documented here, including growing participation in international conferences, new emerging mechanisms for communication, and increased (but still minimal) engagement of developing countries demonstrates an increase in the community's global reach. This expansion suggests a potential enhanced capacity for policy influence within the community, however analysis of the community's actual influence on policy is beyond the scope of this study.

Changing investment landscape

CCS is a set of technologies that involves large infrastructure investment; CCS is not a technology that individual entrepreneurs can easily advance on their own without considerable collaboration and integration with existing energy technology organizations. The technology's dependence and necessary integration with existing energy systems limits the potential for individual 'entrepreneurial activity' that Hekkert *et al.* (2007) define as one of seven critical functions of innovation systems. The interests of large-scale, established organizations and governments instead dominate the investment landscape. Throughout the past decade changes in the level of investment in CCS from several key national governments and some private companies have drastically changed the innovation environment for CCS. Understanding how this investment landscape has changed over time provides additional valuable insight into the development of the CCS community.

Public investment in CCS

Government investment in CCS is critical to its advancement. Given the scale and cost of CCS equipment and infrastructure, the lack of a price on carbon to incentivize development of carbon management technologies, and multiple other characteristics of energy technology innovation that necessitate public support (Gallagher *et al.* 2006; IEA 1998, 2003; Loiter and Norberg-Bohm 1999), significant government investment is key to the future of this technology.

A recent study by Tjernshaugen reviewed the levels of CCS public investment in 2005 within the countries that were investing the most in this technology

(Tjernshaugen 2008). This study, which compiled data from several sources including IEA and government publications, and personal communications with government officials, found that the United States, the EU – including the United Kingdom, Norway, Germany, the Netherlands, Italy and France – as well as Canada and Japan, were the countries with the most intense investment in CCS research, development and demonstration (RD&D). Combined European and North American government budgets for CCS RD&D in 2005 reached approximately US$140 million. This study quantified CCS investment by dividing RD&D budgets by GDP to develop an indicator of political commitment to the CCS strategy. The result showed that Norway stands out as the national government with the relatively greatest commitment to CCS. Using this metric, Canada was second after Norway, followed by the Netherlands and the United States.

This assessment of 2005 levels of public investment can be compared with more recent data on public investment. The IEAGHG maintains a publicly accessible database that records total annual public investment of CCS R&D in countries throughout the world. Data from 21 of the 28 countries in that database is presented in Table 8.4 (the database contains no data on the other seven countries from 2002 to 2009). This database is not comprehensive as data from some countries that have been investing heavily in CCS (most notably Australia) is not included. In several countries, the increase over time in levels of public investment is striking, most noticeably in the United States. CCS investment in the United States, however, has been heavily influenced by the American Recovery and Reinvestment Act of 2009, so the data therefore does not necessarily suggest a consistent trend in energy policy.

The data included in the IEAGHG database is collected from an annual questionnaire that the IEA Secretariat solicits from appropriate offices within national administrations. As noted by the IEA Energy Technology R&D Statistics Service, even the best organized data gathering cannot remedy the non-availability of data, which inevitably leaves gaps in Table 8.4. The lack of data from some countries, including the United Kingdom, the Netherlands, Australia and Germany may be due to communication challenges between the IEA and national administrations. Some degree of ambiguity about what these numbers actually represent is also acknowledged because the categorizations of R&D (research and development) budgets and RD&D (research, development and demonstration) budgets are both used to describe these numbers. Regardless, the data in Table 8.4 does show a clear trend of increasing investment made by governments from 2002 to 2009 in CCS activities (as of May 2011 the 2010 data had not yet been uploaded into this database).

Another source of CCS public investment data comes from the Global CCS Institute which issued a 2011 report reviewing public financing of CCS over time (GCCSI 2011b). This report reveals that globally since 2005, US$25 billion in direct government funding have been announced, and funding that exclusively targets large-scale demonstration projects accounts for 80 per cent of all direct funding announcements since 2005. In addition to support for large-scale projects,

TABLE 8.4: Total government RD&D budgets in CO_2 capture and storage in million US$ (2009 prices and exchange rates)

Country (Total per year)	*2002*	*2003*	*2004*	*2005*	*2006*	*2007*	*2008*	*2009*
United States	–	–	44.704	68.997	144.824	126.173	195.025	3,989.23
France	4.285	4.831	8.996	14.726	37.971	40.236	42.922	–
Canada	–	–	17.114	12.17	13.074	14.825	18.774	118.648
Japan	–	–	–	–	28.545	14.705	40.65	131.905
Norway	–	–	10.152	9.337	18.475	13.598	14.47	27.008
Korea	–	–	1.074	3.308	3.338	5.231	–	–
Italy	0	0	0	0	1.5	3.658	–	–
Spain	–	–	–	–	–	1.715	3.204	3.833
Switzerland	0	0	0	0	0.797	1.019	0.924	0.921
Denmark	–	–	–	1.673	2.734	0.854	–	–
New Zealand	–	–	0.061	0.06	0	0.664	0.641	0.625
Austria	–	–	–	–	0.118	0.104	0.697	–
Australia	–	–	–	–	–	0.088	–	–
Portugal	–	–	–	–	–	0.035	0.025	0.053
Turkey	–	–	–	0.223	0.164	0.012	–	
Germany	–	–	6.59	5.867	0	0	3.992	7.639
Ireland	–	–	–	–	0.389	0	0.469	–
Czech Republic	–	–	–	0	0	0	0.544	0.783
Slovak Republic	–	–	–	–	–	–	0.275	0.276
Netherlands	–	–	–	11.446	4.314	–	–	–
United Kingdom	–	–	8.816	11.237	–	–	–	–

Source: IEAGHG database.

more than US$2.4 billion in government funding have been committed to support CCS research activities and pilot-scale demonstrations since 2005 (GCCSI 2011b). Of government funding directed to large-scale CCS demonstration projects, 77 per cent was allocated to large power generation projects; the remainder was mainly allocated to projects in the oil and fertilizer industry (14 per cent), coal gasification (5 per cent), and others (4 per cent including aluminium, cement, and iron and steel industries etc.) (GCCSI 2011b). The United States has allocated the greatest amount to specific projects at US$6.1 billion, representing 70 per cent of the government's total programme commitments to CCS, with another US$2.6 billion still to be allocated to specific CCS projects (GCCSI 2011b). A total of about US$17.4 billion of funding have been announced by the European Commission and the national governments of Norway and the United Kingdom; however, most of those funds have still not been distributed, with US$14.7 billion (84 per cent) not yet allocated to specific projects (GCCSI 2011b). At this stage, there is therefore limited clarity on how large-scale CCS demonstration projects will be supported by governments in the operational stages (GCCSI 2011b).

Globally, new funding announcements specific to CCS rose from less than US$15 million in 2005 to around US$10.5 billion in both 2008 and 2009, while new announcements of financial support for CCS slowed significantly in 2010 to around US$850 million. A significant share of these new commitments (more than 35 per cent) was stimulus spending in response to the global financial crisis. For example, the United States American Recovery and Reinvestment Act 2009 committed more than US$3.1 billion to fund CCS projects, while the European Energy Programme for Recovery committed US$1.3 billion (GCCSI 2011b).

It is important to acknowledge that there are some inconsistencies in the public funding data reported by the Global CCS Institute and the IEAGHG database. For example, as described by *The Global Status of CCS 2010* (GCCSI 2011b), since 2005, the United States federal government alone has committed close to US$9 billion in programmes that provide direct support to CCS. However, the amount of US$9 billion does not match the IEAGHG database's reporting of total amount of the United States government RD&D funds supporting CCS since 2005 which is about US$4.5 billion (Table 8.4). The complexities associated with different categorizations and self-reporting are the likely causes of these discrepancies. The public funding data synthesized here shows growing government investment in CCS. While many factors contribute to this growth in public investment, the CCS community's growing size is likely to be enhancing the community's capacity to influence policy and is likely to be one reinforcing contributor to investment decisions. Decision-making about public funding allocations involve complex national budgeting processes; the budgeting process has been characterized as a mechanism through which political interest groups 'bargain over conflicting goals, make side-payments, and try to motivate one another to accomplish their objectives' (Wildavsky 1975: 4). Generally, the executive arm of government has primary responsibility for the formulation of budgets and their implementation. However, in practice the spending of public money depends on a range of other factors and includes a range of interested actors, including citizens, users of services, companies, voluntary agencies etc. (Norton and Elson 2002). Given the broad range of actors within the international CCS community and the complexity of processes determining budget allocations, it can be assumed that multiple members of this community engage in policy processes and influence public budgetary decisions in different ways. A more detailed assessment of which specific interactions have been particularly influential would be helpful, but is beyond the scope of this chapter.

Private investment in CCS

While public investment data is available and several organizations have compiled and synthesized this data in different ways, private investment in CCS is more difficult to quantify and assess. Levels of private investment can be inferred to some extent from public-private partnerships in large-scale CCS projects and increasing numbers of private sector attendees at the GHGT conference series. Given the high cost and the financial and environmental risks associated with CCS investments, the vast majority of CCS projects around the world have relied on a combination of public and private funding. To some extent, the magnitude of private investment in CCS technology is connected to national-level CCS investment as well as the private sector's perceptions of the emerging CCS legal and regulatory framework (Gibbins and Chalmers 2008a; Robertson *et al.* 2006). In the past decade, there has been significant progress in addressing some of the most difficult CCS legal and regulatory challenges, such as long-term liability (Wilson *et al.* 2007). However, progress has been largely limited to countries and regions that are part of the OECD, in particular Australia, Europe, the United States and Canada, with less progress has taken place in non-OECD regions (GCCSI 2011b).

A report released by the Climate Group and Ecofin Research Foundation in September 2010 explained that private sector capital providers are ready to invest in first generation industrial-scale CCS projects, but only if three prerequisites are addressed: (i) a performance guarantee should be established across the entire generation and capture chain; (ii) projects should include sponsors with a strong reputation and track record; and (iii) CCS must have a route to being competitive without public funding (The Climate Group 2010).

While a systematic assessment of private investment in CCS is not possible due to limited access to private sector financial information, one rough approach to assessing macro-level private investment in CCS is to compare the total projected cost of proposed projects with the amount of public investment in the same set of projects. The difference between the total projected project cost and the public investment can be assumed to be the level of private investment. For example, in the United States ten large-scale power plant CCS projects are currently proposed or under construction (Table 8.5). The total cost of these ten projects is estimated to reach around US$21.5 billion (MIT 2011). The United States government has provided strong financial support for these projects through direct project grants, the Clean Coal Power Initiative of the Department of Energy (DOE), the American Recovery and Reinvestment Act, DOE loan guarantees, and state tax relief (MIT 2011). In aggregate, committed public funding toward developing these ten projects is at around US$9.5 billion. For these projects to move forward, the private sector will have to contribute the rest, which will require a total of approximate US$12 billion and amounts to about 55 per cent of the total costs of these ten projects.

TABLE 8.5: Large-scale power plant CCS projects in the USA

Project name	*Leader*	*Size MW*	*Capture/ storage*	*Start up*	*Location*	*Anticipated private funding*[1] *(million US$)*	*Current public funding (million US$)*
AEP Mountaineer[2]	AEP	235	Post/Saline	On hold	West Virginia	326.8	341.2
WA Parish	NRG Energy	60	Post/EOR	2017	Texas	167	167
TCEP	Summit Power	400	Pre/EOR	2014	Texas	1400	900
Trailblazer	Tenaska	600	Post/EOR	2014	Texas	900	2600
Kemper County	Southern	582	Pre/EOR	2014	Mississippi	1997	403
HECA	HEI	390	Post/EOR	2014	California	1962	338
FutureGen	FutureGen Alliance	200	Oxy/Saline	2015	Illinois	1200	1000
Sweeny Gasification	ConocoPhilips	460	Pre/Saline and EOR	TBD	Texas	3512	588
Taylorville	Tenaska	602	Pre/Saline	TBD	Illinois	504	2996
Antelope Valley	Basin Electric	120	Post/EOR	On hold	North Dakota	100	400

Source: MIT 2011.

Notes

1 Current public funding commitment for each project as detailed in an MIT-maintained database of CCS projects is listed in the column farthest to the right. The public funding sources include: governmental grants, the Clean Coal Power Initiative, the American Recovery and Reinvestment Act, governmental loan guarantee and the state tax relief. The anticipated private funding is estimated by subtracting the current public funding amount from the total projected project cost.

2 The Global CCS Institute will provide AU$4 million (US$4.01 million) to support the initial engineering and characterization phase of AEP Mountaineer project.

The large scale of investments required for CCS prevents many potential small-scale investors, and the reliance on public-private partnerships also adds an important organizational requirement to participation that favours large established companies. Companies from the oil and gas industry have dominated private CCS investment in large part due to the strategic opportunity associated with their technical capacity in sub-surface geological engineering, and the economic incentive presented by enhanced hydrocarbon recovery, in which pressurized CO_2 is injected into depleted reservoirs to recover any remaining resources.

For the purposes of both public and private investment in CCS, one of the most influential recent developments has been the establishment of the Global CCS Institute in 2009, a non-profit organization established by the Australian government (GCCSI 2011a). The Institute is a non-profit entity owned by its members (both public and private), with the Australian government initially having committed AU$100 million of annual funding to the organization for a four-year period. The Institute works collaboratively with a number of strategic partners active in the CCS arena, including the Carbon Sequestration Leadership Forum, the International Energy Agency, the World Coal Institute, the International Aluminium Institute, the Carbon Capture and Storage Association, the Asian Development Bank, the World Bank, the Climate Group, the Clinton Climate Initiative and the Coal Industry Advisory Board (GCCSI is also a member of the IEA CCC (GCCSI 2011a)). The Institute has built a collaborative international membership, bringing together a partnership of government, industry, non-government bodies and research organizations from around the world with a shared commitment to accelerating the commercial deployment of CCS (GCCSI 2011a).

The increase in private sector investment is also reflected in the increasing numbers of private sector attendees at the annual GHGT conference (see section headed 'Participation at the GHGT Conference Series'). This increase in private sector engagement and investment may also be a reflection of the private sector's awareness of increased government support and an associated maturation of the technological community. As the community becomes more established and the perceived need for climate mitigation technologies increases, the co-dependence of public and private interests is solidified. If public funding allocations are consistent and growing in magnitude, the private sector is likely to respond by increasing its own engagement within the community in order to strategically position itself to take advantage of emerging opportunities. The private sector may also be investing in CCS to hedge against a future scenario in which CCS will be mandated by governments and to thereby reduce the future cost of investment.

Another interpretation is that the increase in private sector involvement in the CCS community is a manifestation of 'path dependency' (Arthur 1989; Praetorius and Schumacher 2009; Torvanger and Meadowcroft 2011; Unruh 2000), i.e. the scale of investment requirements and the sunk costs associated with the amount of money already invested in CCS may be creating a difficult 'lock-in' situation and justifying a change of course becomes increasingly difficult (Shackley and Thompson 2012). For those governments and private companies – including technology suppliers – that have already invested millions or billions of dollars to advance CCS, ending their support for this technology may be difficult even if perceptions of the relative challenges and the potential of CCS become increasingly pessimistic over time. The strength and the recent growth of the CCS community, therefore, contributes to this challenging 'lock-in' situation, because as the appeal of attending the international conferences to engage with the community grows, more individuals and organizations are represented at meetings, which leads to strengthening of networks and relationships within the community.

Discussion and conclusions

This chapter has applied theories of epistemic community and innovation systems to help with the study of the CCS community. These theories have facilitated the analysis of the expanding and evolving community's potential to influence policy and the communication mechanisms involved in technology advocacy and the building of technology legitimacy, as well as the mobilization of resources from governments and industry. In addition, the analysis has drawn on the Triple Helix concept for analysing the sectoral composition of the CCS community. It highlights the roles of government, industry and academia for successful innovation, although the case of CCS shows that attention also needs to be given to the distinct contribution of NGOs.

The empirical details presented in this chapter describe some of the ways in which the international CCS community has been changing over the past decade. The community has been expanding its size, its international reach and its mechanisms for communication among government, industry and academic representatives. In addition, public funding for CCS has been increasing and has been accompanied by private funding, which has enabled and supported the community's expansion. Given its growth and broader reach, the CCS community has clear potential to increase its capacity to influence policy and investment decisions, although this analysis has not directly measured the community's capacity to do so. Despite this growth, obstacles to CCS advancement remain, including high project costs and public opposition, which have limited project developments and resulted in uncertainty about the technology's future.

The empirical data in this chapter supports the notion that the international CCS community can be considered an epistemic community (Stephens *et al.* 2011). Epistemic communities have been defined as thought communities made up of socially recognized knowledge-based networks (Antoniades 2003). Through cognitive and practical levels of actions, epistemic communities influence politics and have the ability to control knowledge diffusion in social interactions by demonstrating 'cognitive authority' regarding the knowledge relevant to policy (Haas 1992). Given the limited knowledge of the general public as well as policymakers about CCS, the CCS community in its various communication platforms supports knowledge diffusion and demonstrates cognitive authority. And the recent growth of the community contributes to legitimizing this cognitive authority and expanding its potential influence and political reach.

Epistemic communities have the power to impose particular discourses and particular worldviews on societies (Antoniades 2003), and the potential for epistemic communities to impose a technocratic perspective has been recognized as they have been described as a specific technocratic type of discourse coalition (Stone 1996). The multiple communication mechanisms that have recently emerged within the CCS community enable international exchange of information and nurturing of a shared perspective on the potential and challenges of CCS technology, both within and outside the community. The shared perspective informs a discourse within the community that is communicated externally to policymakers and the public in various ways.

At a practical level, epistemic communities often play a significant role to governments, by helping to decipher their environment and define their interests. In terms of innovation systems dynamics, a key function is to build legitimacy of the technology among policymakers and other actors. In the CCS arena, there has been an observed correlation between the growth of the CCS community, particularly the growth of private developer interests charted in this chapter, as well as increased institutional attention (de Coninck and Bäckstrand 2011), and the rise of CCS on national and regional energy policy agendas (Langhelle and Meadowcroft 2009). While this discussion has focused on the international CCS community and its potential to influence national level policy, it is important to recognize that many members of the community and some subgroups within the community have their own mechanisms for influencing the policy process, and much of that influence may be also at local or regional levels and/or international or transnational levels (Antoniades 2003).

The growth of the community over the past decade clearly reflects the powerful attraction of the technology among actors (government, industry, academic and NGOs) who are increasingly concerned about, and being forced to face the need for, climate change mitigation. The high level of public investment and interest, coupled with private investment, demonstrates a widespread attractiveness of the technology. But controversy exists over whether this is related more to its potential to maintain the status quo of the energy system regime (Corry and Riesch, Chapter 6, this book), or whether the appeal derives from its potential to contribute to a shift to a more sustainable energy system (Unruh and Carrillo-Hermosilla 2006).

As public awareness about CCS increases, the CCS community's capacity to engage not only with policymakers but also with the public, and to build broad legitimacy for the technology, will likely be critical. Historical examples of energy technology development, such as nuclear energy, have shown that public resistance can have a major influence on the deployment of a technology (Anshelm 2000; Kasperson *et al.* 2005; Lovins and Sheikh 2008; Zimmerman 1982). But discrepancy between the perceptions of energy technology experts and those of the general public is common (Nader 1981). Previous studies have recognized that experts tend to exhibit overconfidence about and a positive bias towards the particular technology they are working on (Hultman and Koomey 2007). Recent research confirms a high level of confidence in CCS technology among CCS experts (Evar 2011; Hansson and Bryngelsson 2009). More generally, it is recognized that positive biases and overconfidence are often perpetuated by communication within groups that tend to focus on factors promoting success (Buehler *et al.* 2005), whereas pessimistic scenarios are often downplayed (Newby-Clark *et al.* 2000). This relates to Haas's (1992) description of the evolution of an epistemic community, or a knowledge-based network of recognized experts who 'not only hold in common a set of principled and causal beliefs but also have shared notions of validity and a shared policy enterprise' (Haas 1992: 16).

In addition to considering the dynamics of the international CCS community, this analysis has relevance to our understanding of the social dynamics in other technology communities. All emerging technologies have a community of individuals and organizations who often become technology advocates, and it should be recognized that technological advocacy is an essential part of the innovation process (CSLF 2010; GCCSI 2009; Hekkert *et al.* 2007). Awareness of this advocacy tendency within a technology community may help actors consider how to most effectively respond to public concerns and opposition in a way that promotes mutual understanding of the broader social context for the emerging technology (de Coninck 2010).

Although this assessment of the international CCS community provides some insights that are relevant to other technologies, the international CCS community is unique in many ways. The mutually reinforcing qualities of a growing community, increased funding and the political salience of a climate mitigation technology for coal constitute a distinct situation. It could be argued that compared to other energy technologies with climate mitigation potential CCS has among the highest levels of powerful, entrenched actors involved in a focused way in its advancement. The CCS innovation system therefore has an advantage in terms of resource mobilization, as compared to other low-carbon energy technologies. While the CCS situation may be unique, the interdependency of the development of all energy technologies is critical to consider. The changing international CCS community represents a changing innovation environment that has complex direct and indirect implications for the advancement of other technologies with potential to contribute to a transition to a more sustainable energy system.

9

UP AND DOWN WITH CCS

The issue-attention cycle and the political dynamics of decarbonisation

Simon Shackley and Benjamin Evar

Introduction

Carbon dioxide capture and storage (CCS) has enjoyed something of a meteoric rise in attention within climate and energy policy circles since about 2003–2004 in both individual industrialised countries and, to some extent, globally. This has been reflected, and may be to some extent stimulated, by the growth in attention devoted to CCS by the specialised media. Prior to 2000, CCS was a fairly esoteric subject and received relatively little funding in comparison with alternative, carbon-mitigating energy-generation technologies. A variety of indicators can be used to measure the shift in attention since that time, including publications in popular and specialist media, mentions in government speeches, changes in R&D funding levels, as well as new policies and regulatory measures. Such indicators reveal that while attention to CCS on the whole has been rising, key developments (or a lack thereof) since 2008 also show an increasing lack of confidence that the technology will deliver on CO_2 mitigation hopes, within the timeframe proposed by policymakers. This is, to some extent, linked to the failure to generate a strong commitment to global carbon mitigation and technology transfer through the UNFCCC during recent Conferences of the Parties in 2009 and 2010 (COP-15 in Copenhagen and COP-16 in Cancun, respectively). Spokespersons such as Jeff Chapman, Chief Executive of the Carbon Capture and Storage Association (CCSA), have also noted the lack of commitment, following an encouraging start, towards CCS in recent years:

> CCS has not been advancing as quickly as it needs to. CCS technology is proven, but the policy and incentives framework is uncertain. … Without greater certainty, investors, backed by pension funds and global money markets that will finance CCS, [it] will not deliver.
>
> (Chapman 2011a)

In this chapter we ask how the lack of confidence in CCS delivery – given the promising messages and signals from scientists, industry and government just a few years ago – can be explained. We investigate some of the potential forces behind this development using Anthony Downs' concept of the 'issue-attention cycle' (Downs 1972) as a tool for considering the changing fortunes of CCS as an issue in the media and public policy arenas.

In the next section we present the key ideas behind the issue-attention cycle, then we move on to apply these ideas to CCS through qualitative analysis and interpretation of key events, developments and documents. In the section following that, we provide more quantitative evidence through an analysis of the articles published on CCS in the UK environmental professionals' journal *ENDS* from 2008–2011. In the penultimate section we propose a revised theoretical apparatus through which recommendations can be presented to the CCS policy community, and end with a set of conclusions and implications for policymaking in the final section.

The issue-attention cycle

Downs (1972) suggests that public attention to 'ecological matters' over time follows a cyclical pattern as follows (see Figure 9.1):

1. *the pre-problem stage*, when 'some highly undesirable social condition exists but has not yet captured much public attention' (p. 39)
2. *alarmed discovery and euphoric enthusiasm*, when

   > the public suddenly becomes both aware of and alarmed about the evils of a particular problem. This alarmed discovery is invariably accompanied by euphoric enthusiasm about society's ability to 'solve this problem' or 'do something effective' within a relatively short time
   >
   > (p. 39)

3. *realising the cost of significant progress*, which consists of a 'gradually spreading realisation that the cost of "solving" the problem is very high indeed … doing so would not only take a great deal of money but would also require major sacrifices by large groups in the population' (p. 40)
4. *gradual decline of intense public interest*, as people are discouraged by the high costs to themselves of responding, while 'others feel positively threatened by thinking about the problem; so they suppress such thoughts. Still others become bored by the issue' (p. 40) (and the media lose interest); and
5. *the post-problem stage*, when 'an issue that has been replaced at the centre of public concern moves into a prolonged limbo – a twilight realm of lesser attention or spasmodic recurrences of interest' (p. 40). Programmes and institutions are often established during earlier stages of the cycle, which persist in the post-problem stage and can have some modest successes towards tackling

> the original problem. Furthermore, 'problems that have gone through the cycle almost always receive a higher average level of attention, public effort, and general concern than those still in the pre-discovery stage' (p. 41).

Downs observed the cycle at work in environmental problems that affect a relatively small number of people, where existing arrangements provide benefits to a majority, and where the problem has no intrinsically 'exciting' qualities to the general public (or has lost these). The costs of responding to such a problem are felt particularly by powerful interest groups and minorities, who therefore have a strong incentive to attempt to block a clear policy response. And because the benefits of a clear policy response are often dispersed, powerful minorities can frequently block action. Downs also suggested that invisible environmental problems would be more susceptible to the erosion of attention, and climate change certainly falls into a category of environmental problems that are arguably difficult to experience directly. Environmental hazards and events that might be climate change related are frequently impossible to attribute unequivocally to anthropogenic climate change. Climate change mitigation is therefore, arguably, one example where public attention has waned significantly since Copenhagen COP-15 in 2009 (Spence *et al.* 2010).

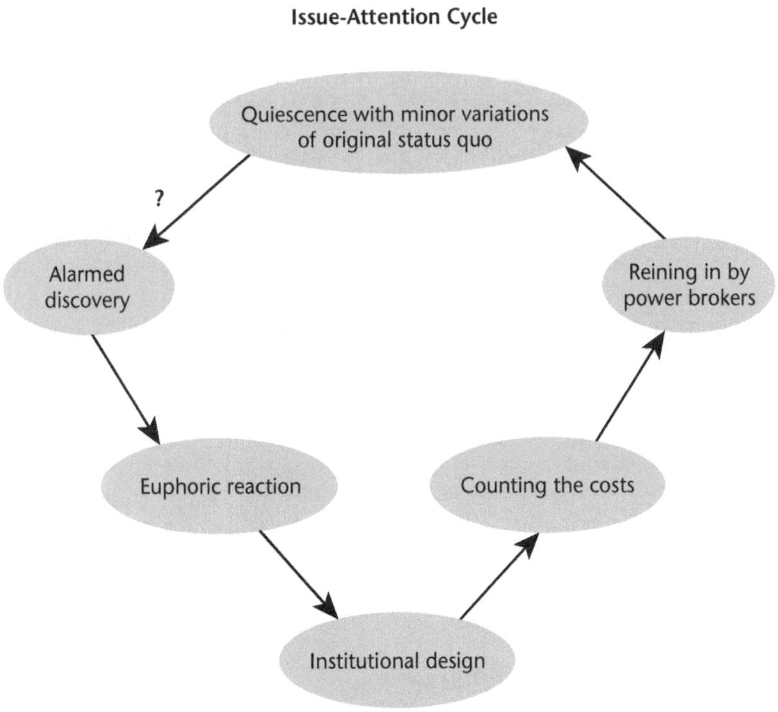

FIGURE 9.1: The issue-attention cycle (after Downs 1972, modified by O'Riordan 2009).

While some writers have stressed the emergence of environmental concern as a long-term trend during the last half-century, initially in the West but now internationally (Inglehart 2008; Mol *et al.* 2009), other scholars have maintained that the issue-attention cycle is a better analytical construct for understanding such changes. Tim O'Riordan, for example, writes that:

> Downs placed his finger on the political pulse when he noted that reactionary forces would look critically at the mounting political and economic consequences of measures, and would lobby to rein in the emerging but hesitant political enthusiasm. This did, of course, happen in the early 1970s and still occurs every day, even in the modern era.
>
> (O'Riordan 2009)

Consequently, O'Riordan finds little evidence that the environmental movement has had much pervasive influence on the political and economic status quo, four decades after Downs wrote his (seemingly prophetic) account of the ups and downs of 'ecological matters'.

In his 1972 article, Downs refers mostly to the public as the subject that provides or withholds 'attention'. However, it is a relatively modest extension of his ideas to apply this framing to the policy community as well (whether conceptualised as an elite, network, nexus or coalition). We also extend Downs' model from an 'ecological issue' proper to consider reactions to a proposed technological solution, which is intimately connected to the science that underlies a distinct environmental issue (climate change). The 'lay public' is rarely involved directly in policymaking, but rather participates via voting for political representatives in addition to lending support on occasion to local and national campaigns orchestrated by non-governmental organisations, one-off campaign groups, newspapers and so on. In this chapter, we therefore assume that it is the *policy community* that is being attentive or otherwise, not the lay public directly (though the perceived responses of the electorate to policies is, of course, an important consideration on the part of the policy community). Other authors have made similar points (Howlett 1997; Howlett *et al.* 2009) proposing a 'political' cycle, in which attention originates in the political leadership and later catches public attention. They also note the importance of exogenous events, such as energy crises, affecting attention and policy outcomes. In the case of CCS, it would moreover be difficult to apply the issue-attention cycle to the lay public directly; recent survey findings suggest that some populations in the EU have heard a great deal about CCS, particularly in Norway and the Netherlands (Pietzner *et al.* 2011), but that less than 10 per cent of the general population 'know what it is' (TNS Opinion & Social 2011).

Criticisms of the issue-attention cycle

Since its publication, Downs' model has received a relatively high amount of attention, both citing it as a useful conceptual framework and criticising it for lacking a rigorous empirical basis. Attempts to statistically test for the hypothesised relationship between public attention and government (in)action have found only a weak correlation, whether the time period under consideration was limited to a decade (Peters and Hogwood 1985) or a month (Howlett 1997). Extending the subject to the policy-nexus side-steps this particular criticism, though a methodological challenge remains of how policy attention can be reliably measured.

The dependent variable in the aforementioned studies, government change, was defined either loosely as government initiations and changes, or regulatory announcements (e.g. as recorded in the Hansard system in the UK), and purposefully excluded budgetary amendments and publications of policies. While such data may be desirable to test for changes related to established technologies, we consider a mix of media receptivity, public acceptance, cost estimates, budget announcements, policy developments and regulatory changes, because CCS is only just emerging. Constraining our analysis to regulatory changes and government initiations of e.g. R&D programmes would miss much of the relevant activity.

A related criticism of Downs' analysis is that it implies a static system, with the issue-attention cycle functioning as a political thermostat, preventing any proposed change that would likely have a substantial effect from taking place. Such a functionalist type explanation finds change hard to explain and O'Riordan proposes some sources of change that might move things beyond the confines of Figure 9.1, including the possibility that natural systems are near to reaching global tipping points, acknowledgement of the vulnerability of socio-economic systems to such changes, and changing debates on definitions of welfare. 'Disruptive technologies' – technologies that do not fit within the (currently) dominant socio-economic frame of society – are another potential agent of change that can re-structure the prospects for policy innovation.

Applying the issue-attention cycle to the CCS policy community

During the mid-2000s, dramatic decarbonisation, by way of alternative technologies, appeared to be an unstoppable policy, political and economic issue in government legislation, reports, political and planning statements, as well as in the media. Relevant interests seemed to overlap sufficiently and were backed by robust science and technological assessments and capabilities, to the extent that CCS appeared to be well on its way as a favoured techno-policy choice in the climate change and energy communities in several countries, and generally within the EU and North America. Some countries were earlier to embrace it than others (Table 9.1) perhaps because of the importance of fossil fuel supplies, the salience of climate change as a political issue and/or their technological capabilities for developing a CCS value-chain.

Yet, as we shall argue, such apparent coalitions of policy actors can dissolve fairly rapidly as circumstances change and interest wanes. While there are clearly important differences between countries, we maintain that there is evidence across the countries in Table 9.1 of a shift from 'alarmed discovery' (stage 2) to 'realising the costs' (stage 3) and, in some cases even towards 'gradual decline'. As an example,

TABLE 9.1: Time of political engagement with CCS

Very early engagement with CCS (late 1990s)	*Early engagement with CCS (early 2000s)*	*Somewhat later engagement with CCS (2005–2008)*	*Latest engagement with CCS (2008–2011)*
Norway	Australia	Germany	China
	Canada	Netherlands	South Africa
	USA	UK	UAE
	Japan[1]	EU	

Adapted from Langhelle and Meadowcroft (2009) and updated for the period 2008–2011.

Note

1 Japan does not appear in Langhelle and Meadowcroft's original list of countries with early engagement.

in all the countries listed in Table 9.1, projects have been announced that have failed to materialise close to schedule or at all.

CCS on the up in the policy arena

CCS rose rapidly up the policy agendas of several countries from the mid-1990s to the late 2000s (Table 9.1). The main CCS 'champion' in many national contexts from the early to mid-1990s had been the national geological surveys, which identified assessment of CO_2 storage as an important function of (and opportunity for) publicly funded research and advisory organisations. A key document in the EU and UK context was the Joule II report on geological CO_2 storage capacity published in 1996 (Holloway 1996), in which several prominent European national geological surveys participated. Also in 1996, Statoil initiated the Sleipner offshore CO_2 storage project, which has since inspired a number of commentators and organisations in the field, to suggest that large-scale storage would be possible. A few other large projects followed suite in the late 1990s to mid-2000s (Weyburn-Midale, In Salah, Snøhvit) along with a larger number of smaller projects (e.g. CRUST K12-B, Netherlands; Mountaineer, USA; SECARB, USA; Zama, Canada) (Shackley and Gough 2006). Project announcements and updates have since become easier to review with institutions tracking CCS activity worldwide, including the International Energy Agency's (IEA) Greenhouse Gas Programme[1] and the US Department of Energy.[2]

In the 2000s, CCS became widely evaluated by the energy, environment and finance ministries in a number of European countries (Norway, UK, Netherlands, Germany, Denmark etc.). During this period, CCS began to be included in several major economies' energy and climate change policies (e.g. the 2002 Norwegian Ministry of Petroleum and Energy Report on Gas Technology, and the 2003 UK Energy White Paper). At the Gleneagles G8 summit in 2005, UK Prime Minister Tony Blair succeeded in putting CCS on the 'international stage' and the UK government announced the competition for a CCS demonstration facility in November 2007.

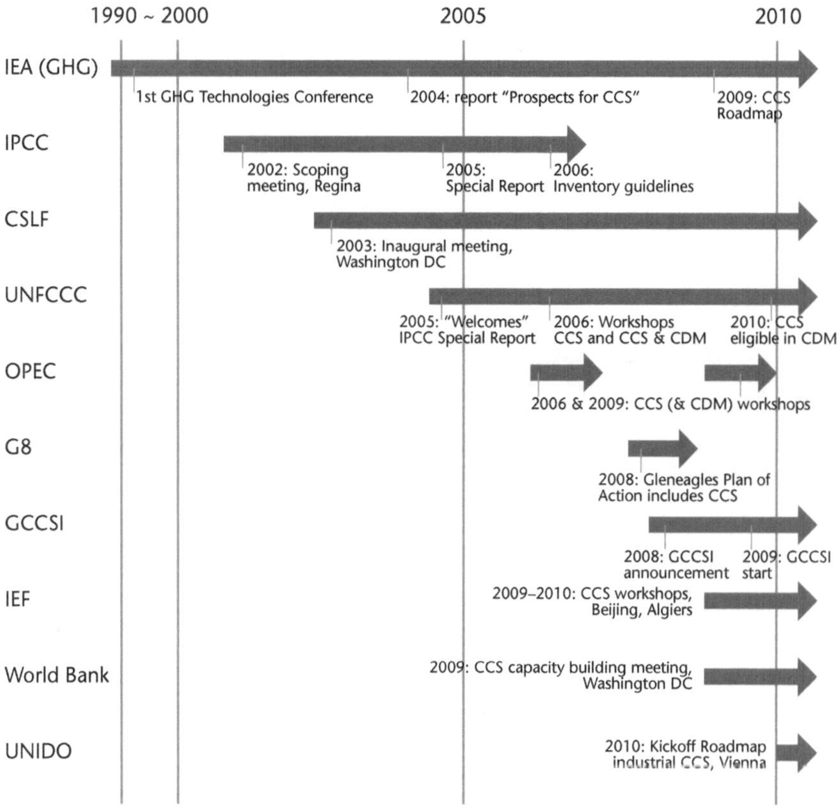

FIGURE 9.2: Time line for involvement of international organisations in CCS issues and policy formulation (source: de Coninck and Bäckstrand 2011).

Note that the G8 Gleneagles Plan was first devised in 2005, not 2008.

De Coninck and Bäckstrand have documented the rise of CCS on the agenda of international organisations and a summary of their findings is shown in Figure 9.2 (de Coninck and Bäckstrand 2011). In only a few years, several international organisations involved with climate change and fossil fuel policy issues, carved out a central position for CCS in their energy technology strategies. As an example, in 2009 the IEA developed a CCS Technology Roadmap with a global ambition of 100 projects by 2020 and over 3,000 projects by 2050 – requiring funding of $4 billion per year for early projects from 2010 to 2020, and a total investment by 2050 of $5.8 trillion (IEA 2009).

A review of public sector R&D funding over this period also suggests that CCS rapidly became another tool in the box of emerging clean energy technologies. IEA data for the US and Norway show the rapid rise in funding from 2004 to 2009 (see Figures 9.3, 9.4 and 9.5 overleaf). In the US, much of this funding has been channelled into support for new projects under the federal government's FutureGen programme, as well as private-led operations with supplementary Federal financing.

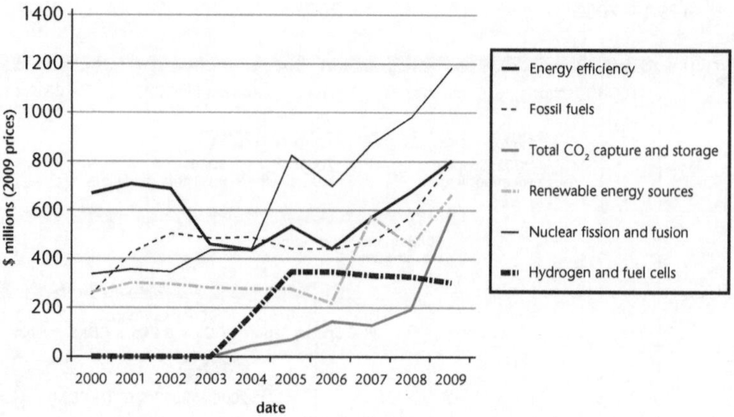

FIGURE 9.3: Expenditure on energy R&D in USA 2000–2009, at constant 2009 US$ (excludes American Recovery and Reinvestment Act 2009 spending).

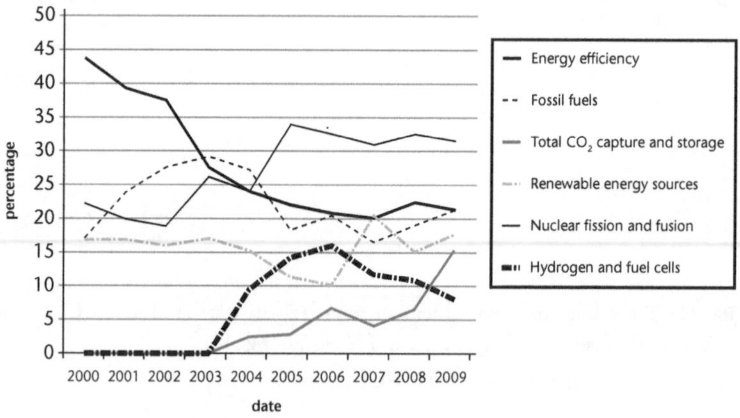

FIGURE 9.4: Share of expenditure on energy R&D in the USA as a percentage of total from 2000 to 2009.

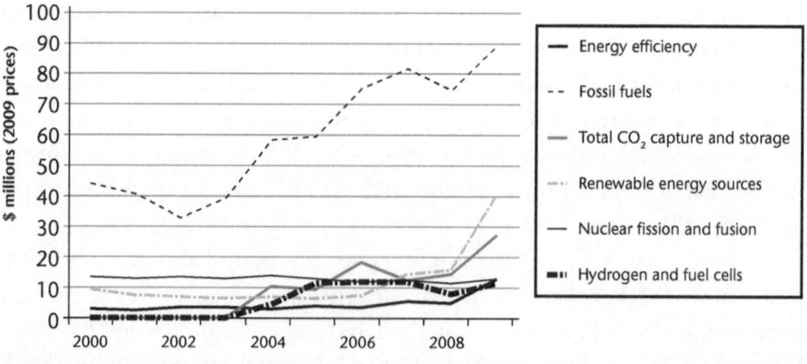

FIGURE 9.5: Expenditure on energy R&D in Norway 2000–2009, at constant 2009 US$ and exchange rates.

The rise of CCS advocacy groups and increased funding over this period cannot be divorced from the equally meteoric rise in media attention devoted to climate change and carbon reduction between the mid-1990s and the COP-15 in 2009. The publication of scientific reviews by the IPCC and economic assessments such as the Stern Review in Britain, the Garnaut Review in Australia, and the Princeton University Pacala/Socolow climate stabilisation wedges in the US brought perspective to the decarbonisation debate by focusing on the required scale of technology deployment. In the 1990s, it was widely believed that the scale of greenhouse gas reduction required from 1990 to 2050 to avoid dangerous climate change was on the order of 60 per cent (IPCC 1990; RCEP 2000). Advances in climate modelling, and a more rapid than anticipated growth in global emissions, led to that number increasing in about 2005–2006 to an 80 per cent reduction (in industrialised countries) by 2050. Government policies from the 1990s promoting renewable energy had mixed but, in most countries, limited success and globally there was a realisation that complete reliance on renewable energy technologies for electricity supply would be challenging due to intermittency in flows, energy storage complications, grid limitations and competition for land (IPCC 2007; Giampietro and Mayumi 2009), as well as public opposition e.g. to onshore wind farms and bioenergy crops (Devine-Wright 2011).

The promise (and criticism) of CCS is that it allows the status quo to remain largely untouched – the reliance on large, centralised fossil fuel power plants embedded within an electricity grid. End-of-pipe modifications to existing infrastructure was an attractive alternative to large energy companies as well as to many governments, compared with overhauls in transportation networks and the introduction of large-scale renewables and micro-grids, which appeared to require extensive socio-technical and behavioural changes (Unruh and Carrillo-Hermosilla 2006). Faced with an inexorable demand for electricity, a need to replace old power plants and a mixed reaction to renewables and nuclear, CCS began to appear as an attractive alternative, especially to policymakers and politicians who had been accumulating political capital from promulgating carbon reduction targets (Winskel, Chapter 12, this book).

CCS under greater scrutiny

We suggest that focus on CCS has now largely moved beyond early enthusiasm and cost assessment, and towards increased stakeholder scepticism (see Figure 9.6). This greater attention to the costs has come about as large-scale demonstration projects have come closer to being realised, raising a plethora of difficult-to-answer questions about design and engineering, planning and logistics, storage reservoirs, legal and insurance issues, regulatory affairs (such as specific public health concerns discussed in Evar and Shackley, Chapter 10, this book) and public acceptability. We observe a greater number of industrial, political and advisory actors beginning to review CCS in the critical light of having to make a project 'work' and within budget. More small-scale demonstration projects have also been undertaken to

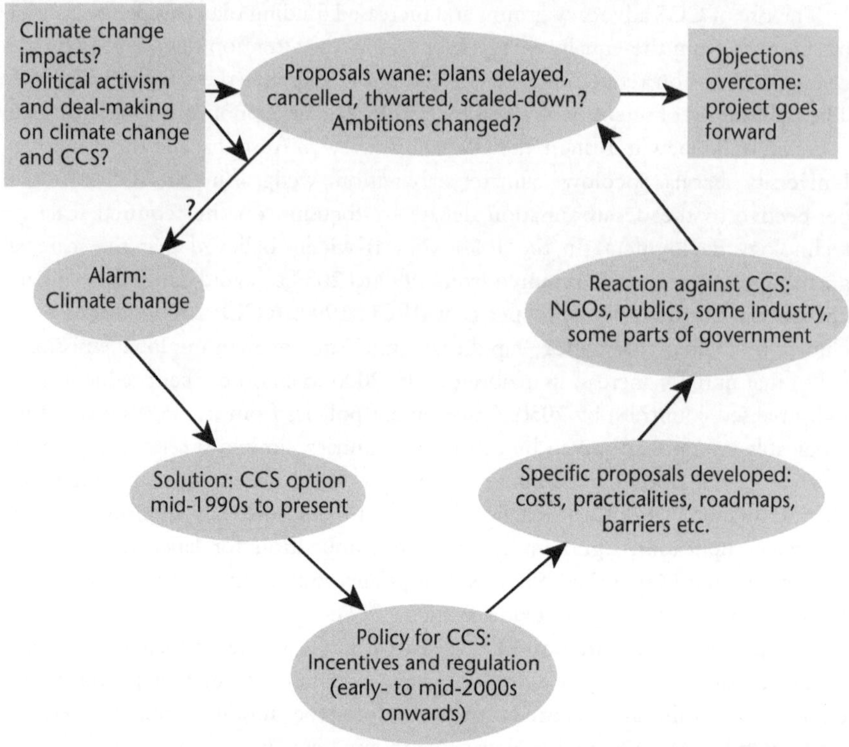

FIGURE 9.6: Application of the issue-attention cycle to CCS.

move towards the goal of commercial maturity and these projects are coming nearer to deployment. There is, therefore, better information available to use in evaluating technical, economic and socio-political feasibility. The results of this new wave of evaluations – not merely by the dedicated CCS community which has evolved over the past 15 years or so, but also by the wider energy technology, policy, commercial and financial investment worlds – has, arguably, led to a more sceptical stance towards CCS.

There is also an important difference between decarbonisation options that are themselves electricity generation technologies, such as renewables and nuclear power, and CCS, which is an add-on to an existing energy generation technology. Unlike renewables, nuclear and energy efficiency technologies, CCS stands or falls entirely on its carbon mitigation potential because it adds to the cost of electricity generation with no other benefit than carbon abatement. The main, and increasingly important, exception to this is where CO_2 is used in enhanced oil recovery (EOR). In the USA, there is 35 years of successful experience with CO_2-EOR using historical sources of CO_2. Approximately 50 million tonnes of CO_2 are used for EOR in the USA annually (Brinckerhoff 2011). The US industry is prepared to pay up to $38 per tonne of CO_2 (Pollak *et al.* 2011), well above the carbon price of $10–$20 per tonne CO_2 on the EU Emissions Trading Scheme. Experience with CO_2-EOR has been restricted to onshore extraction, and the cost of offshore applications is likely to be considerably higher. While some constituencies are considering using CO_2 for both on- and offshore EOR, it remains unclear whether this has sufficiently large benefits over other EOR methods to cover the additional cost of CO_2 capture and transportation and thereby to incentivise CCS (Dooley *et al.* 2010). Current industrial needs are far less than the amounts envisioned stored under CCS and there are currently no other commercial markets for large amounts of CO_2 (Brinckerhoff 2011).

As a profitable investment opportunity, CCS is therefore risky compared to energy generating technologies because it ceases to make sense (and provide a financial return) if carbon mitigation is downgraded as a political and policy concern (and carbon reduction targets are relaxed or pushed further into the future) and/or if alternative, cheaper or more publicly acceptable decarbonisation options arise in the future. If carbon mitigation continues to be a less germane policy consideration in the future, renewable energy and nuclear power plants would still be generating power and returning at least some revenue, while energy efficiency would still save end-users money. CO_2 capture plant could, arguably, be turned off, and CO_2 storage abandoned to save on operational costs, but that might be technically and politically contentious (as well as having incurred a high capital cost). CO_2 EOR, with its additional revenue stream from oil recovery, is therefore a more attractive proposition to investors and it is not surprising that this appears to be the dominant CCS coalition in the US rather than the coalition driven by climate change concern (Pollak *et al.* 2011).

As part of the increased scrutiny, the 'real-world' costs of CCS now appear to be in the upper range of, and even above early estimates. In the period 2000–2006, costs of $20–$80 per tonne CO_2 avoided were widely quoted, e.g. by the IPCC in its Special Report on CCS (IPCC 2005). Cost estimates at this time were more or less the exclusive preserve of CCS specialists, whether in companies or academia. Such optimism by early promoters, developers and enthusiasts is well known from other emerging technologies. By about 2008, however, some experts were putting the 'initial project' costs at $90–$100/tCO_2 avoided with cautiously worded expressions of uncertainty by industry and consultants (*Economist* 2009a, 2009b). And the higher estimates were now being expressed by officials in companies and consultancies beyond the CCS arena. The consultancy McKinsey for example, suggested in a November 2008 report that early demonstration (2015) project costs would result in a carbon abatement cost of $90–$135 per tCO_2 avoided (2008 prices and exchange rates). These costs might come down to $45–$60 per tonne CO_2 avoided by 2030 due to learning effects (McKinsey and Company 2008). However, CCS specialists have tended to raise the lower cost estimates in the 2005 IPCC range without extending it beyond its higher estimates. Hence, in its most recent update, the IEA has quoted a range for CO_2 avoided costs using CCS of $43–$80/tCO_2 depending on the type of combustion (IEA 2010b). It would not be unreasonable to expect this figure to continue to rise, as the history of technology development shows that real world applications of early-stage large-scale technologies generally escalate costs beyond the projections (Collingridge 1992). One potential source of such cost increases will be discussed in the Mongstad case study that follows this chapter and builds on the analysis presented here.

Early project failures began to take place in 2007, with the UK government's decision not to fund the DF1 Peterhead project and cost overruns in the US FutureGen 1.0 project in 2008. Questions began to be raised regarding the funding of future demonstration projects, pending results from pilot studies (GCCSI 2010), and the achievement of economic feasibility within a foreseeable time-frame (Balagopal *et al.* 2010). The new infrastructure required for CCS also began to be discussed more widely in stakeholder circles, alongside issues of regulation and public acceptability. Continued hostility to CCS projects within the Clean Development Mechanism (under the Kyoto Protocol of the UNFCCC, mediated by Brazil with support from other nations), and the failure of COP-15 to agree upon a clear pathway to future carbon reduction objectives, hindered the flow of carbon mitigation funding to developing nations and the establishment of potential export markets for capture technology. On the other hand, COP-17 managed to reach agreement on inclusion of CCS as an eligible technology within the CDM and a proposal from the UAE is rumoured to be under development.

From 2008 onwards, local public hostility to CO_2 storage projects began to appear in the Netherlands, Germany, Denmark and the US, and caught the industry by surprise. CCS advocates had thought of themselves as 'progressives', furthering the cause of carbon reduction, often having to make strong arguments in its favour within their own organisations, and were unprepared to be portrayed as regressive representatives of distrusted multinationals acting in their own commercial self-interests. The main suspicion has been that companies with considerable vested interests in the fossil fuel and power generation sectors would try to use CCS as a way of reconciling fossil fuels and carbon mitigation and hence to justify the construction of new coal-power plants in the near future, and never install CCS in the long run.

NGOs found themselves divided over such concerns, with Greenpeace taking an overtly critical (Greenpeace International 2008) to mixed stance, WWF UK reluctantly accepting the role of CCS alongside other low-carbon technologies (Allott and Kaszewski 2008) and Bellona continuing its strong support for CCS (Stangeland 2008). The very vocal and media-savvy 'climate camps' in the UK captured the limelight in 2008 with their protests over new proposals for coal-powered plants, such as Kingsnorth (Corry and Riesch, Chapter 6, this book). In 2011, Friends of the Earth Scotland issued a statement that was critical of CCS, reversing an earlier more positive stance towards the technology, stating:

> CCS is still largely theoretical and unproven at any commercial scale, and to spend so much time and effort working towards something that may never actually happen risks creating a significant and dangerous diversion in the short-term from the important aim of harnessing and developing Scotland's renewable, clean and truly sustainable energy sources.
>
> (Anon 2011)

In Germany, local protests led state-level politicians to oppose the federal government's plans for demonstration projects in 2010, especially after Greenpeace obtained and published a list of potential CO_2 storage sites. In summer 2011, the German 'climate and energy camp' in Brandenburg devoted itself to a critique of CCS, and in a press release stated:

> the energy companies will be liable for the extremely risky underground CO_2 dumps only for their first 30 years of operation. Afterwards it will be the public that will have to bear the risk of possible damages from these dumps for many centuries.
>
> (*InfoshopNews* 2011)

The environmental NGO community in Europe appears to have become more critical of CCS as delays in project development have extended and as concerns over both carbon and fossil-fuel lock-in have gathered momentum.

Public opposition to CCS has since led to something of an impasse on new coal-fired power plant development creating a headache for governments (with uncertainty on CCS also delaying new gas-fired CCGT in Norway for a decade). Instead of emerging as a 'silver bullet', CCS now seems to have further convoluted a Gordian knot of fearsome proportions. These developments led John McElroy, Director for Policy at RWE Npower, to state in mid-2011 that: 'Certain (EU) member states are no-go areas (for CCS projects). It's disappointing we haven't got to grips with that even though we've been discussing it for 10 years' (Carr 2011). The tone of the communiqué of the Carbon Capture and Storage Association (CCSA) is also one of increasing frustration. A letter to the UK's (then) Minister for Energy and Climate Change, Chris Huhne, in June 2011 from the CCSA's Chief Executive Jeff Chapman noted that:

> the CCSA Board is concerned that industry does not yet see firm Government signals about the long-term viability of a CCS market, particularly when compared to the signals given for nuclear and renewables ... Without clear recognition from Government that CCS will be part of the generation mix, we are concerned that industry's appetite for continued commitment will shrink. We would urge you to say that roll out (of CCS projects beyond the 4 demonstration projects planned for the UK) should be planned on the basis that Projects 1–4 will be successful and will yield valuable learning through planning, design, build and operation.
>
> (Chapman 2011b)

There has been a huge slow-down in planned demonstration projects and few major projects have been developed anywhere since 2006, despite numerous feasibility studies and expressions of intent by governments to resolve the fossil carbon 'problem' accompanied by companies' intentions to develop CCS. The following is a list of some noticeable delays and failures in policy and project development in recent years:

- Barendrecht, the Netherlands: a CCS project of Shell, which was abandoned in 2010 due to public opposition. As a direct consequence, the Dutch government has decided that, for the time being, all CCS projects in the Netherlands will be conducted offshore (at greater expense to developers).
- Beeskow, Brandenburg, Germany: Vattenfall has faced well-organised local opposition towards this German project (Donath 2010). In an unprecedented development in German regulatory culture, the proposed CCS law (2011) permitted states to veto implementation of the national law. The two States with the greatest geological storage opportunities – Schleswig-Holstein and Lower Saxony – had indicated their intention to veto implementation. Public protests in Schleswig-Holstein erupted in 2010 in response to the power company RWE's proposed CO_2 pipeline through the state. Vetoes by these two States would effectively put an end to onshore CCS in Germany more generally given lack of many other large geological formations for CO_2 storage. Offshore CO_2 storage could still be contemplated (beyond the State-regulated zone which extends 12 miles from shore), though there is limited capacity within German territorial waters. In 2011, Vattenfall announced that it was abandoning its proposed project at Beeksow. Meanwhile, the Bundesrat rejected the CCS law in September 2011, meaning that the transposition of the EU CCS Directive into German law has been thrown into new uncertainty.
- The US: objections to CCS have emerged from local communities, e.g. in Greenville, Ohio and Long Beach, California (Bradbury, Chapter 4, this book; Hammond and Shackley 2010).
- Mountaineer, WV: in July 2011, American Electric Power (AEP) announced that it was cancelling its CCS activities at the Mountaineer Plant in West Virginia. AEP's work began in 2009, with the intention of building a $668 million CCS facility that would capture and store more than one million tonnes of CO_2 per year, the costs to be subsidised by the US Department of Energy and the State treasury (Wells and Elgin 2011). However, the state decided not to meet the funding committed by the DOE. At the time of cancellation, the project had already cost $100 million. AEP's CEO, Michael Morris, explained that 'at this time it (the CCS project) doesn't make economic sense to continue' and a company press release cited 'the uncertain status of U.S. climate policy' (Wells and Elgin 2011: 1).
- Mongstad, Norway: project costs have soared since its announcement in 2006 and an investment decision for a demonstration project has been pushed from 2012 to 2018 (covered in detail in Evar and Shackley, Chapter 10, this book).
- The UK: in 2007, BP and partners cancelled their DF1 Peterhead project in Scotland, citing a lack of government commitment to funding, and moved it to Abu Dhabi, where it was announced in 2011 that the project would be suspended pending a more favourable economic context. Incidentally, the project was named DF1 as it was intended as the first of ten proposed CCS projects globally, none of which have come to fruition as yet. The government instead wanted to continue with a 'CCS competition', with up to three

demonstrations being proposed. Later that year, the scope was limited to a single post-combustion capture plant. This reduced the number of entrants to four (BP Alternative Energy, E.ON UK, RWE/Peel/Dong and Scottish Power). Over the following years, all entrants except for Scottish Power (Longannet plant) withdrew from the competition. Despite the plan to announce a winner by the end of 2009, the government announced in October 2011 that the Longannet plant would not receive funding, citing higher than anticipated transportation costs. £1 billion is still available for another project to apply for.

- The UK: the Energy Act of 2010 provided a financial support mechanism for the CCS sector with a proposed levy that would allow electricity providers to charge consumers the costs of implementing CCS. This would constitute a largely risk-free funding source for industry developers. However, in 2011 the new coalition government announced a fundamental Energy Market Reform (EMR) process, which looks likely to retract the levy. Whether the EMR is beneficial for CCS or otherwise remains to be seen, but the loss of the levy is surely a blow to the sector.
- The UK: in 2011, the UK's Technology Strategy Board turned down a proposal from the Yorkshire region to develop a large CCS network by creating one of six new UK-wide Technology and Innovation Centres (TIC) (*Yorkshire Post* 2011). Asked his opinion on the significance of this decision, one prominent academic in the CCS field stated: 'It means that TSB have come to a view that CCS in general doesn't have enough readiness and enough platform technologies to be a primary driver of UK global industry and science innovation'.
- The EU: the CCS Directive, published in 2009, contains a requirement that before responsibility for a storage site can be transferred by the company to a competent authority, the stored CO_2 must be 'completely and permanently contained' (see Box 9.1). On a strict interpretation of the precautionary principle (which is fundamental to European Union law-making), Article 18 implies a zero-risk threshold of CO_2 leakage from a storage site. Some in the CCS community believe that the Article puts such a high barrier for projects to satisfy, that it effectively prevents them from happening. This concern has not disappeared with the publication of Guidance Documents detailing the Commission's interpretation of appropriate evidence of safe storage (EC 2011a).

Downs' framework suggests that the time delays, rising cost estimates and inadequate funding witnessed in the CCS arena should be expected when great hopes are pinned on an emerging technological solution to solve an environmental problem. Rather than see increasing degrees of roll-out of demonstration projects in the near-term as some have proposed (Gibbins and Chalmers 2008b; CCSA 2011), the issue-attention cycle suggests that scrutiny is more likely to increase and delay projects as resource expenditures and actual applications are contemplated.

The effect of the greater involvement by many stakeholders is to contain rather than promote CCS, at least until its position alongside other energy technologies and low-carbon options is better ascertained and more accurately costed. Meanwhile, this largely precludes acquiring the all-important experience with coal power plants at large-scale in the near-term.

BOX 9.1 ARTICLE 18, PARAGRAPH 1, OF THE EU CCS DIRECTIVE, 2009/31/EC

Where a storage site has been closed pursuant to points (a) or (b) of Article 17(1), all legal obligations relating to monitoring and corrective measures pursuant to the requirements laid down in this Directive, the surrender of allowances in the event of leakages pursuant to Directive 2003/87/EC and preventive and remedial action pursuant to Articles 5(1) and 6(1) of Directive 2004/35/EC, shall be transferred to the competent authority on its own initiative or upon request from the operator, if the following conditions are met:

(a) all available evidence indicates that the stored CO_2 will be completely and permanently contained;
(b) a minimum period, to be determined by the competent authority has elapsed. This minimum period shall be no shorter than 20 years, unless the competent authority is convinced that the criterion referred to in point (a) is complied with before the end of that period;
(c) the financial obligations referred to in Article 20 have been fulfilled;
(d) the site has been sealed and the injection facilities have been removed.

Source: EC 2009.

The news for CCS development is not all negative however, and positive developments include the announcement in May 2011 that the Canadian Boundary Dam project would proceed in Saskatchewan, with operations commencing in 2014 (Cansolv 2011). Meanwhile, funding is being made available by the EU, both in its CCS demonstration programme (worth €180 million per facility) and from the auctioning of 300 million EU ETS Emission Reduction Units (ERUs), the so-called NER300 funds, for supporting both CCS and renewable energy projects across the EU. The UK government has also committed £1 billion to support a first CCS demonstration project and indicated that it intends to support three more projects to follow. The forthcoming EMR is also intended to be designed to support CCS through a 'contract for difference' and through 'capacity payments'.

Hence, our argument is *not* to deny that large quantities of public funds (and smaller amounts of private funds) are still being channelled into CCS technology and project development, but rather to assert that we are still in the early stages of technological innovation and demonstration, rather than deployment. A great deal of uncertainty remains with regards to the outcomes of technology development and project management despite projections of imminent success from the CCS community in the mid-2000s, and we are a long way from where the IEA suggests that we should be in its 2009 CCS Roadmap (IEA 2009). This is not merely explained by a tendency towards inflated hopes and hyped expectations on the part of scientists, technology developers (Bazerman 2006; Borup *et al.* 2006; Hansson, Chapter 5, this book) and industry, but also unfulfilled promises of funding, inadequate policy frameworks and financial incentives by governments, poor stakeholder management and public engagement activities, and possibly inadequate strategic management at the project level (Evar and Shackley, Chapter 10, this book).

Analysis of UK CCS articles in *ENDS*, 2008 to 2011

In keeping with Downs' hypothesis that public scrutiny increases as evidence from cost and feasibility studies accumulates, we assessed the portrayal of CCS in articles from the UK's leading environmental professionals' journal, *ENDS*. We selected the journal because of its broad professional readership, and its emphasis on cost assessments, technical viability and detailed policy developments.

The search covered three years, from 22 February 2008 to 22 February 2011. The titles of each article were scored from between 1 and 5, where 1 represents a very negative perspective on CCS, and 5 a very positive perspective. We also scored the contents of each article on the same five-point scale. There is inevitably a degree of subjectivity associated with the scorings, but we each scored a selection of articles individually and compared our results to ensure consistency (intercoder reliability in these cases was close to 100 per cent). Articles with only a passing mention of CCS were not included. In total, 191 articles were scored, the number increasing from 43 in 2008 to 53 in 2009 and 88 in 2010. The results are shown in Figure 9.7 for the titles and in Figure 9.8 for the contents of the articles.

Overall, increased negativity was observed in the tone of reporting over the three-year period, both in the headlines and, more strikingly, with respect to the actual contents of articles. While a negative trend emerges, there is some degree of variability, which can be put down to the UK's general election of May 2010 and to the pronouncements regarding the EMR in 2011. This variability shows that while professionals' perspectives on CCS are coloured by the technology's role in the political context, the overall development has been an increased scepticism, as technical and economic details have emerged from a wider set of stakeholders and received increased attention.

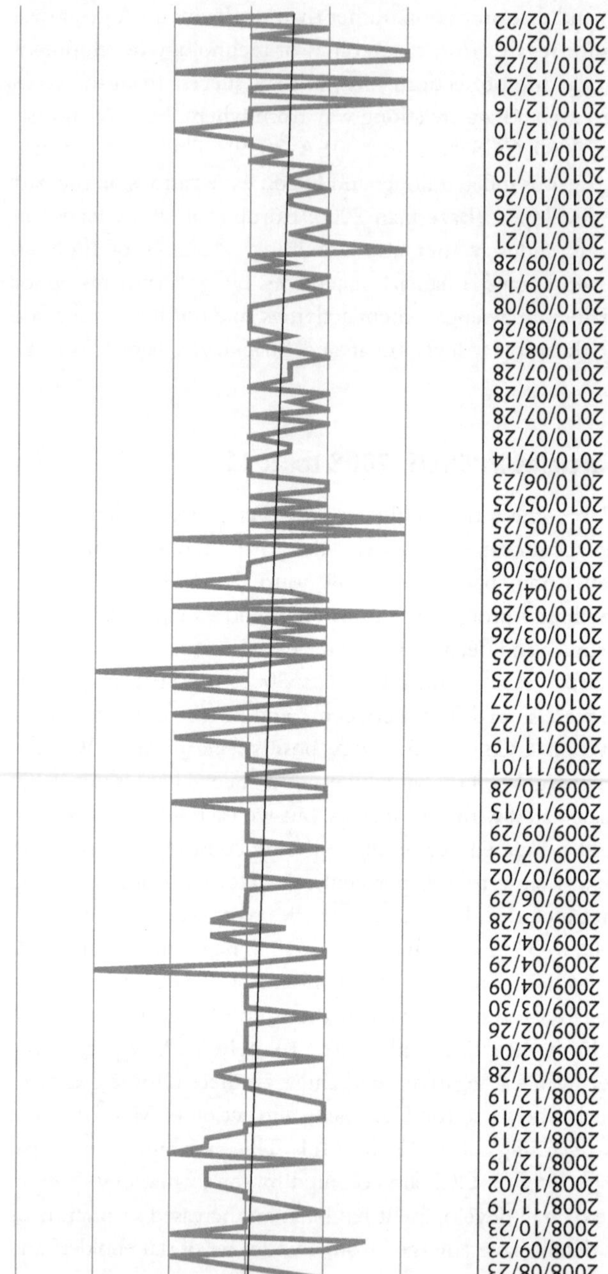

FIGURE 9.7: Analysis of titles of articles published in *ENDS* referring to CCS, February 2008 to February 2011 (5 = very positive view on CCS and 1 = very negative view on CCS).

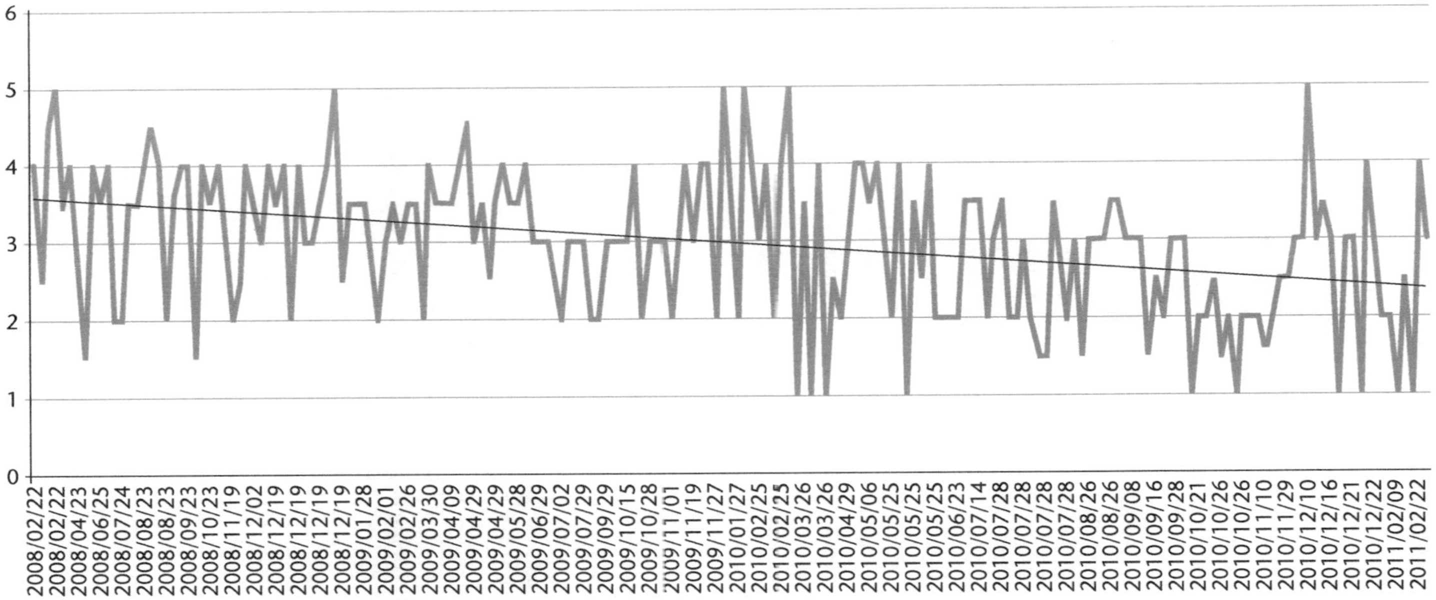

FIGURE 9.8: Analysis of content of articles published in *ENDS* referring to CCS, February 2008 to February 2011 (5 = very positive view on CCS and 1 = very negative view on CCS).

From the under-critical to the over-critical phase in the turning of the issue-attention cycle

We now turn to the question of *how* the growth of critical scrutiny and scepticism during the course of the issue-attention cycle can be explained. Downs explained this in terms of increasing awareness of the economic costs relative to the benefits. We supplement Downs' idea by drawing upon Collingridge and Reeve's (C&R) model of under- and over-critical science for policymaking (Collingridge and Reeve 1986) and Collingridge's previous work on 'error-costs' (Collingridge 1982). This work suggests that the error-costs – that is the cost of making a wrong decision on, e.g. the choice of CO_2 capture technology – increase dramatically as new technologies move closer to demonstration. Issues such as health, safety and local environmental impacts are also paid more attention. Key to the error-costs argument is the notion of technology 'lock-in' (Arthur 1994) – that is, the degree to which options are consciously and unconsciously limited as specific technological development progresses. Because of the risks of heightened error-costs as development progresses, those involved in funding, managing or otherwise scrutinising projects will ask more probing questions as to the technical feasibility, possible problems and potential for cost-increases. Other groups in society who are less enamoured with, openly sceptical of, the proposed technology, or have a preference for an alternative option, will intensify their scrutiny as deployment begins. Hence, as CCS technology progresses around the issue-attention cycle, contending values and technical uncertainties come to the fore.

C&R express this relationship between science and technology policy in a simple model, where political consensus on an issue will result in under-critical acceptance of the knowledge and technologies underpinning it, while political dissensus will result in over-critical scrutiny. Since C&R first proposed their model several decades ago, debates over the role of technical and scientific knowledge in policymaking have received much greater acknowledgement (see, for example, work by Ezrahi 1990; Wynne 1996; Porter 1995; Jasanoff *et al.* 2002; van Zwanenberg and Millstone 2005; Beck 1992). It has been argued that what are essentially political debates, often rooted in conflicting values, masquerade as technical disputes. Hence, technical controversy is frequently a key dynamic by which political dissensus is expressed, within government and beyond.

If political consensus surrounding the need for deep carbon cuts persists, then the under-critical model might be most appropriate to explain developments, as demonstration and deployment gather pace and learning accumulates. In a political context of dissensus, on the other hand, there is a strong incentive to deconstruct knowledge, which does not support the chosen position, and for policymakers and other stakeholders to employ experts who can assist in this deconstruction. In that case, increasing amounts of knowledge are generated, and the associated uncertainties are brought to the fore and emphasised by critics, fuelling dissensus yet further. While such scrutiny arguably results in more robust knowledge constructs, it never entirely eliminates uncertainty nor the scope for different interpretations of evidence.

The types of knowledge that are produced and used within the under- and over-critical models also vary and here we draw upon the idea of Mode I and Mode II knowledge production (Gibbons *et al.* 1994). The suggestion is that, largely disciplinary-based, Mode I knowledge establishes the technical credentials for CCS (i.e. the geological and engineering case), but as CCS gets closer to deployment the demand from policymakers and industry shifts the knowledge required, or towards Mode II knowledge production. Mode II is more problem-focused knowledge, which has to draw upon a range of disciplines, and concerns wider problems of assessing how demonstration projects can be taken forward to deployment.

In the over-critical phase, there is thus *diversification* of knowledge-constructs: from technical and scientific critiques, to social science critiques, and a wider range of participants in both discipline-specific (Mode I) and the wider field (Mode II) of knowledge production. The increased scrutiny results in more robust knowledge constructs but there is less synthesis and integration between findings and claims, so the emergence of dominant ideas is messy, slow and confusing, and consensus is elusive. If there is sufficient policy demand for the technology, companies will begin to become more involved in this type of knowledge production, forcing 'real-world' integration as more reliable information on costs and risks is advanced. Under such a condition project development may 'escape' the issue-attention cycle (see Figure 9.6).

Is there presently enough evidence of the reception of CCS knowledge to suggest that either an exclusively under-critical or over-critical model is in operation at this time? We do not believe that there is, partly because the evidence discussed in this chapter can always be interpreted differently, e.g. as anecdotal with other contrary examples of CCS success, such as announcements of funding support for demonstration projects throughout 2010 and 2011. Yet, while each demonstration project needs to be assessed on its own terms as it emerges, the scientific, technical and public policy disagreements that have emerged since COP-15 do appear to be gathering momentum, leading to explicit or implicit questioning of the core premises underpinning CCS as a carbon mitigation technology. While there continues to be massive support particularly from global energy-focused institutions, professional and trade groups, this support has met with growing scepticism not only from NGOs and public interest groups, but also increasingly from academics (Tsouris *et al.* 2010; Page *et al.* 2009). This is, in many ways, entirely to be expected from the perspective of science and technology studies.

Conclusions and implications

The advocacy coalitions that emerged around climate change, energy policy and carbon reduction, in the 2000–2008 period, are now beginning to look more vulnerable. The relatively low cost estimates of responding to climate change predicted by policy planners such as Stern and Garnaut have met with economic recession and more detailed (and usually higher than anticipated) cost assessments of alternative energy technologies and other low-carbon mitigation options. The liberalised energy markets in Europe and the US have not managed to deliver cheap and clean energy innovation to the degree previously envisioned, and governments now appear to be contemplating continued or more active intervention in energy markets once again (the UK Energy Market Reform, the American Recovery and Reinvestment Act etc.). Hence, it is largely now acknowledged that alternative energy development requires a much more interventionist stance from policies such as price incentives, government-issued carbon bonds, a high carbon floor-price, or mandates such as emissions performance standards on CO_2 emissions.

If climate change remains a relevant public policy issue and is identified as a growing problem, it is reasonable to assume that some mitigation solutions will need to be found, whether on an international scale, through regional partnerships or at the national level. CCS remains a key part of the carbon mitigation 'toolkit' and two major questions therefore emerge from this analysis: how do we make 'good' decisions about technology development where the costs as well as the science and technology are uncertain, and what level of error-costs and precaution are appropriate where alternative mitigation options (e.g. behavioural change) may require far more significant socio-economic changes in the near-term?

We concur with O'Riordan that a modified version of the issue-attention cycle continues to be useful in understanding the dynamics of environmental debate and controversy – in our case regarding CCS. We have applied the cycle to the domain of the policy community (rather than using it to explain lay public attention), and elaborated on Downs' cycle by expanding on the theoretical explanation of why and how the issue at stake is subject to more scepticism and loses out on more favourable forms of attention. We refer to this reworking as a 'positive-negative issue-attention cycle', since it is not necessarily the case that the issue becomes subject to *less* attention, rather that the attention goes from being positive to more negative.

In the years since Collingridge and Reeve developed their 'under-critical/over-critical' theory, science and technology studies has illuminated the extent to which scientific and technical knowledge for policy interacts with, and partly constitutes, policymaking, while such knowledge in turn is also being partly constituted by policy processes. Rather than science (for policy) and policymaking being two separately constituted arenas, they are often mutually constitutive. Scientific and technical knowledge can possibly come to play a larger role in maintaining political consensus than C&R envisaged, and further theoretical developments could seek to combine insights from case studies on science and technology policy into an updated version of the issue-attention cycle.

Acknowledgements

We would like to thank Dr Florian Kern, of SPRU, University of Sussex, and Dr Nils Markusson, University of Edinburgh, for comments on earlier drafts, responding to which has greatly improved the chapter.

Notes

1 Available at: www.ieaghg.org/index.php?/RDD-Database.html.
2 Available at: www.netl.doe.gov/technologies/carbon_seq/global/database/index.html.

10

TECHNOLOGY MANAGEMENT IN THE FACE OF SCIENTIFIC UNCERTAINTY

A case study of the CCS Test Centre Mongstad

Benjamin Evar and Simon Shackley

Introduction

This chapter builds on the theoretical model from the preceding chapter by Shackley and Evar, and applies concepts of uncertainty management from the science and technology studies literature to a case study of the CCS Test Centre Mongstad (TCM) near Bergen, Norway. The TCM follows in the footsteps of Norway's endeavours into CCS technology piloting and demonstration, begun in 1996 with the Sleipner offshore CO_2 separation and storage project, and continued with the Snøhvit and Kårstø projects, as well as the now cancelled Halten facility. It is distinguished from similar projects worldwide as one of few that aims to separate CO_2 from several types of flue streams including a gas-powered cogeneration plant and an oil refinery.

In 2010, an inquiry was launched into the health risks posed by airborne degraded amine components, a by-product in the post-combustion capture process. This chapter addresses the source of this concern, its political and scientific context, and the implications for technology development. Specifically, the chapter aims to explain how strategic technology management could be informed by insights from the science and technology studies and policy science literatures, specifically the neo-incrementalist school.

The chapter starts with an overview of neo-incrementalist concepts, a short account of the political context for climate change mitigation in Norway, and a summary of budget increases and delays at Mongstad. The emergence of a health concern related to CO_2 capture at Mongstad is thereafter covered, with a focus on the implications of differing interpretations of scientific uncertainty for risk management. The chapter concludes with an analysis of how neo-incrementalism would help to inform strategic technology management in the presence of scientific uncertainty.

Incrementalism and the management of technology under scientific uncertainty

The concept of incrementalism has its genesis in the policy science tradition with Charles Lindblom's work. Lindblom observed that the 'rational-comprehensive' or 'synoptic' ideal – in which policymakers examine and compare all possible options before deciding on an optimal approach – was a cumbersome prescription and an inaccurate description of how decision-making takes place under uncertainty and under pressure to establish agreement between political actors. He explained that observers of the policy process instead are well aware that decision-making in such cases follows a set of 'successive limited comparisons', in which choices are made incrementally to reach agreement on the basis of imperfect assessments. Under this approach end goals are still defined, but critically, are open for re-evaluation as new information emerges. And rather than assume that there is a direct connection between a policy and a final objective as the rational-comprehensive method does, a test of 'good' policy under the method of successive limited comparisons is more modest. It simply asks that competing interests can agree on discrete steps rather than end goals: 'In an important sense, therefore, it is not irrational for an administrator to defend a policy as good without being able to specify what it is good for' (Lindblom 1959: 84).[1]

Lindblom's description of the policy process was subsequently applied in the science and technology studies tradition by neo-incrementalist academics such as Collingridge (1982, 1990, 1992), Collingridge and Reeve (1986), Woodhouse (Weiss and Woodhouse 1992; Woodhouse and Collingridge 1993) and Morone and Woodhouse (1986), as a *prescriptive* concept for the management of 'big science and technology' projects. When the cost of committing errors is relatively large, rather than adopt strategies for technology development that aim to meet pre-determined end-goals, these academics have suggested that technology selection and application should follow the path of successive limited comparisons. Such advice follows Lindblom's observations on successful policy development, formulated as 'disjointed incrementalism' in his follow-up article from 1979:[2]

1. Limitation of analysis to a few somewhat familiar policy alternatives;
2. Adjustment of objectives in light of the policies potentially available, rather than considering ends in the abstract;
3. More preoccupation with ills to be remedied than positive goals to be sought;
4. A sequence of trials, errors, and revised trials;
5. Exploration of only some, not all, of the important possible consequences of a considered alternative;
6. Fragmentation of analytical work to many partisan participants in policy making.

Applied to technology development, these guidelines recommend that early analysis is limited in scope; that end-goals are updated by information from early trials; that such trials do not seek to perfect a pre-selected solution, but instead consider alternative routes to solve a problem; that trials are continuous and therefore relatively small in size; that all possible risks are not exhaustively analysed in trials; and that different actors are provided with a chance to inform analysis, so that the solution(s) arrived at have an increased chance of being implemented. These prescriptions are discussed in relation to the TCM case study towards the end of the chapter.

The political resolve behind CCS in Norway

Norway's investment in CCS research has served as an exemplar to members of the European Union and internationally since Statoil began storing CO_2 from natural gas extracted at the North Sea's Sleipner field, in the Utsira geological formation in 1996. While Norwegian domestic electricity is almost exclusively generated from hydroelectric dams, oil and gas exports in 2009 accounted for nearly a quarter of the gross national product (22 per cent) and nearly half of all income from exports (47 per cent). Estimates by the government indicate that such supplies will remain plentiful in decades to come as new fields are developed (NPD 2010).

At the same time that the country is committed to its economic reliance on oil and gas exports, a strong political commitment to climate change mitigation has developed since the early 1990s. In 1991 a CO_2 tax was levied on mineral products and the hydrocarbon extraction industry, which applies to 68 per cent of CO_2 emissions. A national emissions trading scheme (ETS) was first launched in 2005 and merged with the EU ETS in 2008. Also in 2008 a national carbon reduction framework committed Norway to achieve zero net emissions of CO_2 by 2030, partly by investing in afforestation projects in developing countries and partly by delivering efficiency measures to domestic infrastructure. This framework also supports development of CCS technology domestically. A special CCS reserve for new gas-fired power plants is also provided in the country's ETS allowances, which follows the implementation of an emissions performance standard (EPS) in 2000 on new gas-fired power plants (IEA 2011a).

These policy instruments indicate that CCS occupies a central role in Norwegian energy policy, by helping to legitimise continued hydrocarbon extraction and export. In a study of the Norwegian political environment, Andreas Tjernshaugen and Oluf Langhelle suggest that CCS has emerged 'as an important compromise solution to the country's climate policy dilemma' (Tjernshaugen and Langhelle 2009: 111). However, they point to a number of incidents since the later 1990s, which indicate that strategic technology development largely has been left to stakeholders outside the political system, despite the government owning 70 per cent of Statoil – the Nordic region's largest oil and gas operator (NMTI 2006). This in turn has resulted in politicians often being caught by surprise when a compromise solution on technology development has failed.

As an example, in 1998 Norsk Hydro (now Statoil) announced that a pre-combustion CO_2 capture plant would be built in the wake of political controversy over intentions to commence gas delivery from the continental shelf for domestic use. The CCS plan was shelved the following year and rebranded as a long-term research project. However, it soon became a basis for demands to use best available technology to capture 90 per cent of emissions from the intended gas plant. Since such an option was not available, but had instead been based on projections of potential CO_2 capture, the controversy escalated and led to the resignation of the prime minister in 2000.

In 2006, under the Stoltenberg government, CCS again arose as a key compromise solution between industry and environmental groups and plans were formed to develop the TCM by 2012, as a stepping stone to large-scale capture. The centre would be charged with developing conceptual studies of CO_2 capture for an oil refinery and a cogeneration plant, and testing capture technologies through laboratory work and pilot tests. The Mongstad location was chosen for the many emission points connected to the refinery including a catalytic cracker plant, Norway's largest single source of CO_2 emissions.

In what follows, an account of rising cost estimates and delays at Mongstad will be discussed with a focus on the emergence of a health risk concern and the relationship to technology development. The chapter will specifically discuss the high economic and political error costs associated with a technology development strategy that has been subject to much scientific uncertainty, where end-goals have apparently not been updated regularly, where alternative technology options have long been limited in scope, and where a focus on the potential risks from one option has stalled development and subjected the project to political controversy.

Mongstad budget and timeline

At the time of its announcement in 2006, Prime Minister Jens Stoltenberg called the TCM Norway's 'moon landing', referring to the 1960s space race:

> Our vision is that within seven years we must have in place the technology that makes it possible to clean the emissions of greenhouse gases. There will be an important breakthrough for reducing emissions in Norway, and when we succeed, I think the world will follow. This is a major project for the country. It is our moon landing.
>
> (Stoltenberg 2007)[3]

To cement its support for the project, the government initially agreed to take ownership of 80 per cent of the TCM (since reduced to 75 per cent with the inclusion of additional partners) through its company Gassnova, while Statoil agreed to fund the remaining 20 per cent and to manage day-to-day operations (MPE 2011).

The plant was being designed to capture 100,000 tons of CO_2 annually from 2012 for five years, as a basis for a potential decision to scale up to capture two million tons of CO_2 annually. The intention was to perform small-scale trials with a chilled ammonia solution by Alstom at the catalytic cracker plant and an amine solution by Aker Clean Carbon at the combined heat and power plant. Results from these tests would constitute the evidence basis for a feasibility study into a full-scale demonstration capture plant (MPE 2007). Statoil was charged with covering any cost increases for such a plant beyond the projected budget at the time of an investment decision. However, until such time the government would act as the principal funder of research and development at TCM. Early estimates in 2006 suggested that costs would amount to 1,200 million NOK (€150 million; US$220 million), while the large-scale demonstration plant was estimated to cost around 25 billion NOK (€3.2 billion; US$4.6 billion).

Changes to cost estimates were first announced in 2008 when the government budget announced that the TCM would be likely to cost closer to 2,000 million NOK rather than the original estimate of 1,200 million NOK. Separately, additional planning and preparation costs towards full-scale capture between 2010 and 2012 were estimated to reach 1,500 million NOK (NMF 2009). The plans were now also officially delayed and 2009 was deemed to become the most costly year for development. The budget therefore suggested to more than double the allocation for 2009 from 920 to 1,882 million NOK. Cost reductions from further development of amine technology were also curtailed and now estimated to be between only 0 and 10 per cent, far below what was projected by Aker in 2006 (MPE 2009b).

Despite the projection of significant developments in 2009, allocations for the year were subsequently reduced from the proposed 1,882 million NOK back to just 900 million NOK. Allocations towards full-scale planning and preparation were also reduced by 320 million NOK, citing technology and judicial complexities that would require the development of new solutions in years to come (MPE 2009a). Finally, the total cost estimate increased again, now to 5,200 million NOK (MPE 2009b). In April 2010 the government concluded that the original 2012 deadline no longer was prudent and instead moved it forward to 2014 (NMF 2010). The year after, in 2011, the deadline was again moved forward two years to 2016 and the total cost estimate now stood at 6,452 million NOK (MPE 2011).

These significant changes have not gone unnoticed in the Norwegian press. The environmental NGO Bellona, which has been a strong supporter of CCS, has responded to the delays and cost increases with particularly sharp criticism. Bellona was very critical of the April 2010 announcement to delay the investment decision for large-scale capture until 2014, with one leader calling it 'the worst [environmental policy scandal] in my ten years in the Norwegian green policy debate' (Fouche 2010). Following another round of budget changes, Bellona called for Statoil to make all information pursuant to decision-making at Mongstad public. The NGO said that they believed Statoil was creating 'unnecessary uncertainty' around an issue that researchers as well as Aker Clean Carbon believed was manageable (Kristensen and Kokkvoll Tveit 2010). The topic of this uncertainty will be discussed below.

CO_2 capture technology choices

In response to criticism from Bellona and others, Statoil and Gassnova have explained that the revised cost estimates and development trajectories for TCM have arisen from knowledge gaps around the capture technology, specifically related to health risks. A 2008 TCM impact assessment pointed out that a number of research projects had been initiated to address the risks from such uncertainties, including the health-related effects of atmospheric amine emissions and any degraded components (TCM 2008).

These uncertainties were commented on extensively in Statoil's TCM Master Plan released in February 2009 (StatoilHydro 2009), which noted that collaborative research projects were addressing the health risks posed by amine degradation. Because post-combustion amine capture was considered to be the most mature available technology based on evidence from other gas separation and CCS pilot projects, the report concluded that only post-combustion technologies, and specifically amine and chilled ammonia capture, were advanced and promising enough to take forward in a qualification phase. However, the report also noted that the TCM would test amines in a relatively novel fashion, in which some of the solution would be allowed to escape to the atmosphere following the CO_2 absorption/desorption cycle.

Pre-combustion and oxy-fuel technologies had already been ruled out in 2006 because of their relative complexity compared with an end-of-pipe solution, and were now ruled out for the TCM project going forward, as both were deemed to cause significantly more disruption and reconstruction of existing and new plants than post-combustion capture would. It is noteworthy that this reasoning was maintained despite the TCM's goals to specifically 'identify, develop, test and qualify possible technological solutions' and to 'reduce costs and risks associated with building and operating full-scale CO_2 capture'[4] (MPE 2009b). One would therefore assume that pre-qualification of specific technologies would go against the grain of the project's aims. Amines were noted as being particularly promising because of their use in the Sleipner and Snøhvit projects, but neither of these projects emits a comparable amount of amines to the atmosphere. The same report noted that the CO_2 captured from flue gases at Mongstad was likely to contain nitrogen oxides and oxygen and that these compounds could contribute to the formation of secondary chemicals such as nitrosamines and nitramines, formed during the degradation of amines. Both nitrosamines and nitramines, it was noted, were known to be carcinogenic to animals and humans, but their potency and longevity in the environment was uncertain.

In light of these uncertainties, the Statoil Master Plan therefore proposed two possible development trajectories. Under the first one, CO_2 would be captured from the cogeneration plant using available amine technology with the aim of the fastest possible start-up. The refinery's catalytic cracker plant would be developed in a second phase based on the gradual development of the chilled ammonia technology. Because of a CO_2 concentration three times greater than in the cogeneration plant, capture at the cracker plant was thought to be able to reach upwards of 85 per cent. This would make the less pollution-sensitive chilled ammonia solution more appropriate than an amine-based technology. The second option proposed the simultaneous development of technologies for the two plants based on a phased technology plan, with a minimum of one and a half years of successful operation of the TCM before scaling up the technology. It is noteworthy that neither of these plans proposed to test a suite of technologies in addition to the amine and chilled ammonia solutions, despite previously acknowledging that both would present clear challenges to scale-up, and that amine development would possibly include some health risks.

Amines and uncertain health risks

An early environmental risk screening study by the Norwegian Institute for Air Research (NILU) from March 2009 (Knudsen *et al.* 2009) also noted the uncertainties attached to the formation of nitrosamines and nitramines from amine degradation. In an attempt to quantify this risk the report referred to a safe exposure threshold:

> It was discovered that the long-term risk threshold for exposure of the general population by nitrosamines through inhalation is 4 ng/m^3 (nanograms per cubic metre) nitrosamines in air, corresponding to a 10^{-6} [or one in a million] lifetime cancer risk.
>
> (Knudsen *et al.* 2009: 14)

Based on this figure the report noted that, 'the risk related to the studied amines themselves seems to be sufficiently low' (p. 15). However, no mention was made of the source material for this threshold.

In September 2010 the TCM consortium filed an application for an emission permit with the Climate and Pollution Agency (KLIF) (TCM 2010). The report commented that amine solutions were known to release carcinogenic substances, principal of which were nitrosamines, during CO_2 capture and subsequent amine degradation. Because safe exposure levels for nitrosamines did not exist under Norwegian law, this report now referred to a US Environmental Protection Agency (EPA/IRIS 2011) safe exposure limit of 0.07 ng/m^3 corresponding to a 10^{-6} lifetime cancer risk for N-Nitrosodimetylamin (NDMA), a relatively potent nitrosamine, as a reference chemical to compare with release scenarios for other nitrosamines and nitramines. The TCM report noted that the use of this exposure threshold followed the precautionary principle as NDMA, being a potent carcinogen, has a relatively low acceptance value and none of the solvents and emissions in the calculated scenarios would exceed this limit. A set of worst case scenarios were also developed for this limit, which showed that nitrosamines could only constitute a health problem if very strict precaution was assumed.

The numbers and scenarios in this application appear to have been based on a NILU report prepared one month earlier in August 2010 (Flatlandsmo Berglen *et al.* 2010). That report noted that all uncertain parameters in the emissions scenarios were scaled to the worst possible values, including the potential for degraded amines to form nitrosamines. It therefore assumed a high level of 10 per cent nitrosamine formation, and the possibility of direct emissions of nitrosamines when amines reacted with nitrous oxides, to produce a worst-case annual concentration of 0.11 ng/m^3. In comparison, two scenarios that used parameters thought to be more likely resulted in nitrosamine concentrations between 0.008 and 0.029 ng/m^3, far below the EPA limit of 0.07 ng/m^3. As the NILU report noted: 'Worst case emission using feed gas from CHP (over a year) may exceed the EPA/IRIS limit value. All other combinations of scenario and feed gas are calculated to be below the limit value' (p. 61).

The NILU report also pointed specifically to the knowledge gaps around formation rates of nitrosamines. While 2 per cent and 10 per cent values were assumed in the release scenarios, the report acknowledged that the toxicology literature used values ranging from near zero to 30 per cent. This uncertainty in addition to others, led to the conclusion that it would be problematic to rely on a worst-case scenario study of nitrosamine emissions. In other words, the known uncertainties were too great to construct a set of reliable scenarios as a basis for informed decision-making.

Uncertain health risks investigated

In November 2010 KLIF published a letter in response to a request from the Ministry of the Environment, which had asked that the health risks of amine emissions be reviewed more thoroughly. KLIF's letter pointed out that both the TCM impact assessment from May 2008 and Statoil's Master Plan from February 2009 had noted that there were significant knowledge gaps associated with the health risks of amine emissions. The letter also noted that revised estimates were now necessary because (a) new evidence on amine formation had been published under the IEA Greenhouse Gas R&D Programme; (b) the August 2010 report had erroneously assumed that the US EPA figures were recently updated, but in fact dated back to 1986; and (c) worst case scenario calculations were likely to overestimate the risk. However, KLIF also acknowledged that there were no EU reference numbers on safe exposure levels of nitrosamines and nitramines in air and that US EPA figures would therefore continue to be valid (Jablonska 2010).

In a follow-up letter from 1 March 2011 KLIF noted that new evidence from Statoil indicated that nitrosamines constituted a much lower health risk than previous studies had assumed. This was based on new evidence showing that nitrosamine formation was far lower than the worst-case scenario assumptions of 10 per cent (in actuality between 0.6 per cent and 1.1 per cent) and evidence showing that nitrosamines would not be long-lived in the atmosphere. Atmospheric emissions of nitrosamines were therefore deemed to constitute a less severe health risk than previously thought. However, the letter also noted that Statoil maintained that there was significant uncertainty attached to the formation, degradation and health risks of nitramines (rather than nitrosamines). Although Aker Clean Carbon, the developer of the amine solution, had previously claimed that flue gas from the plant could be purified to eliminate both nitrosamines and nitramines, they had not submitted evidence in support of this claim. KLIF therefore asked the Norwegian Institute of Public Health (NIPH) to prepare an evaluation of the health risks associated with nitramines (Jablonska 2011).

A summary of evidence prepared for the NIPH report and released in April 2011 again referred to NDMA as a reference chemical for evaluating carcinogenicity and safe exposure levels to other nitrosamines as well as nitramines, because of the relatively high potency of NDMA. The NIPH assessment also included a table of competing safe exposure levels developed by other national and international institutes of health. NIPH remarked that a figure from the Canadian Federal health department (Health Canada) on safe exposure to NDMA in drinking water was probably more realistic than the US EPA figure, because the 'US EPA risk estimate seems conservative based on the available data' (Låg *et al.* 2011: 23). The Health Canada figure was therefore converted to the equivalent air emissions using the EU Regulatory Framework for the Management of Chemicals (REACH), and yielded a 0.313 ng/m^3 safety threshold for nitrosamines in air, corresponding to a 10^{-6} lifetime cancer risk. The report also drew on an inhalation study of NDMA to conclude that the safe exposure level should not exceed 0.3 ng/m^3, more than four times higher than the US EPA figure of 0.07 ng/m^3. Despite Statoil's reservations, NIPH also maintained that NDMA was an appropriate measure for evaluating safe exposure to nitramines, because nitramines in general were known to be less potent carcinogens than nitrosamines (Låg *et al.* 2011).

It is noteworthy that all of the safe exposure figures mentioned in the report had been available for several years. Little in the way of new data were thus brought into the discussion to argue for the application of a different safe exposure level, and no reference was made to ongoing research projects under the TCM consortium investigating the likely emission levels of amines and derivative compounds. The decision to refer to a far higher safe exposure level therefore appears to have been based on a revised interpretation of appropriate precaution, by reference to competing measures of exposure data, rather than the availability of new and compelling evidence.

Implementation strategy revised

To address the uncertainties surrounding the potential health risks, the 2011 TCM budget introduced the possibility of testing alternative technologies alongside the amine and chilled ammonia solutions, and estimated that this further technology qualification phase would cost 2,400 million NOK (MPE 2011, 2010). In early March 2011, the Ministry of Petroleum and Energy prepared a comprehensive report on Norway's involvement with full-scale CO_2 capture, with specific mention of the TCM. In one section of this report Statoil noted that health risks associated with amines now appeared to be greater than in 2009 when the Master

Plan was submitted to the government. The company therefore recommended a three-year conceptual study and maturation phase to test alternative technologies, and to ensure that at least one option would qualify for full-scale capture. This should be followed by a two-year planning and preparation phase before making an investment decision in 2016 (MPE 2011).

Statoil's suggestion thus ran counter to the company's evaluations from 2008 and 2009 that only the two pre-selected capture technologies should be matured. The report estimated that the additional qualification phase would increase the 6,500 million NOK cost estimate with an additional 2,900 million NOK (rather than the 2,400 million NOK proposed in the government's budget). With a total projected cost now at 9,400 million NOK, the cost estimate had increased nearly eight times since first being announced in 2006.

Bellona again voiced its criticism and was reported saying that Statoil was using concerns over nitramines as an excuse to delay its action on CO_2 processing for other reasons (Bjørnestad 2011). The following day the chairman of Bellona Europa said in an press release that 'Statoil is stabbing CCS in the back' and that uncertain health risks related to nitramines were insufficient reason for postponing the investment decision until 2016 (Saether 2011). The view was shared by SINTEF, the largest independent research organisation in Scandinavia, which said that the uncertainty around health risks could be addressed without postponing the investment decision (Røkke 2011).

Statoil responded to the criticism, calling Bellona's claims 'unacceptable' and denied misinforming the Ministry of Petroleum and Energy, and instead referred to the reliance on worst-case scenarios to inform decisions (*CCJ* 2011b). A few days later Bellona again wrote that developments at Mongstad were being 'needlessly postponed'. The group questioned the validity of a test sample used to inform the TCM emission permit application and asked that the results from an amine degradation test referred to in the NIPH evidence basis, should instead be used in the permit application. Bellona also referred to the KLIF letter from earlier that month, which noted that the scientific basis Statoil had used for dimethylamine emissions was incorrect, according to a press release from Aker Clean Carbon (*CCJ* 2011a; Aker 2011).

Why Statoil had not considered revising its development strategy at an earlier stage when these same health-related uncertainties were already present in the 2009 Master Plan, even in the 2008 impact assessment, was not specified. While Statoil maintained that uncertainties were now greater than previously assumed, there seems to be no published material substantiating this claim. In fact, the NIPH had since, in its April 2011 review, assessed that the risks of nitramine and nitrosamine emissions were less grave than previously thought. It is therefore unclear why there was more cause in 2009 for excluding a suite of alternative technologies by use of a narrow selection process, than there was in 2011.

This confusion was noted by members of the Norwegian Parliament's Energy and Environment Committee in a June 2010 hearing, well before the NIPH assessment was completed. Committee members pointed out that the government had been made aware of uncertainties related to the health risks of amine emissions as early as 2008. They were therefore unsure why such considerations were not taken into account at the time. The Committee also noted that Statoil had used the Master Plan to argue both for the project's viability in 2009 and the need to delay the project in its 2011 assessment (The Norwegian Parliament 2010). The opposition Christian Democratic Party noted that the petroleum and energy minister had waited until the previous month to inform Parliament that the project would be delayed and would face cost overruns, information that Statoil and Gassnova had already informed him of in 2009 (Sandelson 2010b). Following the hearing the Christian Democratic Party therefore proposed a motion of no-confidence in the minister, which was supported by the other opposition parties (Sandelson 2010a), and the minister officially resigned in March 2011.

A further twist in this story came to light after the hearing, in December 2010, when the newspaper *Aftonbladet* obtained documents from the online media organisation Wikileaks, which reported that Statoil representatives had informed US authorities of expected delays at Mongstad two years prior to the hearing. According to these notes, the directors of Statoil chose to 'distance themselves from government politicians that have publicly announced the completion date for Mongstad has been set to 2014, dismiss political guidelines, and state the project will only be continued if it makes business sense' (Tancau 2010). Responding to the leaks, the opposition Progress Party's energy and environmental policy spokesperson said he believed that 'Statoil never intended to realise "the moonlanding"' (ibid.).

The NIPH report became the topic of intense debate again in September 2011, when the new Petroleum and Energy Minister, Ola Borten Moe announced that he would not be bringing plans for full-scale capture forward because of concerns over health risks. The environment minister was quick to announce that these views were not representative of the government and Bellona criticised Moe for ignoring 'the facts on the table' as presented in the NIPH report (Bellona 2011c).

These references to continued scientific uncertainty and the controversy over transparency in government, indicate that developments at Mongstad have been embroiled in disagreements centring on competing interpretations of science, in particular uncertainty, and the proper relationship between technology and policy. Tjernshaugen and Langhelle (2009) suggest that CCS has arisen over time as a 'political glue' to bind different parties together in the Norwegian government. What this chapter suggests based on the Mongstad case, is that scientific claims and technological development trajectories cannot be understood separately from the political manoeuvring of these actors, and that interpretations of scientific precaution, even when presented in the light of scientific 'facts', must be understood simultaneously as value statements. In this light it is therefore not surprising that the NDMA analogue for nitrosamines and nitramines has been presented variously as evidence that exposure is 'safe' and that knowledge of the safety thresholds for each compound is still uncertain. In line with the previous chapter's identification of increased scientific and technological uncertainty as a source of coalition destabilisation, the furore in the wake of these health-related uncertainties appears to have caused a fracture in the Norwegian CCS coalition.

A neo-incrementalist approach to technology development

Bellona's criticisms were admittedly stated from the position that rapid development of CCS would be favourable to the interests of the Norwegian climate change agenda, and in line with the original intentions of the TCM consortium. All the while Statoil has maintained a position that can only be explained with reference to a far more precautionary interpretation of evidence. It is therefore possible to view the TCM case as a development from a generally under-critical assessment of scientific evidence to further the Norwegian climate change agenda, towards an over-critical position, in which different safety thresholds and a reliance on chemical analogues have been used to construct conflicting narratives of (un)certainty (see preceding chapter by Shackley and Evar). Statoil's increased concern with precaution has also introduced more uncertainty into the future of the TCM project, with its attendant higher costs and longer construction timeframe. If these changing views have been based on a genuine concern with the health risks posed by amine degradation, could a greater recognition of scientific uncertainty in that case have informed the planning at an earlier stage?

One option is at hand in the form of a neo-incrementalist approach to decision-making. While a few large infrastructure projects, such as the NASA space programme, may need to be developed on a very large scale from the outset (Schulman 1980), neo-incrementalist writers in the science and technology studies literature have observed that a process of small-scale trial and error has largely been more successful in developing new technologies, particularly when the supporting science has been subject to significant uncertainty (Collingridge 1982, 1992). The incrementalist approach (outlined in the introduction to this chapter) was criticised over the years by scholars in policy science, who interpreted Lindblom's ideas as conservative (because he calls for small steps in development), lacking goal-orientation (because end-goals are revised with information from continuous trials) and opposed to the use of analysis as a basis for decision-making (Howlett and Migone 2011). Such criticism has largely been refuted as misinterpretations of the theory by neo-incrementalists, who have endeavoured to apply Lindblom's formulation beyond policy science, to decision-making in science and technology management (Weiss and Woodhouse 1992; Woodhouse and Collingridge 1993).In this later form, incrementalism was redefined as 'a series of evaluative processes aiding decision-making by limiting analysis to familiar policies and to a subset of possible consequences, employing a trial-and-error approach, and tending towards problem remediation rather than positive goal attainment'. (Howlett and Migone, 2011: 59)

As an example of this application, Collingridge (1982) has noted that learning from nuclear reactor construction was slow to take off in the 1960s, as governments chose to build very large reactors, which took several decades to finish and required large advance security payments by operators. Over the same timeframe, several smaller and safer reactors could instead have been constructed which would have advanced learning for larger projects at a later stage. Other studies of the US space endeavour (Byerly and Brunner 1989), and military research and development (Collingridge 1990) have observed 'that learning is slow and costly when partisans do not press for initial precautions to head off unbearable errors, flexibility to allow error correction, and deliberate preparation for learning from experience' (Woodhouse and Collingridge 1993: 142).

Implications for the case study and conclusions

Drawing on neo-incrementalist literature and the discussion in the previous chapter by Shackley and Evar, we now ask how scientific uncertainty and increased scepticism, a naturally occurring phenomenon in the issue-attention cycle, can be managed as part of a government's attempt to develop technical options for CCS. In the Mongstad case, this literature at first glance suggests that Norway's early policy position to become a world leader in CCS was flawed. It appears to have influenced the choice to pre-select technologies based on limited previous experience in somewhat different situations (where significantly fewer amines are emitted to the atmosphere) rather than to pursue a number of options in small trials. In other words, a very end-goal oriented approach was allowed to lead the way for technology development that relied heavily on upfront analysis, rather than continued input from tests to identify competing options. CCS's status as an arguably emerging technology makes the soundness of such an approach doubtful. While both the amine and chilled ammonia solutions were known to work in CO_2 capture systems, their application in a system where greater amounts of amines are emitted into the atmosphere introduces the variable of public exposure. Neo-incrementalism would therefore advocate several small trials with numerous technologies from the start to complement early analysis.

A related matter was the government's announcement in 2006 of a deadline for the investment decision. Instead of announcing specific dates and using pre-qualified technologies, the planning could instead have accommodated uncertainty as an integral component in technology development. As part of this, neo-incrementalism would also advocate a more deliberate acknowledgement that real-world trials introduce new and unforeseen uncertainties, and that all possible risks should therefore not be attempted to be analysed upfront to guide technology selection. Such an approach would acknowledge that scientific inquiry hardly ever delivers the exact answers that policymakers require to defend their choices in the face of public health risks and other concerns, because the ideal evidence often will be based on very specific studies, with data collected over several decades, and questionable research ethics attached to human trials (such as direct human exposure to toxic samples) (Jasanoff *et al.* 2002; van Zwanenberg and Millstone 2005; Collingridge and Reeve 1986).

The importance of paying heed to the concerns and activities of other actors would also suggest that the approach at Mongstad missed an opportunity to diversify the development of European capture options. Rather than being guided exclusively by a search for the fastest option towards CO_2 capture, the TCM could have explicitly acknowledged that CCS development in Europe is coordinated as a regional project. And since post-combustion CO_2 capture so far has been the dominant approach taken by projects that are part of the EU's Zero Emissions Platform for CCS (ZEP 2011a), the Norwegian approach could have aimed to consider alternative options from the outset.

Announcements of firm deadlines are of course common practice in political arenas and politicians often make grand claims that science and technology will be applied to resolve public policy problems, without commenting on the state of the research underlying their claims. The reference to three different safe exposure levels for NDMA in assessment reports from 2008 to 2011 in the Mongstad case certainly shows how continued scientific uncertainty was used to argue for opposed strategic decisions, in what was from the beginning a highly politicised research project. It is therefore possible to suggest that technology development could not be separated from such agendas. This has certainly been suggested by Maarten Hajer in his analysis of the political framing of desulphurisation debates in Britain and the Netherlands during the 1980s. In his account, political coalitions emphasised and argued against specific scientific findings, with clear implications for emissions regulation and technology development, partly to support political agendas, but also as a result of different approaches to the interpretation of scientific certainty.

As we argue in the preceding chapter, climate change mitigation technologies such as CCS have also become embroiled in an overt political context. However, by acknowledging that the technology is to some extent framed by politics and subject to interest-based manoeuvring, we do not conclude that strategic decision-making for technology is doomed to follow the winds of a dominant political discourse. Beliefs concerning certainty, risk and precaution are to a large extent constructed rather than simply driven by rational calculations of interest and it is therefore pertinent to ask questions about how to best manage this process.

Statoil's decision to change their view on precaution may therefore also have been politically motivated, or based on a genuine concern with health risks. It illustrates how scientific uncertainty can be used to support widely different positions on precaution, with radical implications for RD&D. In this case, neo-incrementalism would suggest that decision-making could have been improved with a more consistent interpretation of precaution throughout. Although neo-incrementalism would seem to advocate a relatively strict interpretation of precaution (certainly in the formulation by Woodhouse), this would have very different practical implications where small-scale trials are developed to identify possible solutions, as opposed to a strongly goal-oriented approach from the start, which locks in to pre-determined options. An allowance for error would be granted in the name of developing the most suitable practice, and failure contained to the outcomes of small trials. Wynne (1992) has similarly noted that if uncertainty is constructed, then responses in terms of precaution similarly function as negotiated concepts. The implication is that precaution cannot be appropriately interpreted without reference to proportionality, a principle that is critical to its interpretation under EU law.

And in the case of CCS, the principle of proportionality will need to be concerned with the effects of unmitigated climate change, as well as perhaps the possibility of CO_2 leakage. In their book *Averting Catastrophe: Strategies for Regulating Risky Technologies*, Morone and Woodhouse (1986: 173), suggest: 'The "How safe?" questions that have become so much the focus of concern are matters of fine tuning; they may be important in the long run, but they are relatively minor compared to the major risks that still remain unaddressed.'

This chapter suggests that while a safe exposure threshold to the carcinogenic effects of degraded amine emissions may be approximated for a risk range between 10^{-5} and 10^{-6}, certitude is an elusive if not impossible aim in applied science, and decision-makers may therefore do well to compare the use of amine emissions for CO_2 capture in small trials over relatively short periods of time to accumulate learning, with the effects of unabated climate change over long periods of time.

Increasing the resources available to projects will not automatically lead to a resolution of such concerns as is underscored in the case of Mongstad, where the majority of costs have been subsidised by the Norwegian government, and which thereby presents an alternative to the general dearth of CCS funding worldwide. Despite this surplus of funds, the government now risks the same political embarrassment that befell the political establishment in the late 1990s when it failed to fit CCS to gas plants. The country's 'success story' with the Sleipner and Snøhvit projects is referred to in much of the support material for CCS globally and it is therefore telling that major delays and cost increases are being faced at a project that was slated to be another 'first-of-a-kind'.

The resolve to restrict tests to the amine and chilled ammonia solutions from the beginning may be said to stem from a poor evidence basis at the time. Crucially, and very relevant to this analysis, the project also appears to have been steered by a rational-comprehensive approach to uncertainty management. In a conference report the Mongstad developers note that the technology qualification approach, following guidelines developed by DNV, was premised on an approach of comprehensive upfront risk identification (de Koeijer *et al.* 2011). In a section covering 'failure mode identification and risk ranking', these guidelines state: 'The objective of this step is to identify all relevant failure modes and threats of concern for the elements defined as new technology in the technology assessment and, for each, judge the associated risks' (Myhrvold 2009: 1531). And as part of this process 'all potential failure modes or threats are then identified and their respective risks are ranked by assigning a probability class and a consequence class based on previous experience and expert judgments' (ibid.).

Attempts to comprehensively quantify such unknown risks from the start have clear implications for views on appropriate technology choices. In contrast, the neo-incrementalist methodology posits that this would be a flawed premise to start from, because it is biased against the outcomes of trial and error testing and because it may minimise concern around 'outlier risks' – those risks that a group consensus process identifies as less severe or less likely. The DNV guidelines do prioritise small-scale trials as a basis for large-scale build-up, as do traditional approaches to risk identification in chemical engineering, but the adherence to a synoptic ideal of risk assessment is a problematic premise for learning processes in emerging technologies. We therefore agree with Ishii and Langhelle that decisions for an integrated carbon capture and storage policy 'cannot be judged a priori, but it may be achieved over time through successive iterations of policy processes and practices' (Ishii and Langhelle 2011: 359).

The TCM project now serves as an example of the increased scrutiny of scientific evidence and technical feasibility that we suggest will befall other projects, as expensive and irreversible investment decisions on emerging capture technologies move closer, and more detailed cost, health and infrastructure assessments are performed.

Notes

1 Ishii and Langhelle (2011) have similarly noted that the rational-comprehensive ideal obscures the reality of environmental policy integration, including integrated policies on CCS.
2 Here quoted from Woodhouse and Collingridge (1993: 132).
3 Translated from the original text in Norwegian.
4 Translated from the original text in Norwegian.

PART III
Innovation

11

INTRODUCTION TO PART III
Innovation

Nils Markusson

Topic/scope

This part of the book covers studies of the processes of CCS technology innovation and the activities involved. It looks at the many factors driving and shaping these processes of technology development and use, and the evolution of the technology as it takes form. The section presents research in the areas of innovation and technology studies.

We can here draw on the insights from the previous two parts. That on perceptions and representations helped us understand how a range of relevant actors understand and portray CCS, which contributed to explaining their attitudes to it and their actions in relation to it. The part on governance looked at how the CCS technological community is composed and organised, how technology can and should be managed and governed, and how the preconditions and context for this have changed. This part focuses on CCS innovation, that is, the processes of change of CCS technology, as influenced by the actors involved with their perceptions of it and as formed by the changing context and by attempts at governing the technology.

The research included here is concerned with innovation processes at different scales, from the micro level of specific CCS demonstration projects, to the co-evolution of CCS with policy and theory for low-carbon innovation in society.

Some of the key questions asked in this literature include:

- How mature is the technology and what has driven its development?
- What have the roles of different actors, like industry, academics, policymakers and publics, been in this process?
- What kinds of knowledge are involved in the development and deployment of the technology?

Relevance to non-academics

Research in the innovation and technology studies fields has been useful in informing public policy as well as the management of companies and organisations on how to support, regulate, use etc. technologies. Much of this work has been driven by the imperatives of national economic advancement and taken up by governments mainly in the area of innovation policy, but has also addressed wider societal needs.

There is also a tradition of critical research, in the sense of a more fundamental questioning of the roles of technology in society and what can and should be done to restrict or reshape specific technologies. This research can serve to inform debates and enhance the reflexivity of the actors concerned. This part seeks to be useful in both of these senses.

Practical questions that will be addressed include:

- What roles do innovation, energy and climate change policies play in CCS technology development and deployment?
- How should learning processes be organised to enhance learning outcomes?
- What can different actors do in the process to promote, stop or shape the technology?

Key research themes (see also glossary in appendix to this chapter)

We will discuss specific theoretical assumptions and concepts below, but it is worth making one basic point already here. The relationship between the categories of 'the technical' and 'the social' have caused much debate (as discussed in Chapter 1). Some of the research in this area accepts a sharp demarcation of a technical sphere – typically including artefacts and engineering expertise. In this view, social things: the public, policy etc. are separate from the technical, impacted by technology and, sometimes, having an impact on technology. In contrast, there is research that questions this distinction, and sets out to study the social nature of technology, how society and social dynamics contribute to shaping the technology as it evolves. Such research sees innovation processes as social phenomena rather than expressions of inherent scientific logics, or at least as a combination of the two.

The point to take from this is that the role of social science is potentially quite different depending on which perspective is chosen. If social things are seen as separate from the technical, the contribution of social science is restricted to the study of mainly publics and policy. With a view of technology as a social phenomenon, the possible and relevant topics of study for social scientists are much wider.

The closely related research fields of innovation and technology studies are characterised by interdisciplinary, social science-based studies of how technology is used and developed. At the most general level, research in this area seeks to answer questions like: what explains the success or failure of particular technologies; what shaped their form; what role do they play in societal affairs?

Innovation and technology studies draw on several disciplinary sources, including economics, sociology, politics, history, management and business studies. This research is thus inherently interdisciplinary, cutting across social science disciplines broadly defined. They are also interdisciplinary in that they seek, to varying degrees, to integrate knowledge across the boundary between social sciences and natural and engineering sciences.

It is also possible to spot differences between innovation studies and technology studies. Where innovation studies tend to rely more on economic traditions of thought and seek to establish universal facts, technology studies are relatively more reliant on interpretative traditions emphasising the actors' own perspectives and stress the contingent, context-dependent nature of innovation.

Systems

Several research traditions have sought to analyse innovation using different 'systems' notions. There are at least two distinct starting points for this work. First, there was the recognition that aggregate patterns of innovation vary across contexts. Seminal work was done seeking to explain variation across nation states (Lundvall 1992; Freeman 1995). The 'innovation systems' approach emphasises the social nature of innovation by drawing attention to the importance of interaction between diverse actors, and draws on institutional economics to delineate the varying institutional setups in different countries shaping innovation processes. The approach has been expanded to include also sectoral and regional contexts, or to focus on specific technologies (Carlsson and Stankiewicz 1991).

In a recent development, the innovation systems approach has been formalised into the so-called functional approach (Hekkert *et al.* 2007; Bergek *et al.* 2008a). The authors have synthesised earlier literature into sets of functions that describe the dynamics of the processes in innovation systems. The exact set of functions varies, but an illustrative example is shown in Table 11.1. The development of an innovation is here conceived as the emergent interplay of structure (actor set-up, networks and institutions) and functions, that is, as feedbacks (or lack of them) within the system.

TABLE 11.1: Functions of innovation systems

Knowledge development and diffusion
Influence on the direction of search
Entrepreneurial experimentation
Market formation
Legitimation
Resource mobilisation
Development of positive externalities

Adapted from Hekkert *et al.* 2007 and Bergek *et al.* 2008a.

The functions approach to innovation systems research has been applied to CCS not least by van Alphen and co-authors, to assess CCS innovation system in different countries, e.g. in Norway (van Alphen *et al.* 2009) and the US (van Alphen, Noothout *et al.* 2010). They have also compared the progress in different countries (van Alphen, Hekkert *et al.* 2010). These studies draw on historical accounts of the emergence of CCS in these countries, as well as on questionnaires to stakeholders asked to assess the strength of each innovation systems function.

A key result of this body of work is that the CCS innovation systems are comparatively well developed in terms of, mainly academic, knowledge creation and diffusion, as well as with regard to the development of visions for the technology that can guide the activities of the involved actors. In contrast, it is clear that the creation of a market for the technology is weak (although secrecy limits the scope for assessing the size of industrial efforts and competences).

Chapter 8 in this book, by Stephens and Liu, also draws on this theory, to help analyse the evolution of the international CCS community. Amongst other things, they found a growing share of industry participation in community forums and growing industry funding.

In international comparison, Norway stands out as a country with a long history of CCS engagement and a more strongly developed innovation system. (Australia may be rather well placed as well, but has not yet been covered in this type of research.)

Moreover, Vergragt *et al.* (2011) applied the functions of innovation systems approach to compare two technical systems: CCS as applied to fossil-fuelled power plants and CCS applied to biomass-based power production and manufacturing (biomass energy with carbon dioxide capture and storage – BECCS). They found that the BECCS innovation system is strong mainly in terms of academic knowledge production, where it overlaps with its fossil-fuelled counterpart.

These innovation systems studies provide a useful assessment of the progress and development of CCS, and provide guidance as to where policy interventions to further develop the technology may most usefully be directed. As the technology progresses, more could be done in terms of studying interactions between the innovation system functions and emerging self-reinforcing dynamics, as well as further studies into interactions between innovation systems.

A second origin of the studies of 'systems' in innovation studies was in response to the recognition of the diverse range of elements that need to be integrated for a technology to function in the real world. Work in the history of technology, especially, introduced the notion of 'socio-technical systems' (Hughes 1983) to discuss the heterogeneous nature of innovators, spanning engineering as well as economics, politics etc. and how diverse elements need to be integrated into a functioning socio-technical system.

Systems approaches to innovation studies have been applied to single technologies, including CCS as discussed below. They have also been applied to the processes of systems undergoing shifts from one technology to another, e.g. Winskel (2002a, 2002b) studied the deregulation and privatisation of the UK power industry in the 1980s and 1990s, and the resulting dash for gas and demise of investment in nuclear power and coal-fired power plants.

In recent years, a prominent approach to the study of system change has been the so-called multilevel approach to transitions (see Chapter 1), again emerging out of historical studies of technology and synthesising insights from a wide range of fields into a compact and concise model (Geels 2004a). A 'transition' is here defined as a major structural shift or transformation in a societal function such as transportation, housing or the energy system. The multilevel perspective sees the transformation of a new societal function ('regime') as a result of contextual change (the 'landscape' of global factors and long-term trends) destabilising the existing regime, and the adoption of new ideas, practices and technologies that have been emerged in marginal settings protected from mainstream market competition ('niches'). Some examples of historical transitions – as studied using transitions theory – include the shifts from sailing ships to steam ships, from traditional factories to mass production, and in the context of energy, from coal to gas systems (Geels 2002, 2005b, 2006a; Correljé and Verbong 2004). The multilevel approach and related transitions research has gained a lot of academic interest and, especially in the Netherlands, some influence on policy (Kern and Smith 2008).

There have been attempts at applying the notion of system transitions to CCS (e.g. Vergragt *et al.* 2011). However, the emphasis of the model on marginal niches as origins of innovation means that the application of the model to CCS is not straightforward, as discussed further by Winskel (Chapter 12, this book).

There is also a growing body of work concerned with the role of CCS in the evolution of the energy system, drawing on systems notions. Unruh (2000, 2002) developed the concept of techno-institutional complexes to analyse how our societies are locked-in to the use of fossil fuels. Here, CCS is seen as an add-on technology that serves to defend the established fossil regime. Markusson and Haszeldine (2010) analysed the proposition of capture-ready power plants to argue that whilst it may mitigate against the risk of carbon lock-in it offers no guarantee against it and might in fact make it worse.

Vergragt *et al.* (2011) sought to develop indicators to assess the effect of adding CCS to a regime with fossil-fuelled power plants, finding that it would be likely to reinforce lock-in. In contrast, Shackley and Thompson (2012) stress that the future of CCS may not be as merely an add-on technology, but may become part of more radical technical change towards a hydrogen economy. Developing a different set of indicators to Vergragt *et al.* (2011), they conclude that CCS might be more flexible than otherwise suggested in this literature.

Learning

Another entry point to innovation studies is the continued interest in the knowledge needed for technology development and use, and in conceptualising innovation as 'learning'. A wide range of theoretical approaches have been developed and used, but for the purposes of this part of the book, it is useful to focus on and contrast two of them: 'learning rates' and 'social learning'.

The learning rates approach builds on the observation of a common pattern of cost reduction as a technology is ever more widely deployed, thought to reflect accumulated experience ('learning-by-doing'). The approach has the advantage of being quantifiable, operationalised as a ratio of cost reductions per doublings of installed capacity. Its wide appeal also steams from wide circulation of stylised, qualitative learning curve diagrams, as an effective way of articulating promises of future improvements in technology performance and costs.

There is a body of research aiming at assessing learning rates for CCS, not least by Rubin and co-authors. An early study (Rubin, Taylor *et al.* 2004) compared capture technology with other environmental control technologies for fossil-fuelled power plants: flue gas desulphurisation technology (FGD) and with selective catalytic reduction, and suggested that the historically observed learning rates of 11–12 per cent for these technologies may come to apply also for CCS. Subsequent work has refined this analysis. In a paper from 2007, Rubin *et al.* analyse capital and operating and maintenance costs separately. They also expand the set of technical analogues to find specific analogues for the different capture approaches (post-combustion, pre-combustion and oxyfuel) and analyse learning for separate component technologies and their integration.

The studies have provided valuable analyses of the learning processes necessary for CCS to develop. Of course, any analogy made with another technology is open for debate as to how good a fit it provides. However, several other authors have also selected FGD as a suitable analogue, including Chalmers *et al.* (2009). Rai *et al.* (2009) draw on an analogy with FGD technology to discuss the preconditions necessary for successful forcing of technology development, and in particular the scope for doing so with CCS.

Learning rates studies have also provided cost trajectory estimates that have been useful for techno-economic modelling of the energy system (e.g. Rubin *et al.* 2007). It should be noted though, that such studies are also dependent on data for early CCS deployment, which are currently very uncertain (de Coninck *et al.* 2009).

The simplicity of the model has a flip side and the assumptions on which it rests have been questioned (Winskel *et al.* 2008). Not all technologies display this pattern of cost reductions over time, and great care should be taken when using it to discuss the future. Also, as a model of learning, is it not clear to what extent wide deployment causes learning-by-doing resulting in cost improvements, as compared to learning-by-research leading cost savings and hence increased market penetration.

Learning rates bring attention to the contribution to innovation from users, with their knowledge and experience from the use setting. This challenges assumptions of technology as being simply applied science, or that it is formed based on engineering ingenuity alone. Such interest in the use of technology has driven much work in innovation and technology studies, and opened up for the study of a multiplicity of actors interacting with technology in multiple settings.

The notion of social learning (see for example Rohracher and Ornetzeder 2002; Munshi 2004) encapsulates this wider perspective. Social learning about new technologies implies a distributed process involving a wide range of actors. This process includes interaction and negotiation about what is being learned (Sørensen 1996). It involves learning about technical facts, but also about other aspects of the technology including the cultural meaning of the technology (Williams *et al.* 2005). Social learning approaches have been used not least for studies of consumer goods technologies with their wider penetration among diverse social groups, but can also be applied to other technologies.

There has not been much work done on learning processes of CCS, other than learning rates research. Chapter 13 in this volume by Markusson *et al.* on social learning in and around CCS demonstrations is a contribution to this, but there is scope for more work. Moreover, there is a gap in the research on CCS-related learning with regard to how companies learn about and manage CCS. The welcome contribution by Bowen (2011) has begun to fill that gap.

The contributions of the chapters in this part

The chapter by Mark Winskel provides a critique of current predominating theories of low-carbon innovation: transitions theory and technological innovation systems, for having an in-built bias towards radical, niche-driven innovation, led by marginal actors. The chapter argues that there is a growing mismatch between this theory and current policy practice, with its strong emphasis on fast results and 'acceleration' of technology development. Winskel marshals evidence for the possibility of major system change that is not the result of disruptive, radical innovation, but instead emerges out of developments within the existing regime. Research in this area needs to re-engage with and theorise the evolution of existing technological systems and the actions and fates of incumbent, resource-strong actors. CCS has emerged in intimate interplay with such actors, including oil and gas giants, who control some of the resources and expertise required, and provides a good test case for theory development along these lines.

In the final contribution to the book, Nils Markusson, Atsushi Ishii and Jennie C. Stephens apply the concept of social learning to the broad learning processes that happen in and around CCS demonstration projects. This way, they are able to broaden our understanding of learning from CCS demonstrations, in terms of the types of knowledge and actors involved. They also show how the specific designs, processes and outcomes of CCS demonstrations are shaped by their contexts, in terms of, for example, the ways that support is built for the project and national and local governance structures. That there can be a tension between the need for knowledge sharing to boost learning and the need for knowledge protection to secure competitive advantage is well known. This chapter suggests that a strong focus on local benefits may drive projects towards a focus on short-term economic benefits, which may thus restrict the broader learning processes that would be required for technology adoption and wide scale deployment.

Appendix 11.1 Glossary

The term	*What it means*
Technology	At a minimum, technology is seen as useful artefacts and the knowledge needed to produce them. Broader definitions include also the knowledge needed to deploy and use the technology, the actors involved in the innovation process (see below), the institutions set up to structure, manage and govern the development and use of the technology, and the cultural meanings of the technology. Note especially that technology is not simply applied science, but draws on a broad range of knowledges and experiences, not least from the context of use and deployment.
Innovation process	Innovation is the process through which an idea is realised – and the solution to a problem or need constructed – and comes to be adopted and used. Most often in market-based societies, this involves commercialisation of a product or service. Sometimes the term 'diffusion' is used to refer to the process of widespread uptake, after the first use. Innovation is not to be confused with the much narrower term 'invention' which is the act of coming up with the idea for a new technical solution to a problem.
Knowledge	The conceptualisation of knowledge varies, from the view of objective facts established by scientific methods, to broader (and sometimes conflicting) views recognising the variety of valid and effective rationalities across different social groups and that knowledge is context dependent. Knowledge is not merely impersonal information, but something that is understood, interpreted and communicated by people. 'Learning' is, therefore, a much more complex process than just a transmission of facts.
System	The term has several possible meanings here. First, it can be used, as in engineering, for a specific technical design composed of multiple components. Second, it can be used to describe a much wider entity, which includes artefacts, but also technology-specific knowledges, actors and institutions, in recognition of the wide range of elements needed to make technologies work in practice.
Institution	The term refers to established rules, routines and practices that structure activity. Here, that activity is the development and use of a technology. Examples of such institutions include design practices and regulatory standards. Note that, unlike in everyday speech, the term is usually not used to mean a specific organisation.
Socio-technical	A term used to indicate the intimate connections between, or inseparability of technical and social aspects of a phenomenon. Just as technical development may influence markets or policy, so can societal developments shape technologies, even down to the minutiae of specific design decisions.

12

CCS

A disruptive technology for innovation theory

Mark Winskel

Introduction

The emergence of carbon dioxide capture and storage (CCS) as a significant part of envisaged energy futures in the UK and internationally has coincided with more urgent imperatives for energy system change. This chapter argues that the CCS case, as a *regime-led* innovation, challenges the prevailing models of innovation theory articulated within innovation studies over the past two decades. In particular, CCS is seen here as a disruptive technology for *constructivist* theories of innovation – especially transitions theory, but also technological innovation systems theory. The emergence of regime-led innovations such as CCS challenges some of the main tenets of constructivist-based innovation theories, especially their emphasis on niche-led disruptive innovation as a primary driver of socio-technical system change. Under stronger and more urgent imperatives for change – environmental, economic and security of supply – policymakers have sought to develop responses by directly engaging with incumbent organisations and networks. As a result, more centrally co-ordinated energy innovation systems have emerged, and these have directed considerable efforts on regime-led innovations such as CCS.

Transitions theory was conceived to articulate a *multi-level perspective* on socio-technical system change, overcoming the limitations of more overtly micro-social agent-based accounts. However, by retaining an essentially constructivist perspective, incumbent interests were cast in secondary terms as respondents to niche-driven change, rather than as resourceful and dynamic agents in their own right. As a result, transitions theory has underappreciated the dynamics and importance of regimes – their variety of resource endowments and interests, and their varied strategic responses to landscape pressures and niche developments. It has also undervalued the role of incremental or end-of-pipe innovation in system change. As drivers of energy system change have become more pressing, constructivist theories' *ex ante* niche-led framings have meant that much innovation studies research is disinclined to interrogate regime-level dynamics in any detail, so limiting its policy relevance.

The chapter first reviews the origins and development of constructivist innovation theories – transitions theory and technological innovation systems – and outlines their main theoretical tenets, especially the primacy awarded to niche-led radical innovation and the secondary role of regimes. It then outlines the emergence of CCS as an important prospective low-carbon technology, mainly with reference to UK energy policy but also with some international context. This highlights a shift from niche- to regime-based energy innovation policy over the course of the 2000s. Then follows a discussion of the co-evolution of policy, strategy and theory in low-carbon innovation over the past two decades. This argues that while energy innovation policymakers, strategists and researchers all underappreciated or underplayed the prospective role of regime-led innovations in the 1990s and early 2000s, innovation theorists have shown a particular attachment to niche-led dynamics. The chapter concludes that the emergence of CCS and other regime-led innovations suggest the need for a reworking of innovation theory in a context of accelerated change, with fuller attention to the inner workings and wider associations of regimes.

Constructivist innovation theory

Introduction

The dedicated analysis of low-carbon energy supply technologies within innovation studies emerged only relatively recently, over the past two decades. Two prominent conceptual frameworks developed over this time: the 'Multi-Level Perspective' on system transitions, also known as *transitions theory* (Rip and Kemp 1998; Geels 2002, 2004a; Geels and Schot 2007) and *technological innovation systems theory* (Bergek and Jacobsson 2002; Jacobsson and Bergek 2004; Bergek *et al.* 2008a, 2008b). While both theories have referenced and been applied to a wide range of technology sectors, they share a normative concern for the promotion of sustainable technologies, and research efforts in both have included many studies of low-carbon technologies. Reflecting their conceptual groundings in the sociology of technology and evolutionary economics, both approaches tend to portray socio-technical change in *constructivist* terms – as a bottom-up and emergent processes initiated in weakly-structured niches;[1] each is now discussed in turn.

Transitions theory: origins and precepts

Transitions theory has been described as 'appreciative theory' (Geels 2002: 1257), in that it assembles research concepts and evidence from a number of disciplinary traditions. For Elzen *et al.* (2004: 284), transitions theory offers a way 'to integrate and position several theories and insights'. Within this, it draws most evidently on 'a combination of sociology of technology and evolutionary economics' that emerged in the 1980s (Geels 2004b: 33), but which have earlier roots in postmodern social theories.

Responding to the abstract, partial and deterministic representations of technological change in neo-classical economics, modern sociology and internalist history of technology,[2] the 1980s saw a wave of *constructivist* contributions that emphasised socio-technical variation and selection in micro-social domains. In the 'new sociology of technology' – a field particularly influential on transitions theory – the main conceptual strands were *social construction of technology* (SCOT) (Pinch and Bijker 1984; Bijker and Law 1992; Bijker 1995) and *actor network theory* (ANT) (Callon 1986; Law 1987; Law and Callon 1992). Both SCOT and ANT drew on earlier constructivist theories developed within the sociology of science, as well as wider social theories of post-structuralism, grounded theory and social constructivism.

The ascendancy of SCOT and ANT in the sociology of technology in the late 1980s and early 1990s attracted some criticism from proponents of more structurally informed accounts of innovation (e.g. Russell 1986; Williams and Russell 1988; Hård 1993; Winner 1993). For these critics, a constructivist approach – though a welcome antidote to deterministic accounts – risked overemphasising micro-social domains as loci of technological change – what Winner (p. 427) described as the 'social determination of technology'. Such criticisms were accompanied by increasing calls for attention to the intermediate *meso* level, where the influence of established organisations and institutions could be analysed, alongside alternative niche innovations (Sørensen and Levold 1992; Russell 1993; Misa 1994).

By the mid-1990s, in recognition of the limitations of micro-level accounts of innovation, 'multi-level' analytical framings began to emerge, mainly from Dutch-based scholars. The first internationally well-known accounts of the Multi-Level Perspective (MLP) appeared in the late-1990s (Kemp *et al.* 1998; Rip and Kemp 1998). From the outset, the MLP was concerned to go beyond micro-level niche dynamics to interrogate the meso-level, where technology diffusion takes place, or is stalled. For Rip and Kemp (1998: 338, 362), 'it is the adoption of novelty which is decisive, not its introduction ... sources of ... innovations are probably less interesting than the conditions of challenge and overthrow'. Geels (2004b) critically referred to three prominent strands of technology studies theory in the 1980s – SCOT, ANT and also large technical systems (LTS) theory as 'point source' theories: preoccupied with the emergence and diffusion of new technologies, and failing to adequately address the replacement of old ones.[3]

The Multi-Level Perspective conceives technological innovation as an outcome of the interplay of social and technical elements across three distinct levels of structuration: micro-level *niches*, meso-level *regimes* and macro-level *landscapes*. Within this, *system innovations* – defined as those innovations that are most significant for overall system make-up and environmental performance – are presumed to originate in niches. Established regimes, which are more stable and structured, are characterised by their resistance to niche-driven change. Therefore, despite its broader framing, the Multi-Level Perspective retained a constructivist perspective; as Paredis (2011: 222) noted, 'a constructivist approach ... is embedded in the transition research field'.

This niche-driven account was associated with a normative concern, shared by many transitions scholars, for the promotion of more environmentally sustainable technologies. Niches were seen as offering crucial 'incubation spaces' where more sustainable technologies could be created and nurtured. Kemp *et al.* (1998) argued that the 'political realities' of liberalised market economies meant that both fiscal measures and central planning were unlikely to be able to bring about the 'regime shifts' necessary for the widespread uptake of more sustainable technologies. Instead, they emphasised the potential of 'modulating the dynamics' of supply and demand, by a mix of technology forcing, network creation and 'strategic niche management'.

As well as its emphasis on niche-led innovation, transitions theory's constructivist origins are reflected in its portrayal of system change as an essentially emergent, unplanned process. For Elzen *et al.* (2004: 284) transitions involve 'no overall rationality ... actors navigate their way through probing and learning', with long periods of experimentation and network building. Elzen *et al.* added that this 'muddling through' character should be matched by an iterative and reflexive policy style, characterised by 'a cyclical process of action and evaluation' (p. 289). The overall policy aim should be to 'bend' innovation dynamics slightly in the direction of desired objectives, rather than imposing any more direct control.

Transitions theory and the role of regimes

Transitions theory's portrayal of regimes has been predominantly in terms of their resistance to niche innovations. Radical innovations, considered essential for any major changes in the character and environmental performance of socio-technical systems, are seen to emerge from niches, and can only diffuse widely by regime challenge and overthrow (Rip and Kemp 1998). For Geels (2004b), 'regimes generate incremental innovations, radical innovations are generated in niches ... [so] system innovations start in ... niches' (pp. 35, 42). As Smith *et al.* (2010: 440) noted, this reflects transitions theory's sociology of science precepts: 'change within the regime ... [is seen as] incremental and path dependent ... in a Kuhnian vein, regimes tend to produce "normal" innovation patterns, whilst "revolutionary" change originates in "niches"'.

A second interest in regimes within transitions theory is in their possible enrolment in support of niche-led innovation (Geels 2004b). For Elzen *et al.* 'the regime and landscape levels not only throw up barriers for niches, they may also provide possibilities to link up [with]' (2004: 287). Elzen *et al.* went on to consider possible regime responses to niche-driven changes, including 'gatekeeping' (screening of niche innovations according to their compatibility with regime interests) and the 'sailing ship effect' (efforts at accelerated development of regime technologies in response to niche-led threats). Again, regimes agents are understood here in essentially defensive, reactive terms.

The early formulations of the MLP attracted considerable interest in wider innovation studies, reflecting a desire for more comprehensive framings than those offered by micro-level approaches such as SCOT and ANT. As the MLP became increasingly established in innovation studies in the early 2000s, however, its niche-driven account of innovation met with some criticism. A notable contribution along these lines, from Berkhout *et al.* (2004), called for greater attention to be paid to the way macro-level landscape pressures place direct 'selection pressure' on regimes (such as through government policy, market structures and public opinion), and also, to the *adaptive capacity* of the regime in responding to such pressures. Noting the need to challenge more 'assertive claims' about the niche-driven account, Berkhout *et al.* went on to outline a range of different possible 'transition contexts', reflecting differences in regime co-ordination and resources.

Soon after Berkhout and colleagues' criticisms, Geels conceded that the MLP 'has a bias towards ... novelty' (Geels 2005b: 85). Subsequent contributions to transitions theory, though retaining an essentially niche-led view, acknowledged more emphatically that niches alone were incapable of system change, and have given more nuanced descriptions of niche-regime-landscape relationships. Geels and Schot (2007) identified different types of 'transition pathway', according to their reinforcing or disruptive *regime-landscape* relationships and competitive or symbiotic *niche-regime* relationships. In one of Geels and Schot's transition types – the *transformation* pathway – incumbent actors are seen as being capable of responding directly to moderate landscape pressures, such that 'new regimes grow out of old [ones] through cumulative adjustments' (p. 407). In another – the *reconfiguration* pathway – regime actors' adoption of radical components from niches eventually triggers a more fundamental change in regime make-up. While both transformation and reconfiguration pathways allow the possibility of more active roles for regimes in system change, Geels and Schot's overall emphasis remained a constructivist one, in that more substantial system change is inherently niche-led, and requires the destabilisation and overthrow of incumbents.

Among transition case research, a number of studies have circumvented the restricted role for regime dynamics in niche-led theory. Correljé and Verbong's analysis of the Dutch transition from coal to gas focused on 'the regime level ... taking into account ... relevant interactions with ... "landscape factors"' (Correljé and Verbong 2004: 115). This allowed for a rich account of regime-level conflicts, power struggles and unpredictable outcomes. Geels has also illustrated the dynamism of regimes in the aviation (Geels 2006a) and hygiene/sewerage industries (Geels 2006b); in the latter case, he observed, 'the transition was not a niche-driven substitution process ... [and was] not disruptive ... existing knowledge remained relevant' (p. 1078). Such regime-led transformation, he noted, 'may culminate in substantial system changes' over time (p. 1081).

Raven (2007) contrasted two possible forms of system change: 'niche accumulation' and the 'hybridisation of regimes'. The latter, exemplified by gas turbine adoption in electricity regimes and also biomass co-firing in electricity/waste regimes, involves add-ons to regimes, rather than their displacement, with incumbent firms having the role of 'driving actors'. Raven suggested that given their tight coupling and high entry barriers, hybridisation (rather than niche accumulation) may be an important form of system change for infrastructure technologies. In a related paper, Raven and Verbong (2009) also considered inter-regime dynamics, and presented a typology of 'boundary-crossing' innovations. Konrad *et al.* (2008) offered a detailed account of prospective inter- and intra-regime dynamics for utility industries. In doing so, they explicitly questioned the regime-overthrow account of transitions, suggesting that this was likely to be especially restricting for 'future-oriented' research: 'we should not presuppose that a regime shift is necessarily the one best way' (p. 1192).

Despite a body of case research illustrating the analytical value of a focus on regime-level dynamics, and some theoretical adjustments to allow the possibility of a more active role for regimes, transitions theory has continued to promulgate an essentially niche-led account of system change. Geels (2010) reiterated the need to destabilise incumbent organisations to allow the acceleration and up-scaling of niche innovations. At the same time, he now recognised that 'incumbent firms *can* overcome the lock-in mechanisms that tie them to existing regimes' (p. 505, emphasis in original), and suggested that transitions theory's account of niche-regime relations could be improved by reference to insights from other research fields, such as strategic management research. Similarly, Geels (2011) acknowledged that many transitions studies have presented regimes as 'too homogeneous or monolithic', with under-attention to their 'internal tensions, disagreements and conflicts of interest' (p. 31). Even so, the regime-led *transformation pathway* was portrayed as an essentially modest 'gradual adjustment' to landscape pressures; more substantial system transition was still seen as requiring regime overthrow and replacement.

Technological innovation systems theory

Less influenced by the 'new sociology of technology' than transitions theory, technological innovation systems (TIS) theory drew rather more on economic and management research fields, especially neo-Schumpeterian 'evolutionary economic' theories of technology variation and selection (Nelson and Winter 1982) which also informed a broader body of innovation systems research initiated in the late-1980s (Freeman 1991; Lundvall 1992; Nelson 1993).

Evolutionary economics (and wider innovation systems theory) were more attendant to the structural aspects of technological change than constructivist sociological approaches – indeed, the concept of 'technological regimes' was introduced by Nelson and Winter (1982). At the same time, evolutionary economic theory developed an essentially constructivist (i.e. micro-level, bottom-up and emergent) view of innovation, within which variation and selection takes place mainly via firm-level market-based decision-making (Nill and Kemp 2009). In introducing their technological systems approach, Carlsson and Stankiewicz (1991: 94) made clear that their starting point was a focus 'on the process of … change at the micro level, including the role of the entrepreneur', and they distinguished their approach from national-level innovation systems approaches by its 'greater emphasis on microeconomic aspects ... rather than on institutional infrastructure' (p. 112).

In addition, because of their normative concerns for sustainable innovation, the focus of many TIS studies has been small, marginalised firms and their supporting networks and institutions. For example, building on Carlsson and Stankiewicz's framing, Jacobsson and Johnson's (2000) influential TIS framework was designed with reference to renewable energy technologies. TIS theory therefore shares with transitions theory an essentially constructivist emphasis on niche-led innovation, despite its differing conceptual origins and its 'functional' rather than 'interpretivist' ontology (Geels 2010). TIS's set of functions for successful system building emphasise micro-level dynamics such as interactive learning and entrepreneurship (Hekkert *et al.* 2007; Bergek *et al.* 2008a, 2008b). Larger firms and mainstream institutions have a significant role in TIS accounts of system change, but mainly as enablers or inhibitors of the emerging technology system under analysis. This lends a rather myopic, introspective flavour to much TIS case research (Markard and Truffer 2008), with the emerging system expanding into a wider socio-technical context which is understood only to the extent that it enables or blocks emergent system growth – what Geels (2007) referred to as a point-source narrative.

Two broad phases of system development are often identified in TIS theory: an initial, *formative* phase characterised by the generation of technological novelty, the trialling and testing of different designs, the establishment of niche markets and efforts to build the societal legitimacy of a new technology; and a subsequent *market expansion* phase, characterised by positive feedbacks between learning-by-doing, scale economies and market growth (Jacobsson and Bergek 2004; Jacobsson *et al.* 2004). Jacobsson *et al.* (2004) suggested that 'several decades' were typically needed for formative phase learning, with often little to show, by way of deployment, over the first few decades. Indeed, Jacobsson *et al.* emphasised the need for a policy favouring of 'variety rather than volume' in early stage innovation – i.e. on supporting learning via small-scale experiments rather than seeking scale economies.

A number of innovation systems-based studies have referenced the archetypal renewable energy innovation success story, Danish wind power (e.g. Jørgensen and Karnøe 1995; Garud and Karnøe 2003). Typically, in such studies, Danish wind power success is attributed largely to bottom-up, distributed and interactive learning. For example, Karnøe and Garud (2003: 22) characterised Danish wind power success in constructivist terms, as 'an emergent transformation ... [involving] networks of interaction ... [with] no central coordination'. Similarly, Jørgensen and Karnøe (1995) highlighted the roles of Danish grassroots activism, counter-expertise and distributed agency, with extended learning-by-experience on small scale units. They contrasted this with top-down science based approaches, adopted in the USA and Germany, which failed in their efforts to fit emerging wind power technology into the established generating and transmission industry. Even though later years saw much greater organisational concentration and technological up-scaling in the wind power industry, Jørgensen and Karnøe maintained that the case broke 'well-established propositions about the development of energy production technology' (ibid.: 58).

Testing the tenets of constructivist based innovation theory

Despite some later theoretical refinements, the essential features of system innovation for constructivist innovation theories are the creation and fostering of radical innovations in niches, and the dismantling (or perhaps enrolling) of incumbent regimes to allow for radical technology diffusion. Incumbent regime organisations, since they are oriented overwhelmingly toward incremental innovations, are incapable of bringing about major changes in system character. The presumptions here about the necessarily radical and disruptive nature of socio-technical system change are worthy of closer inspection.

Necessarily radical? Are major changes in socio-technical systems always associated with radical innovations?

There is some evidence that incremental innovation – gradual improvements to existing technologies – can have substantial effects on the economic and environmental character of energy systems over time. A longitudinal study of energy sector innovation (Burtraw *et al.* 1999; see also McVeigh *et al.* 1999) found that improvements to established (fossil fuel based) electricity technologies tended to be undervalued – in their magnitude and consequence – in forecasts of the growth of more radical (renewables) technologies. Even though the costs of many renewables had reduced substantially over time, their deployment tended to be less than expected. Burtraw *et al.* attributed this to greater than anticipated improvements in the cost and performance of conventional generation technologies – a 'moving baseline' against which renewables had to compete: 'regulatory, technological and market structure changes … reduced generation costs for conventional technologies and … led to a dramatic improvement in their environmental performance' (Burtraw *et al.* 1999: 13). Similarly, taking a long-term historical view, Martin (1996) found more substantial changes in hydrocarbon technologies and a lesser role for non-hydrocarbon systems than had been anticipated, as advances in 'generic technology' such as electronics, computing, and materials had driven innovation in established technologies.

As Meadowcroft and Langhelle (2009b) have noted, 'end-of-pipe' technologies (rather than more wholesale changes) are the established way of emissions control in many industrial sectors. For example, end-of-pipe innovations such as flue gas desulphurisation and catalytic convertors have enabled dramatic changes in emissions from fossil fuel power stations and internal combustion engines. There is also evidence that technological 'substitutions' – the replacement of system parts, rather than entire systems – can have dramatic impact. For example, between 1990 and 2009, improved conversion efficiencies and fuel switching led to a reduction in effective UK electricity sector CO_2 emissions of 36.5 per cent (DECC 2010). Over half of this saving was attributed to increased use of gas rather than coal, with the remainder from improvements in the efficiency of conventional generation technology.

Necessarily disruptive? Do radical innovations always emerge in marginal niches? Does the diffusion of radical innovations require the dismantling of regimes?

A number of significant technological innovations for energy systems, such as steam turbines and nuclear power, were developed within large scale, highly co-ordinated innovation programmes (Hughes 1983; Winskel 2002a, 2002b; Watson 2004). While many incumbent organisations in the electricity sector have shown little appetite for long-term innovation over the past two decades, not all recent regime-led innovative activity in the wider energy sector can be considered incremental – innovations such as deep-water drilling and unconventional gas have had major consequences for global energy systems. In other sectors, such as pharmaceuticals, some incumbent organisations have been able to maintain high levels of innovation via long-term RD&D programmes (some of which may be assigned to 'blue skies' research), by fostering small business units responsible for more radical innovations, by creating defensible intellectual property and by mergers and acquisitions.

The emphasis on the disruptiveness of radical innovation in transitions theory reflects wider debates in innovation and management studies regarding the capacity of incumbent organisations to foster and manage disruptive innovation. Christensen (1997) argued that it was profoundly difficult for incumbent firms to direct sustained effort and resource on disruptive innovation. As Geels (2010: 497) acknowledged, Christensen's account has been challenged; for example, Danneels (2004) criticised it for a selective reading of empirical evidence, for only picking 'winning' disruptive technologies and for overstating the innovative inertia of incumbent firms. Macher and Richman (2004) noted how some established firms are able to develop organisational strategies to benefit from disruptive innovation. In the wake of UK electricity liberalisation, for example, combined cycle gas turbines were first deployed by consortia of outsider and marginalised firms to bypass regime resistance to new entrants. Once the scale and speed of this embryonic independent-led 'dash for gas' became clear, incumbent organisations very quickly abandoned their longstanding commitments to established technologies and invested heavily in the new technology (Winskel 2002b).

The emergence of CCS internationally and in the UK

Introduction

This section presents a short account of the emergence of CCS over the 1990s and 2000s as a significant aspect of envisaged energy futures, first – and very briefly

– internationally, and second – and in somewhat more detail – in the UK. Over this period, CCS has shifted from being an insignificant, unrepresented aspect of international and UK energy and climate policy to become, for the International Energy Agency, 'an important part of the lowest-cost greenhouse-gas mitigation portfolio' (IEA 2010b: 494), and for the UK government 'a crucial part of the solution to climate change, in the UK and globally' (HMG 2009: 67).

As a number of commentators have pointed out, there are still deep technical, economic and regulatory uncertainties and conflicts of interest involved here, and there can be no assurance that the widely shared view in the energy policy mainstream that emerged in the late 2000s on the central role of CCS in energy futures will be maintained. Therefore, this short review is not intended as a unilinear narrative of the 'rise' of CCS, brushing aside its unresolved uncertainties, and the varied set of interests and strategies laying behind its emergence. (It is perhaps worth noting here that CCS is not a uniquely politicised technology – energy technologies and policies are deeply politicised, and visions of energy system change are strongly contested.)

In surveying the emergence of CCS, my aim is not to reach a judgement about its technical, economic and environmental viability, or its appropriate role in future energy systems. Rather, my concern here – while acknowledging its uncertain outlook and the infusion of interests and strategies behind its emergence – is to consider it as a representative case of *regime-led* prospective innovation, in that it is associated mainly with large, incumbent interests in the energy sector. For this chapter, the case illustrates an important shift toward regime (and landscape-regime) dynamics as arenas of energy system change (other examples of regime-led prospective innovations are smart grids, the re-emergence of nuclear power, and large scale offshore wind power). Rather than attempting a detailed account of the internal dynamics and tensions of the CCS innovation system, therefore, my main concern here is describing CCS' emergence into mainstream energy and climate policy arenas. I have relied mainly on official policy documents and secondary sources.

The rise of CCS internationally

Although discussed in the 1970s, the first efforts to develop CCS as a climate change mitigation technology were not made until the mid-1980s, and only started to gather significant momentum in the early 1990s, under the support of major fossil fuel corporations, national governments with fossil fuel interests and the International Energy Agency (Meadowcroft and Langhelle 2009b). Public funding of CCS innovation also began appearing at this time; the European Commission first supported CCS research and development in the early 1990s – although it was a decade later before it became a Commission funding priority (van den Broek *et al.* 2010).

Mainstream climate and energy policymakers only slowly accepted the possibility of a significant role for CCS over the course of the 1990s and 2000s. The

Intergovernmental Panel on Climate Change (IPCC) referred to it as 'limited for technical and environmental reasons' in its 1995 Second Assessment Report; by its 2001 Third Assessment Report CCS had become 'potentially more viable option', but it was only with an IPCC Special Report on CCS in 2005 (IPCC 2005) that it became a mainstream concern for the international climate change policy arena (Meadowcroft and Langhelle 2009b). The Special Report judged CCS to be a technically feasible method of securing up to half of all cumulative CO_2 mitigation worldwide until 2100, 'mainly by applying known technology developed for other purposes' (IPCC 2005: 12, 54). While acknowledging significant economic and other uncertainties, the Special Report also identified the prospect of substantial cost reductions in CCS costs over the coming decade (IPCC 2005: 11).

De Coninck and Bäckstrand (2011) noted that for much of the 1990s and 2000s, a number of mainstream climate change, energy and international relations policy groups regarded CCS as a marginal issue, outside their core remits. Meadowcroft and Langhelle (2009b: 5–7) agreed that during this period, the technology had 'remained the preserve of a narrow group of technical experts … it seems curious that [it] took so long to attract attention'. They attributed this to an unlikely alignment of policy and industry interests: a reluctance among climate policymakers to legitimise 'end-of-pipe' technologies; a cautiousness among energy policymakers to make climate change a central strand of their agenda; and the resistance of fossil fuel interests to support the need for any major (and potentially costly) response to climate change.

Only in the late 2000s, under more urgent imperatives for change and the prospect of stronger regulations, did CCS become a more credible, legitimate and visible part of energy and climate policy mainstreams. Langhelle and Meadowcroft (2009: 262) concluded that 'in a remarkably short time, CCS … moved to the forefront of international discussion as a viable climate change mitigation strategy'. By the end of the decade, International Energy Agency scenarios suggested that without CCS, the overall cost of halving global emissions by 2050 would rise by 70 per cent (IEA 2010b: 494).

Underpinning this broad trend toward greater visibility and legitimacy lay a heterogeneous set of interests and strategies. Langhelle and Meadowcroft (2009) mapped different portfolios of these interests across different countries. De Coninck and Bäckstrand (2011) noted that the multiple uncertainties associated with CCS allowed for widely diverging views on its prospects, with advocacy and optimism among some groups contrasting with cautiousness and scepticism among others. In the face of these deep uncertainties and 'interpretive flexibility' (see Hansson and Bryngelsson 2009), de Coninck and Bäckstrand (2011) cautioned against taking more positive claims at face value, since fossil fuel incumbent interests may 'symbolically' promote the technology's future prospects as part of broader efforts to delay the imposition of any firm climate abatement measures.

Meadowcroft and Langhelle (2009c) noted the ambivalent relationship with CCS for incumbent energy organisations, especially the utilities – in that it is both an unwelcome *cost-increasing* proposition in the short term, and a potentially attractive *sustaining* investment in the longer term. For Bowen (2011), this is itself an over-generalisation, since the CCS field spans both sustaining (post-combustion) and disruptive (pre-combustion/gasification) technologies. In sum, incumbents' interests and engagements with CCS are likely to differ significantly, according to a particular organisation's internal resources, external alliances and strategic outlook. As such, they are a matter for empirical investigation rather than *ex ante* assignment.

The emergence of CCS in the UK

By the late 1990s, the UK energy system reflected a decade of private ownership and laissez-faire governance. For most of the 1990s the UK government dismissed any need for an overall energy policy – system governance, it argued, was best left to the private sector (MacKerron 2009). In a sudden 'dash for gas', investment in the liberalised electricity sector became overwhelmingly oriented to combined cycle gas-fired generation technology (CCGT). CCGT was the only financeable large-scale electricity supply technology in the 1990s and early 2000s; no other technologies were able to attract significant investment. Much of the effort in the UK energy sector during this period was on asset sweating, and on *financial innovation* rather than technological innovation (Winskel 2002b).

This pattern changed only slowly after the Labour Party returned to power in the late 1990s. An early review from the new government on *Energy Sources for Power Generation* (DTI 1998) was concerned mostly with the continued dash for gas, and there was no mention of CCS in the report's discussion of different options for decarbonisation. Two years later, CCS was again omitted from government projections of energy system change to 2020 (DTI 2000). The UK was comfortably on track to meet its modest international and national obligations for greenhouse gas emissions reduction; the weak climate change policy imperative was being inadvertently dealt with by the market. In the absence of any long-term targets for system change, there was no role for long-term planning.

A turning point for UK energy and climate policy was the Royal Commission on Environmental Pollution's 2000 report on *Energy: The Changing Climate* (RCEP 2000). For the first time, an authoritative body recognised that climate change presented a 'radical challenge', and suggested the need for ambitious long-term decarbonisation targets as a central part of energy policy. The RCEP devised four different scenarios consistent with a 60 per cent reduction in UK CO_2 emissions by 2050 – two of which involved a significant role for either nuclear power or CCS. However, while CCS had now emerged on the UK energy policy agenda, the Royal Commission made clear that it saw renewable energy and demand reduction as the priority policy responses to climate change.

The RCEP report marked a watershed in re-establishing the legitimacy of energy policymaking and long-term target setting, and the government commissioned a detailed analysis of the appropriate response. The result was the *Energy Review* from the UK Cabinet Office's Performance and Innovation Unit (PIU) – a quasi-independent body (PIU 2002). The PIU declared that the benign trends that had allowed for market-based system governance over the past 10–15 years had now ended. While it agreed with the RCEP that policy priorities should be renewables and efficiency, the PIU also identified nuclear power and CCS as significant options for the future. At the same time, it noted that CCS faced major uncertainties regarding cost/investor acceptability, safety/environmental impact and public acceptability. In addition, because it was considered a technology system made up of mature components, it was seen as less likely to deliver significant cost reductions in the course of its development and deployment than less mature technologies, especially renewables (PIU 2002: 196).

The government's official response to the PIU report followed on a year later, in the White Paper *Our Energy Future* – the first comprehensive contribution to UK energy policy for 15 years (DTI 2003). The White Paper took forwards many of the recommendations of the RCEP and PIU reports; it accepted the need for long-term targets for CO_2 emissions reductions; the planning of energy system change was now re-established on the policy agenda after two decades of market liberalism – at least as a high-level aspiration. On CCS, the White Paper shared the RCEP and PIU's lukewarm interest, conditional on the demonstration of its technical, economic, legal and environmental feasibility. The government made clear that it did *not* see CCS as a mainstream strand of energy policy and nor, as a relatively mature technology, as a priority for UK energy innovation efforts. At the same time, the White Paper noted a significant and growing level of interest in CCS internationally, for example in the USA and Canada – countries where coal was thought to have a more central part in the energy mix than in the UK. The implication here was that CCS innovation was most appropriately done elsewhere.

In the same year as the White Paper, the House of Commons Science and Technology Committee carried out an inquiry into UK innovation needs associated with a non-carbon economy (HCSTC 2003). The Committee agreed with government that CCS had only a supporting role in UK energy futures, of lesser importance than renewable energy. The promotion of CCS, the Committee argued, 'should not act as a disincentive to the elimination of carbon based fuels' (HCSTC 2003: 42). At the same time, evidence presented to the Committee at this time suggested that some private firms considered CCS likely to have a more central part of energy futures. In their evidence, BP pointed to its significant CCS joint investments with overseas firms and governments. By contrast, the UK's primary funding body for university-based RD&D (the EPSRC) revealed that its spending on CCS in 2002–2003 was just £67,000.

The paltry levels of UK public spending on CCS innovation at this time reflected a wider issue: the erosion of the UK energy system's RD&D capacities over the 15 years since privatisation. Recognising the need to re-establish these capacities, the UK government's Chief Scientific Advisor commissioned an analysis of energy RD&D priorities. The resulting report identified CCS as an area where increased UK research efforts could yield major technical breakthroughs and UK economic benefit (ERRG 2002). Even so, the report supported the prevailing energy innovation policy emphasis on efficiency and renewables. As Scrase and

Watson (2009) noted, the main actors in the re-emerging UK system for energy RD&D of the early 2000s – the Research Councils, government departments and the Carbon Trust – were oriented mainly towards renewables and novel, small scale innovations.

By the mid-2000s, some 'insider' policy groups were presenting the case for a more central role for CCS in UK energy futures. Initially, this was confined to the 'clean coal' policy arena, rather than broader energy and climate policy arenas. The UK government's 2005 strategy report on Carbon Abatement Technologies concluded that, on the basis of energy system modelling commissioned by government, CCS was cost competitive with nuclear power and most renewable technologies, other than onshore wind (DTI 2005). Scrase and Watson (2009) concluded that, inside government, the case for supporting CCS innovation had been won by 2004. However, this victory was a fragile one – government tacit support for CCS innovation had yet to translate into significant levels of funding for demonstration plant.

At the same time, in the wider energy policy arena, the role of CCS in energy system futures was still being questioned. In the mid-2000s, perceiving more urgent policy drivers, especially heightened security of supply concerns, the government commissioned another major policy review. Despite a changing policy context, the 2006 *Energy Review*, like its 2002 predecessor, downplayed the prospect of a major role for CCS in UK energy futures, and questioned its economic and commercial feasibility (DTI 2006: 111). The White Paper that followed (DTI 2007) was similarly unenthusiastic; while acknowledging growing international interest in CCS – reflecting the rapidly expanding role of fossil fuels in global energy production by this time – the government judged that for the UK 'up to 2030, nuclear … and renewables would be more likely to come forward' (DTI 2007: 133). Unless the cost of CCS fell substantially, the White Paper concluded, it was unlikely to become a commercially viable prospect in the UK (DTI 2007: 172).

The House of Commons Science and Technology Committee offered a substantially more positive assessment of the technology's prospects (HCSTC 2006). The Committee judged the costs of CCS were now 'comparable to other low carbon approaches to electricity generation and … there is the potential for substantial cost reductions due to technological development, increased experience and economies of scale' (p. 3). More broadly, the Committee highlighted a confluence of domestic and international forces prompting the need for more urgent policy interventions: growing scientific evidence of climate change risks, rapidly growing emissions internationally; lack of progress in meeting domestic policy targets for emissions reduction; and rising fossil fuel import dependency in a context of more volatile international markets.

The UK Climate Change Act marked the culmination of the gathering trend for long-term policy target setting (HMG 2008). The Act established statutory UK carbon budgets consistent with an 80 per cent reduction in emissions by 2050; with this, UK energy policy had broken decisively with two decades of market liberalism and, after 2008, energy system governance entered a new era of long-term planning. A year later, the government set out the proposed means for achieving its Climate Change Act commitments in a *Low Carbon Transition Plan* (HMG 2009). The Transition Plan offered a more positive assessment of CCS than the 2007 White Paper; its relative contribution was now described in broadly equivalent terms as renewables and nuclear power; all three were seen as being needed 'in some combination' (HMG 2009: 58). In addition, CCS's wider international importance was now more prominently acknowledged – and also, the opportunity that this presented for UK industry creation (HMG 2009: 67). After the high-level targets of the Climate Change Act, and the broad statements of intent in the Transition Plan, more detailed proposals for remaking the institutional and regulatory framework of the electricity sector emerged (Ofgem 2010; DECC 2011).

The late 2000s also saw the remaking of the overall UK energy innovation system. A number of new institutions were now created that reflected the concerns and priorities of incumbent business interests and large-scale technologies. These included, for example, the Energy Technologies Institute (ETI), a public-private joint venture with multinational energy and engineering commercial partners and a large scale engineering focus (including CCS); the Technology Strategy Board (TSB), a government-sponsored business-led organisation whose spending priorities reflect UK competitive advantages, and the Environmental Transformation Fund (ETF), a public fund whose investments focused on accelerated commercialisation (Scrase and Watson 2009).

By the end of the decade, UK energy policy had been transformed and, at the same time, CCS had become established as an important prospect in envisaged UK energy futures. For Scrase and Watson (2009), the emergence of CCS could not be attributed to any changes in its technical or economic character, or any resolution of its major uncertainties. Rather, they argued, its rise reflected the changed perceptions of policymakers regarding the continued use of fossil fuels, both in the UK (in response to growing security of supply concerns) and also internationally (which informed a UK policy aspiration to show international climate policy leadership). Under more urgent domestic and international policy imperatives, CCS had emerged as a significant prospect because it reflected the revised perceptions of policymakers, and the greater engagement of energy incumbent organisations, including power equipment suppliers, construction companies, fossil fuel suppliers and industry associations.

Looking back on UK energy policy developments over the 2000s, MacKerron (2009: 79) suggested that the policymaking style of the early part of the decade represented a radical departure from UK technocratic traditions, 'less incremental … [and] more inclusive'. Rather than any long-lasting change, however, this proved to be short-lived, and for MacKerron, was a 'high point in energy policy development' (p. 83). In the second half of the 2000s a more centralised and authoritarian policy style re-emerged, justified so as to enable the accelerated deployment of large-scale technologies such as nuclear power and CCS. MacKerron concluded that by the end of the decade, faced by trade-off between urgency of response and societal legitimacy, UK energy policymaking had 'largely abandoned the search for legitimacy' (p. 87).

Discussion: the co-evolution of policy, strategy and theory

The social shaping of innovation theory

Innovation policy, strategy and research can be seen as a co-evolving 'triple helix', mutually shaping each other over time (Etzkowitz 2008).[4] For two decades, from the mid-1980s to the mid-2000s, energy systems in many developed economies evolved in an essentially unplanned, emergent manner, under weak top-down imperatives for change and in benign economic and environmental conditions. During this period, technological innovation was a marginal pursuit for energy policymakers and strategists, and energy systems were a marginal topic of interest among innovation researchers.

In a context of relatively low policy-strategy-theory dynamism, energy related innovation studies was driven forwards by a small academic community, from a particular cultural setting and epistemological standpoint: Dutch and Swedish constructivist innovation studies. Constructivist innovation theories assemble an analytical blueprint for characterising major changes in socio-technical systems, and for effecting transitions to more sustainable systems. Essential elements in this blueprint include: radical niche technologies and practices as the source of system-changing innovations; learning processes centred on interaction among distributed agents over long time periods; the breakdown and replacement (or where possible, enrolment) of existing organisations and institutions to allow for the diffusion of niche innovations; and a moderately interventionist, iterative policy style.

As a large-scale technology, drawing partly on components developed from other sectors, fostered largely by regime incumbents, and aimed at adapting rather than replacing established systems, CCS evidently cannot qualify as a system innovation under the constructivist blueprint. Unsurprisingly, then, there are relatively few references to CCS in transitions or technological innovation systems research.[5] When discussed by innovation theoreticians, it has been seen as an essentially defensive response to the decarbonisation imperative – a 'sailing ship' technology, and therefore unlikely to be capable of bringing about system change on the scale required (Geels 2002; Bergek *et al.* 2008b). For Berkhout *et al.* (2004) CCS exemplified regime efforts at 'endogenous renewal' (along with flue gas desulphurisation and fossil fuel steam plant up-scaling) – i.e. a purposeful change resourced from *within* the regime, rather than – as in transitions theory's blueprint – supported by external resources.

After 2005, energy systems faced stronger and more urgent imperatives for change, and more rapidly co-evolving dynamics in policy, strategy and research. In the UK, this led to a profound shift in policy substance and style in the late-2000s, as a substantially more *dirigiste* (centralised and technocratic) style of policymaking emerged than had been considered desirable or legitimate for the previous two decades. By the start of the 2010s, policymaking tenets that had been antithetical in the 1990s and early 2000s – statutory obligations for major system change, long-term planning and deep interventions in market-based decision-making – had become established features of UK energy policy.

As part of these changes, technological innovation became a more serious, central matter for energy policy and strategy, and at the same time energy systems started to command greater interest among academic researchers. In the social sciences, this involved not only the engagement of wider parts of the innovation studies community, but also the significant attention of a number of other disciplines, including business and management studies, psychology and sociology.[6] Within innovation studies, these trends resulted in an opening-up of debate, with greater plurality and heterogeneity of epistemologies and ontologies. While scholars working with transitions theory and TIS framings have, for the most part, retained a constructivist outlook, other, prominent voices in the wider innovation studies community have emerged – offering alternative prescriptions for responding to the accelerated innovation imperative.

Responding to the accelerated innovation imperative

As part of the innovation studies research community's burgeoning interest in energy systems, and especially in how best to respond to the accelerated change and accelerated innovation imperatives, some challenges (both implicit and explicit) have emerged to the constructivist blueprint of niche-led system change. This has been reflected in debate, for example, on the relative merits of *distributed, bottom-up and discontinuous* innovation on one hand, and *concentrated, top-down and continuous* innovation on the other. Some prominent contributions to this debate, and its links to CCS, are now briefly outlined.

In a relatively early contribution, Unruh (2002) identified three distinctive styles of innovation for climate change mitigation: *end-of-pipe* solutions (such as flue gas desulphurisation and CCS); *continuity approaches* (such as large scale, centralised renewables); and *discontinuity approaches* (involving major infrastructure changes); in a later paper, *strategic niche management* was also categorised as a discontinuity approach (Unruh and Carrillo-Hermosilla 2006). Unruh (2002) concluded that under conditions of high levels of carbon 'lock-in', discontinuous strategies were unlikely to be able to bring about rapid change. Continuity approaches, by aligning the interests of dominant corporations and government and using the 'established development path', were more likely to be capable of rapid response (Unruh and Carrillo-Hermosilla 2006).

Yang and Oppenheimer (2007) argued that a highly centralised 'Manhattan Project' style of innovation for climate change mitigation was inappropriate, given the much broader and more diverse socioeconomic context of the climate problem compared to military innovation programmes. Instead, Yang and Oppenheimer suggested that an emphasis on regulatory standards – what they referred to as a *Kyoto Protocol*-style approach – was more appropriate: 'the [climate problem] cannot be simply researched away … using a Manhattan Project as a metaphoric alternative to the Kyoto Protocol … is likely to deflect policymakers in counterproductive directions' (p. 203).

Mowery *et al.* (2010) also noted that the Manhattan Programme had been repeatedly invoked by climate change researchers (rather than the innovation studies research community).[7] Like Yang and Oppenheimer, they argued that this was mistaken, given the heterogeneous, global nature of energy technology development and adoption communities. For Mowery *et al.* the Manhattan Programme analogy was 'at best inaccurate and at worst misleading' (2010: 1021); instead, they recommended a broadly constructivist policy approach – emphasising decentralisation, diversity and long periods of niche-based incremental improvement, and recognising the emergent nature of change, 'making it difficult if not impossible to plan or predict the structure of the overall R&D effort in any detail' (p. 1020).

Some respondents to Mowery *et al.*'s contribution implicitly challenged the appropriateness of constructivist-style innovation as a response to the accelerated innovation imperative. Echoing Unruh (2002), Hargadon (2010) argued that given the high upfront costs and long timescales of technological innovation in the energy sector, the most effective route for system transformation was likely to be through marginal improvements to existing technologies. Hargadon highlighted the pitfalls of a more radical, disruptive approach: 'eschewing the transformational potential of a technology precisely because its technical artefacts, patterns of production and consumption, experiences, labor etc. exist already may preclude the very attributes that enable rapid scaling and broad adoption' (p. 1026). Rather than novelty and disruptiveness, he concluded, energy innovation policy should focus on bottlenecks preventing the improvement of existing technologies.

While advocating a regime-driven 'continuity' response, Hargadon cautioned against top-down Manhattan Programme style innovation – indeed, he highlighted Garud and Karnøe's (2003) account of the historic failings of US science-led top-down efforts at renewables innovation. By contrast, Perrow (2010) explicitly advocated a Manhattan Programme-style approach, particularly for large-scale technologies such as CCS. Perrow argued that while decentralised, learning-by-doing approaches were appropriate for some parts of the energy system, such as transport and energy efficiency, a centralised, top-down approach could make 'a very big dent in the problem' for CCS. A dedicated CCS innovation programme, able to license its results internationally, was seen by Perrow as overcoming any objections to Manhattan Programme-style innovation based on the heterogeneity and global spread of energy producers and users.

Summary

Constructivist innovation theories emerged in response to the limitations of established social science representations of technological change. For transitions theorists particularly, the problem of disciplinary abstraction suggested the need for 'appreciative theorising' – absorbing a broad range of social and economic theories and concepts to achieve more complete understanding. However, as Geels (2010) noted in a survey of relevant literatures, there are limits here; for some theories, Geels concluded – particularly those with 'a default orientation towards stability, equilibrium or incremental change' – the differences with transitions theory were so great that they could not be reconciled (p. 503). Clearly, to retain analytical insight, transitions theory has to offer a distinctive, coherent and *contestable* account of innovation dynamics – it cannot avoid the problem of theoretical abstraction. Like all theories, transitions theory translates reality through a particular epistemology and ontology, and its analytical precepts are socially shaped, in that they reflect aspects of the intellectual and cultural context associated with its inception.

Drawing on post-structural social theories, constructivist innovation theories emerged in the co-ordinated mixed economies of the Netherlands and Sweden in the 1990s and early 2000s; these theories embodied elements of their times and places of origin. These were ambivalent influences: building on normative and deliberative traditions in Dutch and Swedish technology policymaking,[8] constructivist innovation theories also reflected the prevalence of neo-liberal political economy in the 1990s. Kemp *et al.* (1998) made clear that their focus on 'strategic niche management' reflected their assumptions about the limited influence of top-down market interventions and central planning – widely accepted policy norms at the time. Together, these contextual influences informed accounts of sustainable innovation and socio-technical system change which privileged bottom-up, niche-led dynamics.

This constructivist reading resonated with contemporary energy systems of the 1990s and early 2000s. In the UK, for example, there was no significant top-down steerage of energy system development between the late 1980s to the mid-2000s, and incumbent energy regime organisations had a very limited interest in technological innovation in this period. Instead, the re-emergent UK energy innovation system of the early 2000s (under still weak imperatives for change) relied on small-firm entrepreneurship, private capital and limited state intervention; as such, it was readily amenable to a constructivist interpretation. Indeed, UK policymakers showed some interest in constructivist-informed innovation theory at this time – evident, for example, in the UK government's *Renewables Innovation Review* (ICEPT 2003; ICEPT and E4Tech 2003; DTI and Carbon Trust 2004). However, this interest led to only limited policy initiatives, grafted on to a wider system with a continuing primary commitment to short-run market efficiency (Winskel *et al.* 2006).[9]

In its 'home' country, transitions theory also experienced theory-policy translation problems. For Kern and Smith (2008) the limited achievements of the Dutch government's 'transitions programme' reflected an underestimation of established political and commercial influences such as political expediency, investor risk aversion and vested organisational interests; Smith later called for transitions theory to more fully recognise these forces (Smith *et al.* 2010). The contextual embeddedness and limited translatability of constructivist-based innovation precepts is also indicated by difficulties encountered in efforts to apply transitions theory in developing economies (Verbong *et al.* 2010 anon. pers. comm. April 2011). The implication here, that constructivist-based theory resonates across only a limited range of sociocultural settings (see also Heiskanen *et al.* 2009), is a striking one, particularly given that, just as transitions theory was rising to prominence in European innovation studies in the early 2000s, global economic liberalism was encouraging an unprecedented expansion of fossil fuel energy systems in emerging economies.

The dramatic international growth of fossil fuel generation proved to be a significant influence on the reformulation of UK energy policy in the course of the 2000s – alongside an ambitious, long-term domestic decarbonisation agenda, conceived in relatively benign political and economic conditions. As those benign conditions deteriorated, the UK's neo-liberal commitments to energy system governance that had prevailed for a generation were eroded, and by the end of the decade they had been largely displaced. UK energy policy's dabbling in constructivism earlier in the decade was now replaced by a more *dirigiste*, technocratic policy style; ironically, the *Low Carbon Transition Plan* of 2009 was less sympathetic to constructivism than policy contributions earlier in the decade. Though a significant departure from recent experience, the re-emergence of *dirigisme* and technocentrism in energy policymaking was less surprising seen from a longer-term perspective – history reveals a tendency for energy policymakers to invest faith in centralised, technocratic authority under urgent pressures for change.[10]

After 2005, the UK electricity regime became a significant arena for technological innovation for the first time for two decades. Compared to the UK's marginal efforts at energy innovation earlier in the 2000s, the organisations now created to deliver 'accelerated innovation' reflected much more the interests and technological affiliations of established groups. From a constructivist perspective, this emphasis on regime-led innovation represents an uncoupling of policy-strategy-theory co-evolution. Transitions theory's presumptions of limited market interventions and central planning, and its policy prescriptions for 'modulating' market forces and 'muddling through' became increasingly uncoupled with a more directive and interventionist policy style. By the end of the decade, transitions theory's call for 'accelerat[ed] … unlocking of socio-technical regimes' (Smith *et al.* 2010: 445) stood in stark contrast with policymakers' efforts to support and re-energise regime-led innovation.

One strand of the regime-led innovation agenda was the emergence of CCS as a significant prospect for supply side change. Until 2005, CCS had been a rather hidden technology – either inadvertently or more knowingly, a blind spot for the energy and climate policy mainstream in the UK and internationally, and a rather inconvenient proposition for both advocates and opponents of serious climate change mitigation efforts. After 2005, in the absence of any significant curbing of emissions, it became a more visible and legitimate prospect for mainstream policymakers and strategists. CCS was a similarly difficult technology for prevailing innovation theories – ineligible, under their *ex ante* framings, for serious consideration as a system innovation.

Constructivist innovation theories emerged and diffused at a time when academic research largely overlooked energy systems. Increasingly, after 2005, broader sections of innovation studies, and other academic disciplines, directed their attention to energy and climate change problems, and by the end of the decade, these had become substantial interests (and sources of research income) for large parts of academia. This greater academic research mass and heterogeneity brought with it diverging accounts of the appropriate style and substance of energy system innovation, and some of these accounts challenged constructivism's emphasis on radical niches as the primary drivers of system change.

Conclusions

The critique of innovation studies presented here does not amount to a call to abandon the constructivist perspective. Indeed, many insights generated by transitions and technological innovation system theory – on the need to attend to multiple, co-evolving levels of socio-technical structuration, of the benefit of experimentation and design variety in early stage innovation, and conversely, of the pitfalls of more deterministic, technocentric and inflexible approaches – offer important lessons for policymakers as they grapple with urgent, multiple imperatives. Rather, the argument presented here, using the emergence of CCS as an example, is that constructivist innovation theories (and much related case research), have paid insufficient attention to regimes as loci of system change and innovation.

As the chapter recognised, a number of contributions have already demonstrated the insights available from a more robust, circumspect and nuanced attention to intra-regime, inter-regime and regime-landscape dynamics. While some of these have emerged from within transitions and technological innovation systems framings, others have come from different social science research traditions, such as organisational and management studies. Looking ahead, innovation studies research should build on these studies, developing stronger lines of socio-technical inquiry on, for example: regime organisations' varied resources and interests in the face of stronger landscape pressures, on innovation prospects in mature components and systems, on the distinctive innovation dynamics of highly coupled socio-technical systems, on cross-regime complementary assets and skills, on end-of-pipe and incremental innovation, and on the range of sustaining and disruptive technologies and strategies in regime-led innovations. Rather than confirming regimes as the primary engines of energy system change, the overall research mission here is to open-up the regime black-box, recognising the likelihood of their greater engagement in system change given stronger change imperatives and raised policy expectations, and to make the relative roles of different innovation styles a matter for robust empirical investigation, rather than *ex ante* assignment.

As Helm (2003: 400) observed, intellectual commitments and normative assumptions – perhaps undeclared – have greatly shaped energy policymaking over time. He added that such embedded commitments and assumptions present a persistent danger of using yesterday's ideas to address today's problems. Similar dangers apply for theory-making; while energy policy and strategy have been substantially remade in recent times, innovation studies has shown some lock-in to models of innovation and system change conceived in earlier times.

Notes

1 Throughout the chapter, 'constructivism' refers to theories which emphasise emergent dynamics in micro-social, weakly structured domains (rather than more structurally agnostic 'interpretivist' meanings of constructivism). An emphasis on niches and emergent systems lends a strongly constructivist flavour to transitions theory and technological innovation systems; as Geels (2007: 643) stated, 'in niches, structuration is relatively weak, because networks are precarious and cognitive rules are diffuse and unstable'.

2 Rip and Kemp (1998: 330) described this in terms of social science disciplines taking 'cross-sections of the complex whole that is technology'. Elzen *et al.* (2004: 283) also noted that 'single disciplines tend to focus on one aspect of a transition and explain it from that point of view'.

3 While there is some basis for this criticism, it overlooks important differences between the three theories. Unlike SCOT and ANT, LTS offers a detailed appreciation of the dynamics and innovative capacities of incumbent systems, and how these shape the prospects for system change (Hughes 1987, 1989). For example, Hughes (1989: 472) emphasised the role of techno-economic principles such as 'economies of scale, economies of scope, and diversity of load' in system development; elsewhere, he described scale economies and capital interest rates as 'soft determinants' of power system development (Hughes 1983). In a later paper, Geels omitted LTS from a criticism of technology studies for its point source focus (Geels 2007).

4 By 'research', Etzkowitz mainly refers here to science and engineering-based R&D. I am bending his metaphor here to describe the co-evolution of innovation policy, strategy and theory.

5 There are some exceptions. For example, van Alphen and colleagues (van Alphen *et al.* 2009; van Alphen, Hekkert *et al.* 2010) have applied the technical innovation system (TIS) framework to CCS development in Norway and other countries. In doing so, they did not explicitly consider any qualitative differences between niche- and regime-led innovation.

6 For example, the author contributed to a large innovation studies conference at the University of Sussex in 2006, in which energy and environmental transitions were addressed only within a small parallel conference stream (The Future of Science, Technology and Innovation Policy: Linking Research And Practice SPRU 40th Anniversary Conference, 11–13 September 2006). By 2010, Sussex University was able to host a conference solely dedicated to energy transitions (Energy transitions in an interdependent world: what and where are the future social science research agendas? 25–26 February 2010).

7 The use of 'Manhattan Programme' as shorthand for a highly centralised and monolithic innovation style ignores important elements of variety and flexibility that contributed to its success; see Lenfle (2011).

8 Notably, Constructive Technology Assessment (see Rip *et al.* 1995; Kemp *et al.* 1998).

9 The UK's constructivist-informed efforts in the early 2000s also faced largely unacknowledged cultural differences between the 'negotiated economy' Scandinavian origins of innovation systems theory (Maskell 2004), and a UK system shaped by almost two decades of market liberalism. As some innovation systems theorists have acknowledged, the framework reflects its cultural roots in its emphasis on distributed agency and collaborative learning (Edquist 2000; Lundvall *et al.* 2002).

10 For an example of this in the UK, see Winskel (2002a).

Acknowledgements

The author wishes to thank the editors for their comments and suggestions.

13

LEARNING IN CCS DEMONSTRATION PROJECTS

Social and political dimensions[1]

Nils Markusson, Atsushi Ishii and Jennie C. Stephens

Introduction and objective

As the global community raises its ambitions to tackle climate change, carbon dioxide capture and storage technology (CCS) has gained increased interest as a mitigation option (Pacala and Socolow 2004; IPCC 2005; GCCSI 2009). Accordingly, the last decade has seen rapid growth of the knowledge needed to explore and develop CCS technology (Stephens 2006; van Alphen, Noothout *et al.* 2010). However, multiple challenges for technology development remain, including reducing the cost of capture, validating the storage potential of saline aquifers and integrating the components and required infrastructure with power plants or other large-scale CO_2-emitting facilities (IEA 2004; IPCC 2005; MIT 2007; McKinsey and Company 2008; Stephens and Jiusto 2010). It is increasingly acknowledged that further advancement of the technology requires large-scale demonstration, in different contexts and configurations, of integrated systems with power plant, capture, transport and storage (de Coninck, Stephens *et al.* 2009; van Alphen, Noothout *et al.* 2010).

The Zero Emissions Platform (ZEP) proposal on CCS demonstration policy in Europe provides a useful example of the articulated expectations of CCS demonstration. A portfolio of 10–12 demonstration projects across Europe with different combinations of fuels, capture technologies and storage options is proposed. These projects aim to contribute learning about a wide range of aspects of CCS, including technological performance, infrastructure requirements, environmental impact, health and safety, legal and regulatory factors, funding and public understanding (ZEP 2008a, 2008b). To achieve the expected learning outcomes, the proposal includes recommendations to enhance learning through knowledge sharing. The proposed plans, however, appear limited compared to the ambitious aims. Only four categories of knowledge to be shared are identified:

technological, commercial/regulatory, environmental and stakeholder engagement. The report lists relevant stakeholders for knowledge sharing: the public, government and those involved in other CCS projects. However, nongovernmental organisations (NGOs) and other potential stakeholder groups are not mentioned explicitly. This tendency to articulate lofty learning goals without detailing comprehensive mechanisms or plans for achieving the goals is also evident in other reports on CCS demonstration (DNV 2009; GCCSI 2009). Such circumstances highlight the need for a deeper understanding of learning processes surrounding CCS demonstration. The learning required for demonstration projects does not simply involve technical experts learning about the technological configuration of the project. Rather, the situation calls for a broader process of social learning that includes different types of knowledge and expertise and that involves engagement with a wide range of stakeholders (Rosental 2004; Desbarats *et al.* 2010).

Given the multiple uncertainties surrounding CCS technology, its widespread appeal, the high costs of investing in the technology and its perceived value as a critical climate mitigation option, much is at stake with regard to the social, political and technical ramifications of emerging CCS demonstration projects. In the past few years, growing public concern about the safety of CCS has emerged, particularly in Europe (van Alphen *et al.* 2007; Greenpeace International 2008) but also in the USA (van Noorden 2010), and public opposition to specific proposed projects and to the technology in general has been influential (Bielicki and Stephens 2008). CCS demonstration projects and their outcomes are, therefore, likely to be contested, although the extent and type of contestation is likely to vary considerably based on project-specific factors including the project design, the host community, the perceived public benefits of the project and the established relationships among industry, government and civil society associated with each project. The learning associated with CCS demonstration projects is also social in the sense of being an outcome of the associated political processes (Collins 1988).

The specific goals of this chapter are to empirically explore the context-specific social and political complexities of learning in CCS demonstration projects and to broaden consideration of and encourage integration of social learning processes in CCS demonstration projects. These goals are achieved through a comparison of the structure, framing and social context of three cases of CCS demonstration projects: the Yubari project in Japan, the FutureGen project in the USA and the Longannet project in the UK. The specific research questions to be addressed by this study are:

1. What are the major differences in the framing of project goals, structure and expectations and how do those differences impact the potential of the demonstration projects to produce social learning?
2. How are the processes of social learning in current and proposed CCS demonstration projects influenced by socio-political factors, including engagement with different actors, project financing and political contexts?

At a general level, this research also aims to demonstrate the need recognised by others for further social science research into energy technology innovation (Webler and Tuler 2010) and energy policy (Allcott and Mullainathan 2010).

This chapter first provides theoretical background on technology demonstration, social learning and framing relevant to CCS demonstration projects and describes the methodology of the comparative approach. Detailed descriptions of the three cases are followed by a discussion of the results, implications and suggested future research directions stemming from this research.

Theoretical background

The social nature of technology demonstration

The word 'demonstration' when used in conjunction with technology has multiple meanings, including a specific stage in the development of a technology (Sagar and Gallagher 2004) or a project or event having the aim of showing an artefact to an audience (Karlström and Sandén 2004; Shapin 1984; Rosental 2005). These definitions are related, but they highlight different aspects of the concept. The stage of development definition emphasises that demonstration is an integral part of an innovation process. Demonstration provides a link between R&D and deployment/marketing (Sagar and Gallagher 2004) and it contributes to the overall progress (or failure) of the technology's development. The project/event definition of demonstration highlights the role(s) of different actors and thus acknowledges the potential social complexities of demonstration.

Demonstration does not generally mean the communication of existing information, but rather implies an activity that is part of a social process of knowledge production and technology development. For example, demonstration activities may backfire and thereby highlight the need for more development work (Collins 1988). Frequently, tensions emerge in demonstration projects between the need to exhibit success and the need to test prototypes to their limits (Spinardi 2008). This situation may result in disconnection between hype and reality (de Coninck 2010). Demonstration is an opportunity to build new social networks around the technology (Karlström and Sandén 2004) and to expand the set of actors engaged with the technology. Promoting, and even selling, a technology is generally a central aim of demonstration projects (Rosental 2004). Securing government support for the demonstration may also help legitimise the technology (Karlström and Sandén 2004). Demonstration is an interactive process, focused on establishing consensus about the technology's properties and on building support (Shapin 1984).

Demonstration as social learning

Demonstration is often seen as part of a process of technical learning, understood as the establishment of objective facts by the application of scientific and engineering methods and the subsequent communication of these facts by acknowledged experts to a lay audience. Social science has shown that developing and learning about technology is a more complicated social process through which interactions between actors having different types of knowledge and differing claims to expertise will influence both the dissemination and the production of knowledge.

Adopting the notion of 'social learning' (e.g. Wenger 2000), this chapter examines the design and evolution of several CCS demonstration projects. Social learning provides a broad theoretical framework for considering and analysing societal change and has been explored in multiple contexts, e.g. the advancement of green building (Rohracher and Ornetzeder 2002) and the diffusion of agricultural technology in India during the Green Revolution (Munshi 2004). Here, social learning about new technologies implies a distributed process involving a wide range of actors. This process includes interaction and negotiation about what is being learned (Sørensen 1996). The process not only involves learning about technical facts, but also involves learning about other aspects of the technology and about the wider meaning of the technology (Williams *et al.* 2005).

This social learning perspective meshes well with the literature on technology demonstration, where demonstration events are seen as interactions between demonstrators and audiences. These interactions involve a focus on demonstrators' efforts to convince audiences about the properties of the technology. An equally important focus is the active role of audiences in establishing agreement or disagreement about these properties (Collins 1988; Shapin 1984; Rosental 2004). The multiple audiences for CCS demonstration projects include local residents, governments, the media, NGOs and the professional CCS community. Asymmetric power relations between demonstrators and lay audiences have been emphasised (Collins 1988). The social and political context within which each demonstration project is conceptualised, designed and implemented influences both the learning process and the content of the learning that takes place. In turn, social learning processes influence the associated social and political contexts. Taken together, demonstration projects co-produce learning processes and its situated social and political contexts. However, this co-production process is outside the scope of our paper and we here concentrate on the influence of the project contexts on the learning process.

Framing of demonstration

Throughout the development of a demonstration project, from initial conceptualisation through development and design to implementation, the framing of the initiative's purpose, structure and potential learning integrates assumptions that reflect the social and political context within which the technology is developing. In the case of CCS, examples of such assumptions include, for example, the need to mitigate climate change, the need to sustain centralised and large-scale electricity supply, the inevitability of continued use of coal, the need for governments to provide financial support for technology innovation and the trustworthiness of CCS experts (Hansson and Bryngelsson 2009).

The concept of framing has been used to analyse the ways in which these assumptions about matters beyond the immediate artefact (technical system) demonstrated are implicated in technology demonstrations (Rosental 2005). Multiple definitions of framing have been proposed. These definitions reflect various contexts and purposes. We adopt the one proposed by Gitlin (1980: 6), one of the most common citations in the field of framing analysis: 'Frames are principles of selection, emphasis and presentation composed of little tacit theories about what exists, what happens, and what matters.'

Framing is here understood to be strategic (explicit and intentional) action, on behalf of demonstrators or other stakeholders, in the pursuit of their interests (see also Pan and Kosicki 2003). Moreover, the framing may or may not be successful, in terms of being adopted by other actors, and need not be persistent (cf. Reese 2001).

Framing is particularly important in considering learning because framing guides learning and delimits the learning domain. The framing of a CCS demonstration project may change over time, but the initial framing will have a major impact on the design (of both the process and the product) of the project. Framing will also determine how learning occurs and what types of learning result from the project. It is also important to note that, in turn, learning may also result in reframing. However, in this chapter, the focus is on the effects of framing on the actual learning process.

Methodology

This research involves a comparative analysis of the structure, framing and social contexts of three cases of CCS demonstration projects in different contexts, with different scales and structures. The criteria used to select the CCS demonstration projects included (1) strong industry involvement, (2) some degree of public engagement and (3) national-level government involvement and support. Based on these criteria and a desire for geographic distribution, the three projects selected were the FutureGen project in the USA, the Longannet project in the UK and the Yubari project in Japan.

Data collection sought to include perspectives of different project stakeholders, including those involved in the project as well as actors external to the projects. Information on each demonstration project was integrated from available documents (both published and unpublished) as well as from semi-structured interviews with individuals involved in the respective demonstration projects. For the purpose of this exploratory study, only a small number of interviews per case study were undertaken. In the US case study, this information was complemented by informal discussions with additional informants. The interviewees primarily held managerial positions and those interviewed included technical staff in all the cases. In both the US and UK cases, communication staff were interviewed, and in the UK case study, a representative from an environmental NGO (ENGO) was also interviewed.

The interview protocol was structured around several themes including *basic information* about the scope and structure of the projects, the sequence of events so far, as well as organisation and financing arrangements. The topic of *social learning* was addressed in depth, through questions on who was involved and what they contributed to and learned from the project. Attention was paid to both core project participants and the roles of other actors, including governments, local publics and ENGOs. Specific questions addressed arrangements for learning from other projects, information sharing and public engagement. To understand the *framing* of the project, a key interview topic was also the project's rationale and justification. This topic was further probed using questions about risks and uncertainties and about the expected roles of external stakeholders.

The comparative analysis of the three cases focused on variation in structure and objectives of each project. The social learning processes were analysed in terms of who was involved in learning and what aspects of the projects they were learning about. Differences in social learning among the cases were further explored in terms of the impact of the ways in which support was built for the projects, the ways in which the projects were framed and the different governance contexts.

Three demonstration projects

Yubari, Japan[2]

Project scope and design

The Yubari project, which ran from 2002 to 2008, had the ultimate goal of establishing a comprehensive, economically feasible system of Enhanced Coal Bed Methane Recovery (ECBM) via CCS in Japan. It was composed of multiple RD&D tasks. The project is located in the western part of Hokkaido, the northern main island of Japan. It was the first project that actually sought to verify the feasibility of ECBM in Japan and was Japan's second project in the field of CCS RD&D. Japan's CCS RD&D began in the late 1980s (METI 2006) and Japan has assigned a very important role to RD&D projects. Since then, Japan has been one of the leading countries in developing CCS technology (and especially in technology for CO_2 capture).

Despite the ambition stated above, the project was designed only as a first step towards realising ECBM-CCS. The main components were a field test of ECBM where a total of about 880 t of CO_2 was injected and an environmental monitoring programme conducted during the period of project operation to detect any leakage of the injected CO_2. The project was funded by the government (Ministry of Economy, Trade and Industry, METI) totalling about US$17 million, of which 80 per cent was allocated for the ECBM field testing.

As listed in Table 13.1, diverse actors were involved: the Japanese government, private companies, government-related organisations and research institutes, universities and a non-governmental research organisation. The main contractor and coordinator was The General Environmental Technos Ltd (GET). Some public outreach activities took place. The project was subject to an official evaluation process (mid-term and ex-post evaluations) by an external committee set up by METI and consisting of technical experts. Because there had been no protest by the local communities against the project and because no external actors, including Japanese ENGOs, seem to have shown any particular interest in this project, this case study focuses primarily on the participants and on the external evaluation committee.

TABLE 13.1: Project components and the corresponding participants in charge

Project components	*Corresponding participants in charge*
Fundamental research work including laboratory testing of coal samples	• Hokkaido University (studying the mechanism of CH_4 displacement by CO_2)
	• Kyoto University and RITE (studying the effect of CO_2 injection on coal behaviour (swelling, etc.))
	• Akita University (developing a model of CH_4/CO_2 behaviour in coal seams)
Development of monitoring technology	• Nongovernmental research organisation (Kyoto Shizenshi-Kenkyujo)
Investigation of cost reduction potential of CO_2 capture from flue gas	• Waseda University
CO_2 injection micro pilot test in a selected coal seam	• JCOAL
Cost reduction of CO_2 capture from flue gas	• Kansai Electric Power Co. (KEPCO) and Mitsubishi Heavy Industry limited (MHI)
Investigation of an integrated system of ECBM and its commercialisation	• The General Environmental Technos Co. Ltd (GET)

Source: Coal Utilization Technology Evaluation Committee (2005).

The learning process

The project was first discussed in 2000 in the context of utilising CO_2 capture technology developed by KEPCO and MHI, who were also participants of the Yubari project, as a pure technology development proposal. This plan was then proposed to METI and was accepted as one of its CCS RD&D projects without any explicit competition. At its initiation, the project was expanded from a pure technology development project to a more comprehensive set of activities needed for ECBM-CCS (hereafter ECBM). An added component of estimating direct costs for commercial scale deployment of ECBM based on a few commercialisation scenarios was of particular importance. These key decisions were made mainly by the private companies. No external actor (including the local communities) was involved in the decision-making.

The injection site, Yubari town, had experienced a difficult economic recession. The town was chosen because both the expected amount of CH_4 production and the prospects for acceptance by the neighbouring communities were larger than in other candidate towns. Before the actual injection of CO_2, explanatory meetings with the neighbouring communities were held without any trouble. The project was basically welcomed as a positive effort to mitigate climate change. It was seen as offering some hope for economic benefits and new employment opportunities in return.

When the injection started, it was soon found that the CO_2 injection rate and the CH_4 production rate were low. Nevertheless, measurements showed that the latter rate was enhanced fivefold due to CO_2 injection compared with CH_4 production in the absence of CO_2 injection. Because of this development, the mid-term evaluation of the external committee resulted (predictably) in a relatively low score. The report of the committee recommended setting numerical targets for the CO_2 injection and CH_4 production rates. Targets of a CO_2 injection rate of 5 t/day and a CH_4 production rate of 600 m^3/day were set based on computer simulations. The committee also recommended strengthening the capacity of management to coordinate the project components and the internal structure of the project was changed accordingly.

To achieve the new injection rate target, the engineers tried to inject N_2 in the 2006–2007 financial year. This measure was expected to shrink the coal, thereby making more room for CO_2 absorption. The experiments showed that, although N_2 injection actually enhanced the CO_2 injection and CH_4 production rates, the former effect did not last long enough to achieve the injection target value. Overall, the actual injection rate of CO_2 was 3 t/day. This value fell short of the injection target, but the CH_4 production rate target was achieved.

Furthermore, it was confirmed that some inter-project learning had resulted from dissemination of technical knowledge produced by the Alberta (Canada) Research Council's project.

Summing up, the project was able to enhance both the CO_2 injection and CH_4 production rates, based on technical learning. Regarding learning on other aspects, the project administration changed the institutional structure and set indicative numerical targets for CO_2 injection and CH_4 recovery, following advice from the external evaluation committee.

Framing

The framing of the Yubari project is predominantly 'technocratic'. That is, the justification of the project, the overall goals and results and the project's relationship with society have been presented almost exclusively on technical and economic grounds. This is consistent with the overall CCS policymaking process of Japan (Ishii and Langhelle 2011).

The project justification includes two major arguments. The first argument cites the need for government to be involved in advancing technology. The basis for this claim is the assertion that the GHG reduction target of 6 per cent below the 1990 level of the Kyoto Protocol puts a heavy burden on the Japanese government and that, therefore, technological innovation, including CCS, is needed. The Industrial Structure Council acknowledges ECBM as an innovative technology that can be utilised after 2010 (METI and The General Environmental Technos 2008), but that to realise the potential of this technology, the government must invest directly in ECBM R&D projects because the risk and associated costs are too high to be managed solely by private companies. The project itself is justified primarily by emphasising the potential amount of CH_4 production by ECBM in Japan, namely 254 million m^3 and the benefits that can be expected from the project: economically efficient GHG reduction and enhanced energy security through CH_4 production. The economic benefits of ECBM through selling the CH_4 are also emphasised in terms of the advantage of this approach relative to CCS in saline aquifers.

The second argument justifying the project is associated with the assumed public acceptance of ECBM. In this argument, the relative ease of public acceptance of ECBM compares favourably with the risks to marine ecosystems associated with ocean storage.

In this project, the general public is considered completely external. It is assumed that the major societal concerns of the project are associated with environmental impacts and safety issues. The project addresses such concerns exclusively through monitoring of the injected CO_2: 'Monitoring is regarded as essential to demonstrate that the sequestration is safe, does not create local environmental impacts.'[3] Informational meetings with local communities were held twice during the project lifetime but were mainly based on a one-way communication model. No feedback mechanisms that would have allowed learning from these interactions were built into the project. Some minimal efforts to investigate existing public knowledge and acceptance of CCS were made, but these efforts received very little emphasis.

FutureGen, USA

Project scope and design

FutureGen was initially conceptualised by various US energy experts to be the first near-zero emissions, commercial scale power plant (275 MW) that would simultaneously demonstrate CCS, hydrogen production and the advanced coal power plant technology known as Integrated Gasification Combined Cycle (IGCC) (DOE 2004). The project was announced publicly in February 2003 and was presented as the flagship programme for the Bush Administration's strategy on clean-coal technology development and climate change mitigation. Despite almost a decade of planning and political strategising, FutureGen has not yet been built and the project's design and structure have undergone two major 'restructuring' initiatives.

FutureGen is a public-private partnership between the US Department of Energy (DOE) and the FutureGen Industrial Alliance, Inc., a non-profit consortium of some of the world's largest coal and energy companies. In addition to cost and risk sharing (initially the DOE was to contribute ~74 per cent of the $1.5 billion project and the Alliance the remainder), the Alliance has had responsibility for the design, construction and operation involved in integrating advanced technologies into the plant. The DOE has been responsible for a major portion of the financing, for independent oversight and for coordinating participation from international governments.

An extensive competitive site selection process occurred throughout 2006 and 2007, during which the 12 potential sites initially identified were narrowed down to four semi-finalists: two in Texas and two in Illinois. Recognising the potential economic and political benefits to their States of hosting FutureGen, both Texas and Illinois invested in competing for the project by providing technical justification to explain why their State should be selected, and also through public engagement initiatives to raise awareness and acceptance about the project's potential benefits. In January 2008, the Alliance announced that Mattoon, Illinois, was to be the site for the FutureGen plant.

A few weeks later, responding to increasing projected project costs, the Bush Administration's Secretary of Energy, Samuel Bodman, announced a restructuring of the FutureGen project and the withdrawal of the federal commitment to contribute funds to project as it was initially conceptualised. This 2008 restructuring of the project from a research demonstration programme to a 'near term commercial demonstration' programme was justified as a more cost-effective way to advance CCS by redirecting funds to demonstrate components of CCS on multiple existing plants rather than investing in the construction of a new single, large plant. The restructuring also altered the DOE cost-sharing from the original 74 per cent to a maximum of 50 per cent per demonstration project. Then, in the summer of 2009, the Obama Administration's Secretary of Energy, Steven Chu, announced a revival of the initial FutureGen project and pledged government support. In mid-2009, the Alliance purchased a site in Mattoon, Illinois, and groundbreaking for the new power plant was expected in 2010. However, in August 2010 the DOE announced FutureGen 2.0 – another major restructuring of the project. The redefined, limited scope of FutureGen 2.0 includes demonstrating oxyfuel combustion in a nearby existing power plant rather than construction of a new state-of-the-art power plant. Responding to this announced change and to the associated reduced significance and reduced international and national prestige of the project, the community of Mattoon withdrew its involvement in FutureGen. Mattoon expressed disappointment and an unwillingness to participate if the community would only be providing the geological storage location for the captured CO_2. In the fall of 2010, in conjunction with the DOE, the FutureGen Alliance began a new site selection process and in February 2011 the FutureGen Alliance announced its selection of Morgan County, Illinois as the preferred location for the FutureGen 2.0 CO_2 storage site, visitor centre, research and training facilities (FutureGen Alliance 2011).

The learning process

The initial FutureGen project was designed to serve as a backbone that would connect other smaller scale CCS R&D efforts and thereby enable the integration and testing of multiple different technological components. The goal was to encourage researchers and research groups worldwide to come to FutureGen to utilise different slipstreams for their R&D efforts, to take advantage of the potential for testing new capture technologies. With the more limited scope of FutureGen 2.0, this ambition has been scaled back. Another critical aspect of the project is that all information and technical advancements developed within FutureGen are to be publicly available, so that intellectual property issues do not produce limitations on learning.

From the onset, the FutureGen project was designed to have an international influence. The FutureGen Alliance consists of multinational companies, and the project is designed to facilitate learning throughout the global CCS community. Recognising that China is another global leader in coal use and has the world's third largest coal reserve base, FutureGen partnered strategically from the outset with the largest coal-fuelled utility in China, the China Huaneng Group, particularly in reference to the IGCC design component of the project (FutureGen Alliance 2009). Although FutureGen was initially designed to be the first full-scale integrated IGCC/CCS demonstration project in the world, China has developed its own very similar project. With the delays and the restructuring of FutureGen, the Chinese GreenGen project is now further along in demonstrating CCS.

Although FutureGen has not yet been built, the project has become widely known and has received significant media attention, thereby facilitating learning by multiple actors. One particular aspect of the FutureGen case, the competitive nature of the site selection process, has been critical in enhancing social learning about the project, about CCS, and about the need for climate change mitigation technologies. Owing to this site selection process, multiple entities (DOE, FutureGen Alliance and the individual states) were simultaneously investing in and promoting public awareness and support for the project and CCS technology. The FutureGen Alliance invested in a stakeholder engagement team that conducted over 200 interviews and group meetings with local communities to listen to concerns, to address questions and to explain the project (Hund and Judd 2009) and both the States of Illinois and Texas invested resources and instructed relevant agencies to promote and prepare their state for the potential project. This multipronged effort during the competitive site selection process broadened the scope of social learning and awareness about both CCS and the global challenge of climate change. One CCS educational representative from Illinois explained that the FutureGen project is so widely recognised in the state by the general public that it has become a frequent reference within public discourse about CCS.

An explicit mechanism designed to enhance 'learning' in both the initial and current versions of the FutureGen project is a proposed on-site, public-access visitor centre that will welcome local, as well as national and international visitors (FutureGen Alliance 2009). This visitor centre has been planned to involve a public 'demonstration' of all activities conducted at the site. In the original project design, with the power plant and CO_2 storage in the same location, this proposal would have allowed visitors to witness a range of plant activities, from the coal being unloaded to the pipelines delivering the captured CO_2 to the storage injection site. With FutureGen 2.0, plans for the visitor centre are associated with the CO_2 storage location rather than with the power plant (FutureGen Alliance 2011).

Throughout the almost decade-long FutureGen planning process, different kinds of learning have occurred. Key players in the project have been learning about the unstable and increasing projected project costs, the political instability of government support for the project and the ways in which various publics have responded to the project. In addition, the high-profile nature of the proposed project has resulted in social learning and increased awareness about climate change and the climate-mitigating potential of emerging energy technologies, particularly CCS.

Framing

The notion of the US as the home of the first large-scale, fully integrated coal-fired power plant with CCS emerged from an aspect of US national identity where the country is seen as a technologically advanced world leader (Stephens 2009). From the onset, FutureGen was framed as a critical project that would demonstrate to the world how a state-of-the-art, near-zero emissions coal-fired power plant could be designed and built. This framing of international technological leadership is also associated with both political and economic benefits. The strength of such international framing as well as the economic benefits associated with FutureGen can be seen in the logo of the FutureGen for Illinois organisation: 'The world needs FutureGen, FutureGen needs Illinois.' (FutureGen for Illinois 2009).

Given the presentation of FutureGen in public and political discourse, the proposed project has itself become a framing device, i.e. the project signifies US leadership in technological advances on climate change mitigation. The framing also promotes the notion that a zero-emissions coal-fired power plant is possible and thereby supports the concept of 'clean coal'.

Longannet, UK[4]

Project scope and design

The Longannet CCS demonstration project in Central Scotland involves retrofitting an existing, coal-fired, subcritical power plant by integrating a large-scale CCS system. The project, currently in the planning stage, is now the only remaining competitor for UK government funding available to support construction and operation of one CCS facility. The proposed facility would process the flue gases from the equivalent of 300 MW of power production and be up and running in 2015. Because the size of the Longannet power plant is 2,400 MW, the government funding will enable only some of the flue gas to be captured.

The planned project includes capturing the CO_2 using amine post-combustion capture and then piping it offshore to be injected into a depleted gas field in the North Sea. The planned pipeline would include around 100 km of pipe onshore and a few hundred kilometres of pipe offshore, in both cases re-using existing gas pipelines. The power plant is owned and operated by the UK utility Scottish Power (SP). SP, formerly a relatively small UK-focused company, was purchased by one of the world's largest utilities, Iberdrola, in 2007.

The UK government set up the competition in 2007 in response to another planned CCS demonstration project at Peterhead (a project later cancelled owing to industry frustration with the delays in government funding). The government's intention was evidently to avoid the perception that it was going to 'pick a winner' (Scrase and Watson 2009). The competition includes only offshore storage and post-combustion capture for a coal plant, has gone through several stages, and now the one remaining contestant of the original nine is negotiating with government.

The government competition is influencing how the Longannet project is designed. The competition guidelines have specific requirements that address the technology, the type of power plant that is eligible, the timing, the size and the organisation of the demonstration project.

The learning process

Multiple actors have been involved in learning throughout the development of the Longannet project. SP has set up a consortium including companies having expertise in capture, transport and storage. This group also includes other companies involved in other demonstration projects: Aker, which will supply the capture technology, Shell, which is to provide expertise in the gas field work, and National Grid, which will focus on the pipelines.

SP has defined five key stakeholder groups for knowledge sharing: academia, ENGOs, global CCS projects, the industry supply chain and government at local, national and international levels. SP plans to work with these stakeholder categories to facilitate knowledge sharing and communication. In addition, an SP interviewee mentioned the public and financers as important groups with which to communicate.

SP claims not to have learned much from previous CCS projects in terms of technology, but does acknowledge learning with regard to communication and engagement with stakeholders. A central lesson so far has come from the In Salah project where the formation of an alliance with academics had proven very useful.

The company has set up formal collaborations with a couple of UK universities and now enjoys quite extensive academic contacts. Academic collaborators have contributed research-based knowledge as well as training, mainly in science and engineering.

SP also mentioned being impressed with the outreach activities of the Hydrogen Energy project in California. SP seeks to learn from the public protests in other European countries, as well as from debates in Australia over clean coal. The need for early involvement with publics is the key lesson SP has learned from this process.

Contacts with ENGOs are based on personalised, pre-existing relationships. More contact is expected in the future, when permits and construction will be discussed. WWF says that owing to the commercial and competitive nature of the project, the organisation has limited access to project details.

The competition both stimulates and hinders exchange with other CCS projects. On the one hand, SP views itself as advanced and therefore as a resource from which other projects can learn. The government subsidy would also mean that results will be shared (to some extent). On the other hand, until a contract is signed, the competition rules limit what can be communicated. Interaction with companies outside of the consortium has been limited so far.

Government interaction can be characterised as being at 'arm's length' because of the competition. Apart from interaction related directly to the competition, SP has briefed government agencies and regulators on its proposal. For SP, a key benefit of involvement with the government competition has been increased access to UK policymakers.

In the original competition memorandum, intellectual property protection was emphasised, rather than knowledge sharing (BERR 2007). However, knowledge-sharing requirements have been increasingly emphasised and clarified during the competition process.

Although this project has not yet been built, multiple types of social learning have occurred. A process of social and distributed learning by the many actors involved is underway as the government and SP proceed to negotiate multiple issues, e.g. the balance between intellectual property and sharing of results. The actors have also learned how to argue for and promote the project, i.e. they have learned about framing. SP is also 'learning about learning' from other projects by adopting good practices from other projects regarding ways of managing communication with stakeholders, including academics and publics.

Framing

The proposed Longannet project has been supported by an array of stakeholders and has had minimal public opposition. The Scottish government has supported the proposed project largely because it sees CCS as a critical technology in Scotland's climate mitigation efforts. The local council is also supportive. The council is aware of the jobs provided by the large power plant and it also sees the project as part of an emerging potential for local economic growth related to green energy technology, including renewables and hydrogen energy. ENGOs have supported the project, primarily because it does not require construction of a new coal-fired power plant and because integrating CCS into the existing Longannet plant would reduce overall CO_2 emissions.

For SP, a key reason to participate in the government competition has been to secure the future of the almost 40-year-old Longannet power plant. With the recent flue gas desulphurisation technology upgrades and the addition of NO_x reduction equipment, the plant may continue to operate for a few decades if it can comply with future CO_2 emissions regulations.

SP justifies maintaining the Longannet power plant by arguing that the UK faces a risk of power supply shortage during the next decades. It believes that the development of renewables will be too slow to satisfy the increasing demand and that there is no support for new nuclear energy in Scotland. Adding CCS to the plant will also create new jobs and it will contribute to the development of new expertise that can be applied globally. The urgency of the need for CCS and the need for abated fossil fuel capacity is further supported by SP in reference to the critical role of CCS in recent International Energy Agency (IEA) projections.

The company acknowledges risks associated with the technology. Representatives highlight the risk of capture being too costly and they recognise uncertainties relating to geological storage. But, as expressed by an SP interviewee: 'CO_2 going into the atmosphere is most risky' and we need to 'demonstrate, demonstrate, demonstrate!'

The project has produced little controversy so far. No major ENGO has argued against it and local publics have been supportive. WWF, an environmental organisation, emphasises that Scotland is a small country in terms of emissions from coal but that the country can make a difference to international mitigation efforts by contributing to CCS development. At the same time, an interviewee at WWF Scotland says that 'for the demonstration of post-combustion CCS, we favour the demonstration of this on existing plant, rather than new largely unabated plant which would act to increase emissions from the power sector.'

Analysis

The empirical examination of the three cases illustrates the diverse nature of CCS demonstration activities, spanning multiple and varied dimensions, including technology, scale, public engagement and status of the project. FutureGen is an example of a public-private partnership in which industry has been willing to invest because of a large government funding commitment. Longannet is an industry-led investment project initiated in response to a government competition for funding. Whereas both cases include elements of R&D alongside large-scale integrated projects, the smaller-scale Yubari case is closer to a pure R&D project, with no immediate plans for larger scale investment. A summary of key characteristics of the three demonstration projects is presented in Table 13.2.

TABLE 13.2: Comparison of basic case study data

	Yubari	*FutureGen*[1]	*Longannet*
New build/retrofit	(No connection to actual power plant)	New build/retrofit	Retrofit
Capture technology	Post-combustion (chemical absorption)	Pre-combustion/ oxyfuel	Post-combustion (chemical absorption)
Storage option	ECBM	Saline aquifer	Depleted gas field
Integration	No	Yes	Yes
Scale	Small	Large	Large
Stage	Project completed	Site to be identified	Funding not clear FEED starting
R&D ←→ demonstration	RD&D, but no large scale demonstration	Plans for capture R&D	Capture R&D ongoing in parallel

1 FutureGen plans have changed considerably from initial to current plans.

Social learning

Who is learning?

In all three cases, as with most CCS demonstration projects worldwide, academia and research institutions are working with private companies and various 'publics' are involved in or aware of the demonstration project processes to varying degrees. Key differences among the cases are found in two areas: (1) the relationship between industry and government and (2) the involvement of civil society.

The Longannet case has been characterised by an arm's-length relationship between the project companies and the government during the competition for funding. In contrast, close collaboration between government and industry is apparent in the other cases, as the projects are organised as public-private partnerships.

In the Yubari case, minimal outreach to the local community was coupled with largely one-way communication. The project was generally closed and had little interaction with the wider society. The Longannet and FutureGen cases involve quite extensive engagement with civil society: local communities, ENGOs and media. With FutureGen, the competitive selection process mobilised extensive outreach and education campaigns at multiple levels and locations and the two government 'restructurings' of the project have created media attention and frustration within the partnership.

The Yubari case involves the least amount of international exchange. In both the FutureGen and Longannet cases, international exchange of information is planned and the projects themselves have been conceptualised based on international strategy. In both cases, some international learning has already occurred through the participation of multinational companies.

What is being learned?

The three cases differ in their relative focus on purely technical learning versus broader social learning beyond the limits of technical performance and cost. The goals of both FutureGen and Longannet integrate broader social dimensions. These dimensions include, for example, public acceptance and education as well as development of regulation, whereas the Yubari case is more narrowly focused on technical learning. This difference reflects the almost exclusive orientation towards R&D of the Yubari project, but also a more technocratic model of governance in Japan, where civil society input is not generally given much value. As Joseph Wayne Smith puts it, in the Japanese model of technocracy 'the political process is completely degraded into a form of scientific management of public opinion' (Smith 1991: 2). More engagement with civil society, as in the FutureGen and Longannet cases, would likely have required a less exclusively technical framing and a broader scope for learning.

Each of these cases exhibits learning from other, earlier CCS projects, but given the limited number of completed CCS demonstration projects worldwide (GCCSI 2009), the scope for such learning has been limited. Both the FutureGen and Longannet projects have plans for sharing the knowledge produced in the demonstration projects. The Longannet project is also learning explicitly about and applying knowledge-sharing techniques from other CCS projects.

Factors shaping social learning

The building of support

Controversy surrounding each of these three demonstration projects has been minimal. However, public opposition has played a role in the cancellation of several other proposed CCS demonstration projects, including the Vattenfall storage projects in Denmark and Germany (Buhr and Hansson 2011) and Shell's and Exxon Mobil's storage project in Rotterdam (Chazan 2009). With the recently restructured FutureGen 2.0 project design, the community of Mattoon has withdrawn its participation, an indication of local opposition and controversy. In each of the three demonstration projects explored here, alignment of the different interests of various stakeholders was apparent throughout most of the planning process and opposition was minimal.

Initial alignment of interests was constructed in somewhat different ways across the three cases. With Yubari, achieving alignment of interests was facilitated through the shared technocratic worldview among the participants, a worldview that tends to regard the actors who do not share such a view, e.g. local communities, as non-crucial stakeholders. With FutureGen, a close collaboration of private and public interests was achieved. This alignment was not sustained as the costs of the proposed project rose and the government twice announced major restructuring of the project, with minimal participation by the industry alliance partners in these decisions. With Longannet, alignment of interests was achieved as a consequence of the alliance of rather different interests. Multiple framings were aligned in support of the project. Specifically, the ENGO supported the project for different reasons from those of industry or government. The way in which interests have aligned in each case relates to early learning about how projects are planned and designed and this initial alignment also shapes the subsequent learning process by determining whose concerns are taken into account.

Framing

The discourse surrounding these projects was dominated by two frames: the potential value of learning about CCS as an important *climate mitigation* technology and expectations of potential *economic benefits* associated with the projects.

Difference among the cases emerges, however, in terms of their emphasis on particular economic beneficiaries. The framing of the FutureGen project emphasised the benefits to the local economy, to the State of Illinois and also potential economic benefits at the national level associated with the US identity as a technological leader. This framing contributed to and drew from plans to share all knowledge acquired. The emphasis is on technological advances expected to lead indirectly and over time to economic gains. Additionally, the site competition produced an emphasis on local and state-level benefits. These benefits took the form of near-term jobs. Increasing projected costs led to the restructuring of the FutureGen project. The restructuring included a shift from a research demonstration focus to a near-term commercial demonstration focus. The Yubari project is framed in terms of its contribution to national and industrial benefits, with little attention to any local benefits. As with FutureGen, the focus is primarily on technological benefits that will lead to economic gains over time. The focus on costs and revenues increased marginally over the time of the project. Initially, the key themes of the Longannet project were national competitiveness and industry gains, with emphasis on the role of intellectual property protection for the industry consortium that would win the competition for funding. During the competition process, the emphasis on knowledge sharing has increased somewhat. Locally, the project has also attracted support because of its potential contribution to the economy.

Table 13.3 summarises the main differences in economic framing. In the projects, a focus on long-term economic benefits has been coupled with a framing of technological progress and leadership, whereas a short-term focus has involved more expectations about immediate effects on industry and competitiveness. An exclusive focus on national-level benefits went along with an emphasis on long-term knowledge creation, whereas a local framing added the attention paid to short-term effects. The table also suggests a correspondence between the economic framing and the arrangements made to shape the learning processes, with a short-term focus favouring intellectual property protection over knowledge sharing.

TABLE 13.3: Variations in the framing and learning arrangements

	Yubari	*FutureGen*	*Longannet*
Framing			
Time scale	Long term	Long term	Short, but increasingly long-term focus
Geographical scale	National	National and local	National and local
Learning arrangements			
Outreach	Minimal	Local and regional	Mainly local
Collaboration	Minimal	Some, and intent to integrate into plans	Limited so far
Knowledge management	Published results	Intent for open access of information	Intellectual property protection, but growing focus on knowledge sharing

The shifts in the relative focus on research versus near-market activities in the projects illustrate the tension in demonstration projects between the need to learn from testing and development and the need to promote and sell a successful technology. These shifts are further evidence for the continued uncertainty about the maturity of the technology and they demonstrate the need for further learning.

Governance context

Each of the three cases has been associated with high hopes for national benefit and commercial returns on the projects, either in the near or in the distant future. The high levels of government subsidy for each project, ranging from 50 to 100 per cent, have influenced social learning in several ways.

With FutureGen and Longannet, government support has been distributed in the form of competitions between projects/sites. In the Longannet case, the competition limited knowledge sharing with other projects, but the government competition simultaneously helped raise the profile of the project and attracted collaboration interest. With FutureGen, competition also resulted in state-level investment in education, raising awareness and social learning at multiple locations throughout Texas and Illinois in particular. The close collaboration between industry and government and the absence of open competition for funding in the Yubari case reflect a more technocratic model of governance that limits social learning to a sparse, one-way exchange of information with the wider society.

These examples show that the ways in which demonstration is designed, structured and organised have strong implications for the learning that occurs. The combination of market forms of governance and government coordination is clearly important and raises multiple issues, including the balance between protecting intellectual property rights and free knowledge sharing. The governance arrangements observed in these cases in part reflect government intentions regarding learning, e.g. the FutureGen policy of open knowledge sharing. Other factors, however, also influence the governance arrangements. The choice of competitive versus non-competitive funding, for example, represents the need in at least some countries to avoid the political sensitivity associated with 'picking winners'. Moreover, the role of and engagement with civil society influences the extent and type of social learning associated with any project. The Yubari case suggests that technocracy is perhaps the least favourable mode of governance for facilitating broad social learning processes, but additional research could explore this suggestion further.

Practical implications

This chapter has provided a better general understanding of learning in the context of CCS demonstration projects. The analysis also leads us to propose a few practical lessons for those with an interest in CCS demonstration.

First, the analysis shows the scope for considering a wider set of kinds of knowledge than normally considered when discussing learning from CCS demonstrations. Clearly science and engineering expertise is crucial, but CCS demonstrations potentially draws on and feeds into learning in a much wider sense. It is useful to think about this in terms of (1) disciplines, professions and kinds of experts and (2) the types of actors and institutions involved with their varying perspectives and interests. In particular projects, depending on the specific objectives, some of these may be more important than others, but it is worth considering a wide range.

Second, this study confirms that CCS demonstration and the related learning are contingent on the social and political context. This means that CCS demonstrations are not independent of the social and political contexts, but are in part shaped by them. It also means that since these contexts vary, there can be no universal 'recipe' for how to design for learning in CCS demonstrations. Rather, it becomes important to be informed about the context and how to engage with stakeholders. This study provides some guidance about what may be important contextual factors: e.g. competitive versus collaborative structures, consensual or conflictual contexts, etc. The study also sheds light on the processes through which they influence learning may look like: e.g. that learning co-evolves with framing and support building over time.

Third, the study shows that how demonstration projects and policies are designed does indeed impact on learning (and vice versa). Support building processes, communication practices and the very framing of the projects all matter for what learning will happen. The study also shows that there is scope for deliberate ('second-order') learning about how to design effective learning processes.

Finally, we have seen how the term 'demonstration' can encompass quite different things and activities. Many different framings and objectives may be involved. At the very least, this shows that some precision may be needed to avoid misunderstanding, although flexibility in the language used can of course also be useful. Perhaps more interesting is the fact that there may be tensions and contradictions among these objectives. The tension between protection of intellectual property and knowledge dissemination has been identified by others (de Coninck, Flach *et al.* 2009). This study confirms the existence of this tension in practice.

This study also suggests that this tension may have a geographical dimension in the form of a local interest in the direct project impact on short-term economic opportunities versus a national interest in technology leadership and climate mitigation over the longer term. This suggests that increased local public engagement might push projects towards a shorter-term orientation, to the detriment of open knowledge sharing with its mainly long-term benefits. Conversely, a strong long-term focus may make local engagement more difficult. There is perhaps no simple solution to this dilemma, but we have identified choices to be made in project design and potentially conflicting interests to be handled.

Conclusions

This chapter explores social and political aspects of the learning processes related to CCS demonstration projects by comparing three different cases. The comparative analysis has shown that major differences in the framing of the projects, as well as key socio-political factors, can impact the learning associated with CCS demonstration projects.

This analysis represents an exploratory effort to develop a conceptual framework for considering the multiplicity of factors that influence social learning in CCS demonstration projects. The analysis suggests that these factors include the way in which alignment of interests is achieved, mechanisms for communication among stakeholders, project framing, governance structures and the national contexts within which projects are designed and implemented.

Although some of the factors influencing social learning will remain constant throughout the project and the learning processes, others may change as the project progresses. For example, whereas early framing is likely to shape the design and subsequent development of the project, projects may also come to be reframed in the light of insights gained. Initial high hopes for quick commercialisation with Longannet, for example, appear to have given way to more realistic longer-term framing. Similarly, the way in which interests align initially may influence who learns what, but the situation may also change as more is learned about the project. Insights about rising costs with FutureGen, for example, led to restructuring and subsequent withdrawal of support by a local community. Framing and interest alignment can thus both shape learning and be outcomes of it. We propose that such processes of co-evolution of framing, interest alignment and learning represent an important topic for future research on CCS social learning.

The research presented here also suggests that further work on 'learning about learning', sometimes called 'deutero-learning' in the organisational learning literature (Argyris and Schön 1996), should be part of further development of a conceptual framework for analysing CCS social learning.

This analysis highlights a broader range of actors and types of knowledge than is typically considered, and confirms that there is need to direct more attention than is presently paid to the different types of learning that take place in CCS demonstration projects (e.g. ZEP 2008a). Integrating consideration of these factors into planning, design and implementation of demonstration projects is likely to enhance the potential and actualisation of social learning. This preliminary study highlights the value of applying more in-depth social science research to questions of social learning associated with future demonstration projects. Given the unique context for each specific project, however, we caution that those involved in CCS demonstration projects should not look to social science researchers for a recipe for maximising social learning. Social learning involves multiple, complex context-specific processes. Social science research may provide insights into ways of facilitating and enhancing these projects, but no generalisable guiding checklist to optimise social learning can be developed. The initial findings presented here could be applied and further developed by a more comprehensive comparative analysis of emerging and proposed CCS demonstration projects around the world to assess in more depth whether CCS demonstrations are designed to deliver effectively on policy expectations of learning outcomes.

Given recent public protests against CO_2 storage projects in the Netherlands (Vergragt 2009), Germany, Denmark and the USA, it is interesting to note the civil society's support for the three projects explored in this study. These cases were characterised primarily by aligned interests, rather than by dispute or public opposition. As a consequence, minimal learning about confronting and delivering public opposition has occurred. Interest alignment was achieved in different ways in the three projects and the influence of local publics, ENGOs and the private and public sectors was different in each case. Additional case studies on other CCS demonstration projects would be helpful in understanding this diversity. In particular, studies of cases in which serious opposition has emerged would provide different kinds of insights from those offered here. Learning takes place both in consensual and in conflictual processes; in both cases, it is shaped by local contexts and contingencies.

Notes

1 An earlier version of this chapter appeared as: 'The social and political complexities of learning in CCS demonstration projects', Markusson N., Ishii, A. and Stephens, J.C. (2011) *Global Environmental Change*, 21(2): 293–302.

2 This case study is based on interviews with project participants, official documents ((Coal Utilization Technology Evaluation Committee (2005); METI and The General

Environmental Technos (2008); Industrial Structure Council (2009) etc.) and unpublished documents (The General Environmental Technos 2003).

3 IEA RD&D Projects Database, available at: www.ieaghg.org/index.php/RDD-Database.html, keyword Yubari, then see project summary of *CO_2 Sequestration in Unmineable Coal Seams in Japan* (last accessed March 2010).

4 This case study is based on interviews with representatives of Scottish Power and WWF, as well as the company website and news reported in the media. Since this chapter was written, the UK government competition for funding has been abandoned, due not least to disagreements about the costs.

14

CONCLUSIONS

Nils Markusson, Simon Shackley and Benjamin Evar

Welcome to the social side!

Actors, frames, processes and settings

Having presented a range of social scientific analyses of CCS, it is now time to take a step back and revisit the objectives of the book. One of the objectives has been to demonstrate that social science can contribute to our understanding of CCS, and that this includes the important topic of public understanding as well as a wider terrain of analyses on governance, innovation options and interpretations of scientific certainty.

The chapters each accomplish this by providing insights into the ways that *societal actors and groups* are involved with developing, promoting and debating CCS. The chapter by Stephens and Liu (Chapter 8) maps out how the CCS community has grown over time and changed in terms of sectoral composition, and the chapter by Hansson (Chapter 5) discusses the role of CCS experts in portraying the promise of the technology to wider audiences. Bradbury (Chapter 4) contributes significant insights into local community responses to CCS project plans, and Corry and Riesch (Chapter 6) analyse the varying ways in which ENGOs have depicted the technology. Other chapters examine the roles of industrial companies, policymakers, funding bodies and international organisations.

A number of the contributions are therefore chiefly concerned with recognising that arguments about CCS are delivered by actors with different material interests, as well as differentiated interpretations of the relationship between social organisation, science, technology and the natural environment. One route into understanding these differentiated views is to look at how different actors *frame* CCS. The chapters in the first part of the book (Bradbury, Hansson, and Corry and Riesch) all recognise that technological framing is not a matter of simply understanding the engineering or not. Rather, framing refers to the varied foci of actors and organisations, and associated assumptions about options for technological learning, belief in scientific progress, social equity and good governance. Such framings are directly addressed in the chapter by Hansson, which shows that scientists and engineers make subjective choices when they present the prospects for technology development. Markusson *et al.* (Chapter 13) similarly discuss how framing matters in the design and evolution of demonstration projects.

Another angle into social science analysis of CCS is to consider a range of *social processes*. A number of the chapters – particularly in the first part of the book on representations – seek to analyse the processes that shape how CCS is framed (Chapter 6: Corry and Riesch), how expectations are generated (Chapter 5: Hansson) and how such expectations have changed over time (Chapter 9: Shackley and Evar). In other chapters the focus is explicitly on communication processes (Chapter 4: Bradbury) and learning processes across cultures (Chapter 13: Markusson *et al.*; Chapter 10: Evar and Shackley). Finally, decision-making and governance is a key topic of social science research on CCS, represented here not least by contributions from Evar and Shackley (Chapter 10) and Winskel (Chapter 12).

A range of *social settings and contexts* are covered throughout the contributions. Some take a global perspective not least Chapter 8 by Stephens and Liu, which maps the international CCS community and Chapter 9 by Shackley and Evar, which examines how the governance of CCS development has changed in the past decade. Other chapters limit the focus to national contexts, as important arenas of debate and policymaking. For example, Winskel (Chapter 12) discusses how innovation policy, strategy and theory have co-evolved in the UK. Bradbury's chapter (Chapter 4) is even more localised and limits the analysis of CCS perceptions to a distinct set of local communities in the USA, and the chapters by Evar and Shackley (10) and Markusson *et al.* (13) only consider developments in a few specific projects.

We believe that these contributions show how social science analysis of technology provides a better understanding of the wide range of actors involved in assessments of technologies; how varied beliefs shape the options available for future development; how complex processes have interacted to govern CCS; and how such processes are shaped differently across cultures.

The collective endeavour in this book has also led us, as the editors, to reflect more deeply upon the social science of CCS, not only to draw out themes from across the chapters but, we hope, to create at least some novel material below that is 'greater than the sum of the parts'. Throughout the remainder of this section, we will discuss how societal support for CCS has been constructed – with what arguments and by whom – and how CCS as a sociotechnical entity has evolved until now. This leads us to address a question in governance: will CCS be able to attract enough support among actors with divergent agendas, without becoming too diluted and fragmented an endeavour?

Toppling the pyramid ...

What is CCS for? This simple question might appear to have a simple answer: it is all about CO_2 emissions mitigation of course! What other sensible reason could there be to reduce the efficiency of a power plant, incurring a massive inconvenience and a vast financial investment, with no clear revenue stream in sight? In other words, the entire CCS narrative rests upon the problem of climate change. We can think of this as related to a number of other arguments organised as a pyramid (Figure 14.1). These arguments become progressively complex and technical, but ultimately rest upon two basic scientific arguments. The first is that anthropogenic climate change (ACC) is happening and is very likely to result in dangerous levels of climate change (first level in Figure 14.1). The second is that deep cuts in CO_2 emissions are required in industrialised countries (second level in Figure 14.1).

Some countries have already passed laws that inscribe carbon reduction targets into legislation, e.g. the UK's Climate Change Act of 2008, which created statutory carbon budgets consistent with an 80 per cent carbon emissions reduction by 2050, relative to 1990 levels (Winskel, Chapter 12, this book). The inscription of such targets as legal statutes is intended to reassure the wider community – policymakers, industry, the public, NGOs, civil servants – that 'the government is serious about climate change'. However, in reality such laws can be repealed and changed by future administrations. Furthermore, there are no sanctions or penalties in place to enforce targets on national or global levels, and policy-savvy actors know this.

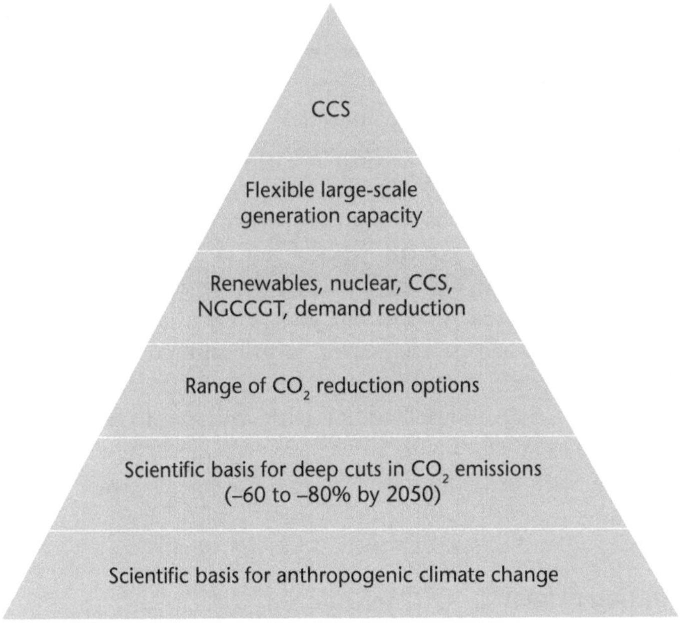

FIGURE 14.1: The CCS rationale pyramid.

Nonetheless, the scientific foundations for the CCS rationale pyramid do look peculiarly solid when consulting technical and policy literature. To our knowledge, never before has such a powerful coalition of scientific and technical agencies, academies, universities, international organisations and governments coalesced to 'resolve' a scientific problem in this manner. Nuclear energy research certainly presents an example of multiple actors agreeing that they had overlapping interests (energy generation, military grandeur, energy security), but environmentally concerned groups have largely clashed with this coalition. In contrast, there is little doubt that several powerful policy actors have accepted the CCS rationale pyramid and its priority of arguments, and Corry and Riesch (Chapter 6, this book) show that this extends to a number of environmental NGOs.

However, scientism – the idea that policies can be founded upon scientifically confirmed knowledge claims – has a dubious record as a governing principle. Science per se is rarely a sufficient basis for political action, regardless of how robust the knowledge claims appear to be. The arguments in the CCS pyramid in Figure 14.1 therefore work well as a set of necessary, but not sufficient, rationales for the uptake of CCS. A social perspective on governance replaces this pyramid with a network of agencies and individuals that take ownership of these arguments and others; a network that must be marshalled and enrolled into a stable coalition or partnership long enough to effect a change in public support of the technology and in government policies.

Companies, government agencies, NGOs and most other policy actors know this well, though they are loath to directly admit it. On the one hand, (Western) society acts as if its commitments are founded primarily upon scientific and technical justification. On the other hand, we all appeal to numerous other forms of justification – moral, cultural, political, commercial, pragmatic, etc. – a veritable hotpot of ingredients. Bruno Latour notes this when he says that 'we have never been modern' (1993). Ultimately our beliefs and decisions about technology cannot be divorced from acts of faith and habits.

It might be helpful here to reflect on the history of flue gas desulphurisation technology (FGD). The first systems, recognisable as FGD technology on power plants, were installed in the UK around 1930 for reasons including the need to mitigate crop damage caused by acidic flue gases, on fields in the vicinity of power plants. FGD has since been seen as a solution to health concerns over emissions from coal-fired power plants, and towards the end of the twentieth century as a solution to regional acid rain problems. Recently, it has again emerged as a means of mitigating the health risks posed by coal combustion. FGD technology has in this way been nimble in finding support across a varied terrain with different interests over time.

... and finding frames

In response to a narrative that is driven exclusively by science and technology concerns, we here provide examples of key arguments for and against CCS that address a number of social concerns, ranging from ethics to economics and the language of risk. The section also elaborates on which actors might advance each type of argument.

Ethical imperatives and equity arguments for CCS arise from concerns over global environmental change, global climate change, fossil fuel extraction and its local and global impacts, excessive and wasteful energy consumption, and the squandering and inefficient use of precious resources. The maintenance of international governance and the management of an orderly transition to a low-carbon sustainable energy future is another powerful ethical imperative. Being seen to lead the debate on issues of global significance is a powerful temptation for politicians. 'Saving the world from climate change by promoting CCS' has attracted the likes of Al Gore, Tony Blair and Jens Stoltenberg.

As CCS has bubbled up to the surface of public debates, officials from companies heavily involved with the technology have framed it as more than just an option for carbon reduction, but as an opportunity for wealth-creation. In a survey of the north-east Scotland based *Aberdeen Press & Journal*, the most frequently positive framing of CCS over a six-year-period was as a new enterprise opportunity, and the combined responses for 'jobs' and 'EOR' (Enhanced Oil Recovery) opportunities exceeded those for 'reducing CO_2' (see Figure 14.2).

CCS has been framed as an economically attractive activity in several ways. First, there is EOR using CO_2. This is a particularly attractive option because oil

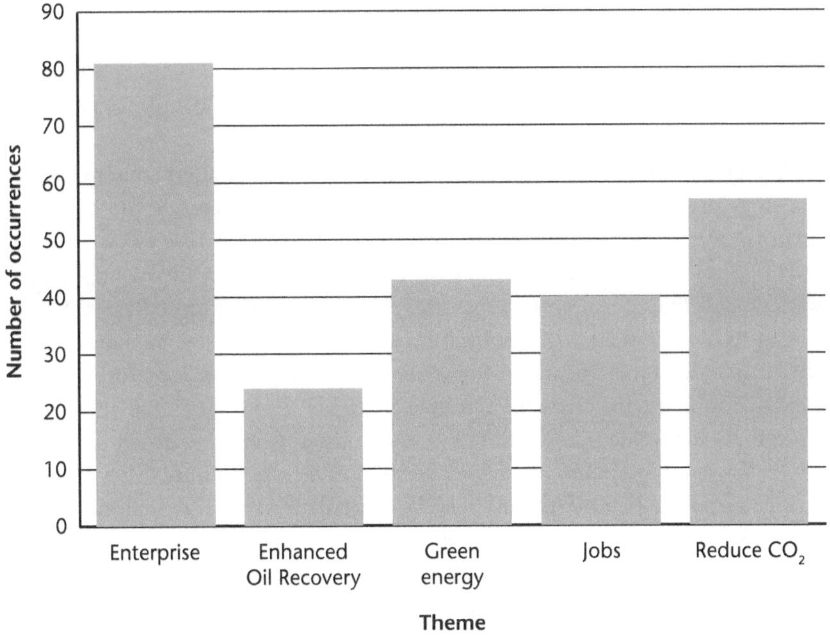

FIGURE 14.2: Positive themes related to CCS expressed in 176 newspaper articles in the *Aberdeen Press & Journal* mid-2005 to mid-2011 (source: Brunsting et al. 2011).

extraction can be a highly profitable (if risky) use of capital. However, one of the real downsides of CCS from a commercial perspective is that it is entirely dependent upon public subsidies that can always be removed by political change. CCS is even more risky in this regard than renewables because, rather than adding to electricity generation capacity, it actually reduces it. In other words, without a public subsidy CCS reduces the return on investment of private developers' capital.

In addition to this commercial opportunity CCS is being promoted by a rationale of security of supply. At least some renewables (wind, biomass, solar) are now seen as being competitive with CCS in this regard because they generate energy. And a change of government – hence of policy – has already been highly significant for the role of CCS in security of supply concerns in Norway, the UK and Australia, further demonstrating the reliance of commercial prospects on views in policy.

At one time, the carbon market appeared as a credible financial mechanism for private developers to argue that government subsidies should be left out of the equation. In the mid-2000s officials from some multinationals were thus arguing against any sort of regulations on the choices of generation in the energy sector, and against forcing developers to decarbonise their energy fleet. Rather, such signals should be delivered by the price of carbon in the EU ETS to provide the dis-/incentives for investment in alternative energy technologies such as CCS. Evidence that the EU ETS market players were instead discounting the likelihood of a high carbon price (Blyth *et al.* 2007) in favour of business-as-usual was itself discounted. Such perceptions have changed remarkably in recent years. Trust in global carbon markets has taken a veritable battering, especially in the wake of the collapse of post-Kyoto negotiations in Copenhagen and the sidelining of environmental concerns in government rhetoric since the 2008 global economic recession.

CCS-EOR therefore stands as an increasingly attractive alternative to insulate CCS from political meddling given growing policy uncertainties and failures to deliver on the promises. And the built-in revenue stream makes CCS-EOR the great hope of the CCS community. However, we expect to see increasing divisions and new fault lines between organisations that, at one time, united around CCS. CCS-EOR will be regarded as contributing to carbon emissions by many NGOs, rather than reducing them, and furthermore as providing justification for continued oil extraction in environmentally sensitive areas.

Another economic justification for CCS arises from the potential exports markets for equipment and specialised services. This opportunity relies upon the presence of political commitments as well, particularly at the international scale to ensure technology transfers and open markets. Interestingly, the international dimension has raised a further question about the scale of action. Does CCS limited to the EU really contribute to tackling climate change, or must the technology simultaneously be used in the USA, China and maybe India to be truly effective? Put crudely, is CCS without Chinese adoption worth bothering about? This question complicates Figure 14.1 because in order to be effective CCS now requires significant 'buy-in' from more than just one country or political bloc. Figure 14.1's implicit scientism has to be broadly adopted worldwide, but there is scant evidence of this taking place post-Copenhagen. Scientific arguments then revolve around whether China will act in its own self-interest as climate change impacts accumulate within its territory. Political arguments for climate change mitigation revolve around brokering a deal with China, but what is the pay-off on either side? Where are the reciprocal benefits for politicians to display to their constituencies from such political brokering? What do the voting public in the West care if Europe's political leaders manage to agree with China that CCS is an attractive priority, when there is an unprecedented squeeze on public finances and environmental issues have all but sunk from public policy debates? These may be uncomfortable questions for the CCS and the climate change communities of interest to address, but realpolitik has never been sensitive to special pleading, however valid the ethical or scientific arguments may be.

Framing also governs the critique of and concern about CCS. Several of the chapters in this book have addressed the differing, sometimes even clashing, interpretations of what risks are involved with establishing test facilities near the public. Practitioners will be familiar with such concerns from publications and meetings that deal specifically with local stakeholders' concerns. Judith Bradbury's chapter (4) is a valuable contribution to this line of research and highlights the pitfalls of poorly implemented public engagement plans. Moreover, Olaf Corry and Hauke Riesch (Chapter 6) note that the way environmental NGOs frame the risks of science and technology has an influence on the lay public's beliefs and willingness to accept local developments.

An interest-based interpretation may conclude that different stakeholders strategically emphasise the concepts and data that further their individual agendas and that they do so knowing that there are alternative and legitimate ways to understand a situation. We do not argue that this is not the case. On the contrary, even local communities with a relatively limited understanding of the technology have been observed to value financial remuneration above concerns related to risks of local groundwater contamination. However, acknowledging that strategic framing is at work does not release practitioners from dealing with the different interpretations of 'the (public) good' that are now increasingly emerging as CCS nears commercialisation. On a purely instrumental level, such interpretations should be given increased attention to ensure that policy is based on sound advice, that regulatory tools do not have adverse consequences, that demonstration projects do not fail because of poor planning, and that private developers direct their resources towards innovations and projects that will yield successful results. An examination of conflicting positions on risk can also enrich our understanding of how conflicting definitions of legitimacy can co-exist and how one notion becomes equated with the 'public good' over time.

As Corry and Riesch point out, CCS debates on risk can therefore be understood as couched in concerns over 'problem definition'. Another example is Evar and Shackley's chapter (10) on the Norwegian government's framing of CCS as a critical national undertaking. While Norway's aspirations have previously been faced with political debates over feasibility, the current bottleneck over scientific and technical uncertainty draws in the question of safety thresholds and differing understandings of precaution. We should only expect such questions to increase as CCS nears commercialisation and demands for 'scientific certainty' increase. Whether they will continue to occupy a central role in national debates will depend on how much and how far the technology is developed and on experience with handling any risks which do appear to arise.

After technology push …

So far we have argued that the case for CCS has been constructed as a more or less tightly coordinated network of different actors and framing-dependent arguments, rather than as a set of straightforward applications of an inherent, scientific logic. But building a knowledge base and social support is not enough in itself for CCS to happen, there is also a need for a social process of technology development and deployment, and we need to analyse the processes through which the network may evolve in this process. Shackley and Evar (Chapter 9, this book) analyse the evolving conditions for knowledge production and decision-making about CCS. Here, we add to this analysis drawing on a debate around technology push and pull, to further analyse the challenges facing CCS when the time comes for visions to meet reality 'on the ground' in demonstrations and deployment.

The analyses in this book tend towards a technology push model, that is, they tend to see CCS innovation as primarily driven by forces internal to the CCS community. We know from work on other technologies that technological innovation is generally the result of the meeting between such technology push and wider demand pull, although the balance between the two can vary. The push versus pull question is closely related to a longstanding debate on the so-called 'linear model of innovation', which roughly argues that a technology begins in the scientific lab and develops through a series of ordered scale-up stages until it is finally realised at full scale, and subsequently diffuses and becomes deployed more widely. Critics of this view have pointed out that knowledge inputs from prospective users may be just as important as the scientific basis, that technologies often encounter setbacks rather than progress steadily ('negative learning'), and that innovation processes are generally more chaotic and unpredictable than the linear model would suggest.

It is also true that innovation processes may be *made* more or less linear, at least to some degree. In some areas, for example pharmaceuticals, new technologies go through a strictly prestructured process of development and testing, creating a more or less linear path from lab to mass market. Even here contingency sometimes plays a role, as in the case of AIDS-related medication in San Francisco where patients demanded access to drugs that were undergoing tests in clinical trials after having extensively researched the topic themselves the topic (Epstein 1998; Collins and Evans 2007). In contrast, much software innovation appears to happen in ways that are much less ordered, and seemingly much more difficult to intentionally order (Williams 2008).

Innovation in the area of energy production is likely to be somewhere in between these extremes, and varied, with utilities typically strongly dependent on external technology suppliers whereas, for example, oil and gas companies have much stronger in-house capabilities. And an issue for CCS is indeed how to integrate the different innovation dynamics along the value chain.

With this background into possible innovation trajectories, we can ask to what extent the technology push model is valid as a depiction of CCS innovation. Drawing on the chapters in this book and other social science contributions on CCS, we can begin to formulate an answer to this question.

The main CCS-related activities to date have first focused on building the knowledge base through academic as well as industrial R&D, as well as several large-scale demonstration projects, and second on developing visions, scenarios and expectations about the future of the technology. Clearly, CCS researchers have had an important role in stimulating the development of CCS visions (Hansson, Chapter 5, this book), and used these to support the formulation of policy and regulations (Stephens and Liu, Chapter 8, this volume). In this sense, the technology has progressed so far in a vein not unlike the technology push model. But, as Shackley and Evar discuss (Chapter 9), the conditions for knowledge production related to CCS appear to be changing, with increasing need for researchers to respond and adapt to external pressures from funding limitations to public concerns in policy decisions.

Academics, governmental and industrial scientists involved in producing knowledge relevant to CCS are undoubtedly central to the social support base for CCS. And the technology or science push model presumes a well-coordinated research community, sharing knowledge about and visions of the technology, and that this community successfully enrols policymakers and technology users over time. The question is whether a coherent CCS community exists in this sense.

CCS has risen fast on the agendas of many organisations over the last decade or so, and the reason for this is not least that it fits with the interests of a wide range of actors. Stephens (2006) described the rather heterogeneous range of CCS stakeholders at the time, and discussed their varying reasons of interest. The promise of CCS has worked to some extent to bring these different actors together. But it remains unclear how stable this new social constellation is, and there is also some evidence to suggest that it has substantive internal tensions. For example (as mentioned earlier), Pollak *et al.* (2011) found a rift in the US between fossil-oriented interests seeking to orientate regulation and policy towards EOR on the one hand, and on the other, climate change actors concerned with ensuring CO_2 not only is placed underground but that we monitor it and know whether it remains there permanently or not. Monitoring, in fact, may not be enough for those concerned about CO_2-EOR increasing the available oil and gas reserves.

The challenge of building a coherent social support base across a social landscape populated with actors that hold very different objectives about the goals of the technology should not be underestimated. The argument that there is only one clear reason to engage with CCS – climate change mitigation – may serve to drum home a message on what CCS is hoped to deliver, but may also fail in attracting the attention of stakeholders who hold other primary interests. And research has shown that this may include the greater public as well as private developers.

... comes the bounded fuzziness of CCS

CCS innovation is not just about forging coalitions among actors who get to accept or reject a given technology proposal, but also about changing the technology itself in response to the requirements of users and other stakeholders. It is a process of reframing and forging of social networks, but this is not just a matter of rhetoric and advocacy. These social manoeuvrings ultimately affect how the technology is designed and deployed. The choices that are made about the technology are social events, shaped, in part, by the character of the social support base for and context of the technology. What is at stake are not just benefits and risks, but the very definition of what CCS is.

We know from studies of this and other technologies that such (re)framings are not just ways of talking about a technology, but that they can in fact contribute to changing the development path of the technology, and that the framing co-evolves with the set of social actors that engages with a technology. One example is the focus in the UK on offshore storage, in spite of the presence of suitable onshore storage formations. This is likely at least in part a result of a fear that the general public may come to see onshore storage as unattractive, probably combined with a hope of EOR and CO_2 storage revenue from depleted North Sea oil and gas fields.

From the point of view of the future viability of CCS, the technology is certainly not exclusively a climate change mitigation option, and its social base may be stronger than this would imply. Looking back at the not-so-very-long history of CCS, we see how the concept of CCS has not been static. In terms of storage, an option that attracted early interest was ocean storage, but this option has since been ruled out (for now). The CCS concept has also been fuzzy and fragmented in other, perhaps more interesting ways. There are the variants of air capture and capture from biomass conversion that do not (on the whole) offer any direct advantages in terms of dealing with the existing assets of the fossil economy, but which have received interest as options in deep emissions-cut scenarios (and which would maybe be prohibitively expensive without fossil CCS providing the rationale for investing in transport and storage).

In contrast, CCS has sometimes also been described as part of 'clean coal', which can be seen to include technologies for improved efficiency in the conversion of fossil fuels to power. Here, CCS is more aligned with agendas seeking to secure the future of fossil fuels. As discussed above, EOR (and enhanced gas recovery – EGR) with CO_2 can be made to work as storage, and if so constitute a version of CCS, but there are clashes between different interests here regarding how to regulate these activities. In terms of securing more or less clean future power supply, CCS is sometimes also seen as a step towards the hydrogen economy, potentially enrolling a new set of actors with an interest in hydrogen-related technologies. Most recently, there are ideas emerging for new productive uses to put the capture CO_2 towards, and the notion of 'carbon capture and use' has been coined, with obvious overlaps with CCS and again with the potential to attract new actors and resources. The CCS concept has thus been somewhat fluid and the different actors have been, and may become, attracted to the technology, depending on what framings are used – and, conversely, the framings will depend on what actors manage to define the technology.

CCS may have to continue for some time to be flexible. We note that concerns have been voiced about the possible rise of an overly coherent and rigid CCS community, giving rise to 'group think' and insensitivity to alternative perspectives and elements of tunnel vision or hype (de Coninck 2010). Based on the discussion here, we would suggest that such group think is necessary to construct a shared vision of what the technology is for and where it is headed. However, it is crucial to maintain a balance between such univocal support and a flexible and sufficiently fuzzy set of framings, to allow new actors to be enrolled into the CCS coalition.

Whilst many members of this coalition have not necessarily been too modest in listing the benefits of the technology (Hansson and Bryngelsson 2009), the stalling of CCS suggests that it may need some reframing, and thus a shift in its support base, although CCS might then come to seem less attractive to actors primarily motivated by the climate change mitigation imperative. And ultimately, the risk is that if CCS tries to please too many masters it loses its identity and stability. Such a support base might build in too many difficult-to-reconcile tensions. The challenge is to achieve the right degree of fuzziness – 'bounded fuzziness' as it were – steering a course between flexibility and stability.

We have here built on the contributions to the book, and developed a social science analysis of CCS evolution. The perspective set out chimes, in broad terms, with the discussion in Chapter 1 about a contrast between a traditional view of technology, emphasising linear processes and the primacy of science and engineering knowledge, and a social science perspective with a broader set of actors, networks, knowledges and interests.

CCS and society

Before we summarise and develop the practical lessons arising from this book, we now want to address whether and how CCS fits with existing theories of the greening of society. This section aims to shed light on the wider societal dynamics and politics involved with a transition to a low-carbon society built on CCS, and the initiatives and social processes that might further or hinder CCS development and deployment under different assumptions. Through this, we hope to learn something about what the emergence of CCS says about the decarbonisation of society, as well as gain ideas on how CCS could be governed and managed.

We have chosen technocracy as the starting point for our analysis, as an implicit default position of many actors involved with CCS today. Three additional theories follow to challenge the assumptions implicit in the theory of technocracy and suggest alternative governance models. We have chosen ecological modernisation and risk society theories as examples of theories of greening with wide currency. They also provide useful contrasts in terms of politics and governance models – not least with respect to state versus market as leading institutions in societal coordination, and in terms of centralised versus decentralised social agency. Finally, we have added eco-socialism to the brew, for a flavour of conflict driving change in the context of a strongly structured society.

Technocracy

Technocracy proposes that decision-making should be heavily informed by technical experts. In its most extreme versions, validated scientific and technical knowledge is held to be the only basis for good policy- and decision-making. Scientists and experts play a key role both in formulating what issues policymakers need to act upon and in providing the assessment and evaluation of the policy options, which could be adopted in response. Policymakers and politicians still make the final decisions on which policy options to adopt.

This sequence of stages is shown in Figure 14.3. Van Zwanenberg and Millstone (2005) provide a useful account of the development of models of science-policy interaction and discuss how Max Weber promoted a model of 'decisionism' for governance, in which policy and political actors identify the problems and options available, and technical experts subsequently are heavily involved in the assessment of the options. Technocracy goes one step further by providing the experts with a key role in problem and option identification and establishment.

In reality, politicians and powerful policy actors are unlikely to ever cede full authority to experts and scientists in identifying problems and possible options. The power to define 'the problem' and 'the potential solutions' is a key part of what constitutes the power of the politician and policymaker. Ceding that power completely to an external agency that cannot be fully controlled would be very risky. Nonetheless, policymakers may find themselves beholden to address problems that emerge through the activism of agents who take their lead or inspiration from science and technical problems, especially when actors unite in what have been called epistemic cultures, communities, advocacy and discourse coalitions (Haas 1992; Knorr Cetina 1999; Sabatier and Jenkins-Smith 1993; Hajer 1995).

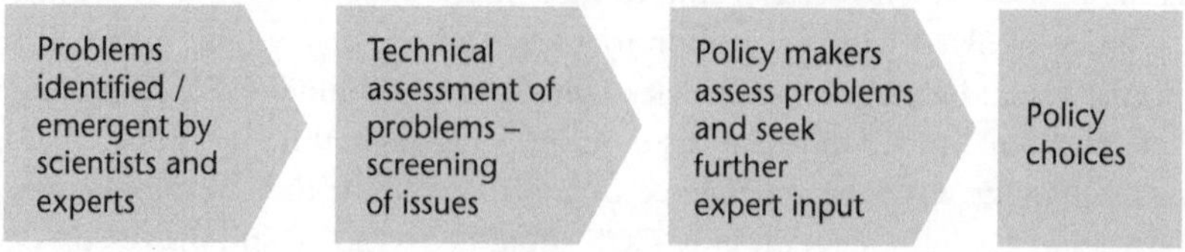

FIGURE 14.3: The technocratic model of policymaking.

In the case of CCS, what does technocracy look like and who is advocating it? The technocratic view defines anthropogenic climate change (ACC) as one of the key challenges of the era, as discussed in relation to the CCS pyramid (Figure 14.1). A panoply of highly skilled experts and academics are involved in the production of the pyramid: climate scientists and modellers, ecologists, computer scientists, economists, technology specialists, land-use change and material consumption modellers, and so on.

The technocrat sees the problem of ACC as too big and significant to leave in the hands of political interests which are uninformed as well as narrow, tied to particular interests and lobby groups, and relatively self-serving. Instead, the issue has to be dealt with by 'the experts', working closely with the politicians and policymakers, preferably through international organisations. A good model here might be the Intergovernmental Panel on Climate Change (IPCC). Here, the decision-making process is controlled by experts and the assessment is 'handed over' to policymakers to determine what policy outcomes should be adopted (even though we acknowledge that international climate change negotiations in reality are somewhat more complex and involve policymakers in the assessment process). Hence, the importance of the IPCC's Special Report on CCS in 2005 in terms of 'validating' CCS for the policy, expert and bureaucratic communities.

A technocratic approach would be to replicate the type of assessment process and mechanisms instituted by the IPCC at the national level and to institutionalise low-carbon policies, measures and practices. This has been attempted in a number of EU countries with legally binding carbon reduction targets and new offices and agencies established within the government; at the EU-level overarching objectives and targets have been established with the EU Emissions Trading Scheme and Climate and Energy package. Attempts to institutionalise low-carbon policy and practice have been less successful in other legislatures, such as the USA, though the State level has been increasingly active (Byrne g 2007; Rabe 2007). Technocracy also comes in more than one flavour. Engineering-focused technocracy will promote picking technological winners and designing policy so as to get them deployed, but with the neoliberalist turn of the 1970s and 1980s the very notion of who is the relevant expert has changed and techno-economic assessments have largely replaced exclusive engineering advice. Hence, the beginning of an era of market-based mechanisms for environmental problems, the crowning glory of which has been the EU ETS. In this respect, economic technocracy shares many of the same features as ecological modernisation, discussed in more detail in the following section.

With specific reference to CCS, the necessary stages to go through at the national level are: expert evaluation of the energy system and options for decarbonisation (usually involving use of science and engineering expertise, as well as techno-economic and economic models such as MARKAL); and assessment of other issues such as health and safety and environmental impacts, planning and regulatory issues. The identification and assessment of issues and potential problems is largely expert-led and the results, conclusions and recommendations are fed to policymakers and politicians who are tasked with creating the necessary legislative, regulatory and incentives framework for CCS to become a reality. As a large-scale, lumpy investment, CCS deployment requires a secure and reliable, i.e. low-risk investment environment. Companies, investment banks, sovereign wealth and hedge funds, would only invest in such a technology if they were highly confident of a reasonable and secure return on their investment. To obtain this level of confidence would require the creation of a policy framework guaranteeing incentives for CCS, such as had been proposed by the out-going Labour government in the UK in 2009 in the form a CCS levy charged on all electricity consumers (subsequently dropped in favour of a more extensive market reform process).

The technocratic approach to CCS deployment has been the predominant practice of governance, with extensive technical, economic, legal and regulatory assessments having been undertaken in OECD countries such as the USA, Netherlands, France, Italy, Germany, UK, Japan and Australia and, increasingly, assessments in non-OECD countries such as China, Brazil, Indonesia and India. Questions of social acceptability have also been addressed, though more frequently by academic researchers than as part of formal governmental evaluations. Wider questions of the type that are increasingly being asked of CCS by environmentalists and scholars – such as whether the prospect of CCS is acting as a 'Trojan horse' for the construction of new coal power plants, which will never be retrofitted with capture technology, have not been addressed in formal policymaking. However, they certainly feature prominently in the national politics of CCS (see Chapter 6 by Corry and Riesch, this book; Markusson and Haszeldine 2010; Shackley and Thompson 2012).

Policymakers, politicians and officials who adopt a technocratic approach are unlikely to be doing so explicitly or 'consciously'. Yet as Majone (1989) points out, all policies do imply some sort of theory, albeit often implicitly. The other theories that we consider in this section are often more explicitly adopted and promoted by supportive stakeholders.

While dominant, the technocratic approach to CCS policymaking appears to have been encountering unexpected difficulties in past years, as discussed in this book by Shackley and Evar (Chapter 9). This might open up opportunities for other approaches to policymaking to be progressed.

Ecological modernisation

A central question in environmental social science is whether and how society can address limitations to ecological carrying capacity by reforming our interactions with the natural environment, and what this implies for governance structures. Ecological modernisation theory, first proposed by Spaargaren and Mol (1992), is one attempt at providing an answer and it places pollution reduction at the centre. Huber (2008: 362) describes the philosophy of the theory in the following manner:

> Moving beyond its old-industrial stage, the modernisation of society now also entails ecological modernisation, i.e. readaptation of industrial society within

> the global geo- and biosphere by modern means such as a scientific knowledge base and advanced technology in order to upgrade the earth's carrying capacity and make development more sustainable.

Although there have been alternative formulations of the theory, these all generally have a positive view on the possibilities to integrate ecological rationality with social institutions, and specifically to 'economiz[e] ecology by placing an economic value on the third force of production: nature' (Spaargaren and Mol 1992: 335). 'Environmental management is seen as a positive-sum game: pollution prevention pays' (Hajer 1995: 3). In a later formulation of the theory, Mol (2000) specifies the connection between environmental protection and technological innovation more explicitly stating that ecological safeguarding and reform arises from changes in science and technology and, further, that: markets have a key role to play in restoring environmental integrity; the state should enable rather than mandate activities and governance should be less centralised; and social movements should be more engaged with the rest of society and not assume the role of an outside critic. With these clarifications, ecological modernisation stands as a highly technologically deterministic theory of sustainable development.

While Huber's description is largely adaptive and suggests that the natural environment is simply another realm that society can and will influence with technology, Mol and others have taken a more explicitly normative view of this relationship and suggested that both the natural environment and the economy will benefit from applications of technological ingenuity to resolve environmental problems. Such benefit may accrue both from avoided damage and from restoration of degraded environmental goods, now to re-enter the economy as natural capital. Not only does this imply that nature can be institutionalised as an element of social interaction. It also suggests that environmental gains can be made from the increased inclusion of environmental capital in the economy, e.g. by assigning clear property rights to commodities and internalising environmental externalities. Setting aside this difference in normative emphasis for the time being, both of these versions suggest that the adoption of new practices and forms of production are fundamental to achieve a more positive society-environment relationship than the status quo.

Implications for governance also vary, with some suggesting that the regulator be left out of the production process as much as possible to give firms the flexibility to create cost-effective and environmentally sound improvements (Porter and van der Linde 1995). In contrast, Mol and Spaargaren and Mol (1992) have suggested that there is a role for the state in ecological modernisation, but that its focus should be on flexible, market-based policies, rather than forcing regulation. There is therefore general agreement that government regulation and an ethically conscious citizenry alone cannot change the problems that arise from societal interaction with the natural environment. Instead, new technologies should lead the way for reformed production and consumption patterns, to relieve the stresses on environmental resource consumption and degradation. To address this, Huber

(2008) develops a framework for technology adoption that prioritises regulation by governments, aimed at stimulating and backing environmentally sustainable innovation activities by industry, to create pioneer markets and access to new technologies in less developed nations.

This version of ecological modernisation, which accords with Mol's later (2000) characterisation of the role of state governance, largely envisions regulation as a vehicle to deliver the signals to industry to engage with innovation on their terms. Regulatory strategies such as command and control, diffusion of consumer information, protection of special rights, and direct state action to force particular technologies are seen as inadequate or misguided approaches. Such approaches either place the role of government or citizens at the centre of social change programmes while even Mol's later characterisations, that specifically include other actors, merely ask that government and citizenry support the role of industry at the centre of social change. Ecological modernisation can therefore be seen as a logical offshoot from neoliberal theory in which state actors provide the right conditions for industry to shape social circumstances to the benefit of the general public (Harvey 2005).

Levi-Faur and Jordana (2005a) suggest that the relationship between government and industry has indeed emerged in this manner and become increasingly institutionalised with the introduction of more, rather than fewer, regulations since the 1980s. A closer analysis of what they call the new order of 'regulatory capitalism' reveals a distinctly supportive role for regulation to legitimise markets transactions and facilitate trust on the part of the citizenry rather than impose direct checks on market power (2005b). Jänicke (2008) argues that this has particularly been the case with the relationship between environmental protection and technology innovation. Just as pro-market policies have emerged with a distinct 'steering' role for governments, Jänicke's more current interpretation of how ecological modernisation has worked out in practice therefore follows Mol (2000), in acknowledging that regulation can provide the conditions for strongly market-based practices to shape social circumstances. The implication is that governments adopting this model of governance should set high-level goals rather than mandate specific practices or channel resources into technology forcing, and thereby induce innovation in the private sector.

Jänicke (2008: 560) specifically notes that regulation is innovation-friendly as long as the resulting instruments meet the following criteria:

- Provide economic incentives;
- Act in combination with one another;
- Are based on strategic planning and goal formulation;
- Support innovation practices and take account of the different phases of innovation/diffusion.

The regulatory instruments of choice under Jänicke's formulation therefore go far beyond trading schemes applied to specific pollutants or efficiency targets.

Somewhat more limited is Huber's assertion that performance standards on specific pollutants and efficiency targets should function as the instruments of choice. Although performance standards are more restrictive by mandating technology use or tolerable emission levels, given a flexible implementation plan they can be seen as following the line of ecological modernisation. In the case of CCS, the Norwegian government's adoption of a performance standard on CO_2 emissions from natural gas extraction activities in the North Sea falls in a grey zone. Though a tax-based standard rather than a trading scheme, the government did not mandate that non-performing practices should be phased out. Instead, the high tax induced Statoil to store CO_2 that it was already separating from its gas stream. The UK is currently considering different ways of implementing a performance standard on CO_2 emissions under its forthcoming Electricity Market Reform, but has yet to specify the conditions for flexible operations of power plants.

The theory of ecological modernisation in technology innovation practices offers helpful insights to comment on the multiplicity of views on CCS promotion and development in countries around the world. The UK government originally issued a CCS competition in 2007 (in line with original formulations of ecological modernisation), but has since been forced to re-evaluate the suitability of this approach, as there has been little response from potential developers. The US has also largely followed a market-based approach to CCS innovation to date, but one that is arguably more in line with Jänicke's views on 'regulatory capitalism'. Under this model, the DOE continues to partially fund a number of industry-led commercial research projects, similar to the European Union's support of several industry-led demonstration projects in Member States. The Australian Barrow Island Act is perhaps the best example of regulations paving the way for industry innovation, with a tailor-made regulatory framework that covers gas processing and legislation to facilitate CO_2 storage from the Gorgon project.

Has this institution-government-industry relationship proven successful when attempting to mitigate global environmental problems? Clearly, some success has been achieved with the mitigation of nitrous oxides, sulphur dioxides and CFC emissions using emissions trading schemes and emissions performance standards. But some scholars have argued that the slow pace of innovation for many technologies that address climate change mitigation serves as evidence that increased efforts to steer innovation are needed, rather than expecting industry to lead the way (de Coninck 2009; McKibbin and Wilcoxen 2009). Increased public ownership of the innovation process may therefore be necessary to ensure that long-term governance goals are met with emerging technologies such as CCS. Similarly, increased sharing of technologies between nation-states and projects may be needed, particularly in CCS where development progresses over long timescales and investments are relatively large and lumpy (van Alphen, Hekkert *et al.* 2010). This is not to say that privately held multinational companies arbitrarily contain their innovation practices within traditional state boundaries. Rather, it suggests that the necessary capital – financial and social – as well as the long lead times for developing large and complex technologies, are not conducive to fragmented and competitive development practices.

Climate change presents a global environmental public good problem for inter-state cooperation, which implies that there are significant barriers to begin the arguably costly process of dramatically reducing CO_2 emissions. This is particularly true for technologies that attempt to reduce emissions without adding a clear revenue incentive, such as additional power generation from renewable technologies. CCS development is also burdened with extremely high up-front costs and a short-term alternative in the form of natural gas. Investments are therefore risky not only because the outcomes of the innovation process are uncertain, but also because interest in its development is premised solely on a belief that other nations will choose CCS as the favoured low-carbon option for fossil fuel reliance, rather than simply switch to natural gas.

One criticism of ecological modernisation as a theoretical framework for CCS development is that it cannot adequately address the global coordination barriers facing large investments into climate change mitigation technologies. Nor does the facilitation of fragmented market-driven innovations incentivise industry to develop risky technologies such as pre-combustion plants on integrated gasification combined cycle power plants, which theoretically involve higher efficiencies and opens the possibility for a hydrogen-based infrastructure.

Risk Society

In 1992, Ulrich Beck's book *Risikogesellschaft* was published in English as *Risk Society: Towards a New Modernity*. It seemed to capture, and certainly contributed to, intellectual debates of the 1990s and 2000s and numerous books and articles were published on the issues raised (Mythen 2004). The *Risk Society* remains one of the key theoretical frameworks for understanding debates over risk in late modern societies. Beck's ideas are numerous and complex and it is only possible to present an outline of his theoretical framework here.

Beck argues that the traditional sociological distinctions such as class, gender and different patterns of social relations, which had been used to understand pre- and industrial society by nineteenth and twentieth century thinkers, are increasingly irrelevant to our society. This is because of the extension of negative risks in late modernity arising from increasingly complex technologies, proliferating unanticipated and unpredictable effects. The complex interaction of human societies with the natural world, which entails many feedback loops, generates these new types of risks. In an earlier era, capitalism led to the proliferation of goods and the key political issue related to distribution of the benefits of industrialisation. There were, of course, negative impacts to health, environment and so on, and these were heaped upon the working class. The wealthier classes were able to use their greater material resources to protect themselves from exposure to such risks. The risks at that time (such as air and water pollution, degraded landscapes) were 'containable' spatially and temporally and hence could be avoided by the wealthy class.

With the onset of the risk society, such insulation from risk becomes increasingly problematic despite significant personal wealth. Risks change from being local, specific and ameliorable in the short term, hence containable, to becoming global, universal and long term. It is no longer possible to protect oneself by utilising personal wealth to insulate oneself from health and environmental risks. Risks from radioactive materials from nuclear power plants, chemicals and anthropogenic climate change are examples of such types of risk. Radioactive materials remain dangerous for thousands of years and, as Chernobyl illustrated, can act at a long distance. CO_2 emitted to the atmosphere anywhere has an impact globally. Beck argues that such 'manufactured risks' have the potential to be catastrophic and invokes the 'worst imaginable accident' to demonstrate this. Climate change and nuclear waste incidents could cause major disasters with long-lasting impacts, thereby posing an inter-generational ethical dilemma. Hence, Beck adopts the view that knowledge of the catastrophic potential of, for example, climate and ecological tipping points, is sufficiently reliable and robust that it can create a universal reaction, such as fear and anxiety (albeit laced with various types of uncertainty). In this regard, he is set apart from some social constructivist thinkers who place greater emphasis on the social construction of risk and the impossibility of making fact/value distinctions. Such thinkers point to the uncertainties and contingencies of assumptions taken for granted by risk assessment experts, a critique that becomes only more poignant when we are forced to balance such contingencies against one another in a distribution of risks over time and space.

Beck (1994) does however argue very clearly for 'reflexive modernisation' in this new era of manufactured risks. Beck distinguishes between 'reflection' and 'reflexivity'. Reflection occurs where, for example, the public are asked to distrust and to question more thoroughly traditional institutions and their expert staff. The heightened focus on risk stimulates a whole new politics that garners its own experts to counter the prevailing system of expertise that has become discredited. Reflexivity, on the other hand, he defines as follows:

> [T]he further the modernization of societies proceeds, the more the foundations of industrial society are dissolved, consumed, changed and threatened. … this can quite well take place without reflection, beyond knowledge and consciousness.
>
> (Beck 1994: 176)

Beck distances himself from post-modern thinkers in that he argues for a 'radicalised modernity' in which many of the self-same tools, methodologies and epistemologies that underpinned modernity are applied with renewed vigour to challenging the knowledge constructs, systems of expertise and institutions of late modernity. But, as noted in the above quotation, such reflexivity may take place at a systemic level and unconsciously.

Beck does not advance an explicit political programme, but does certainly promote an explicit political change agenda, which can be termed 'globalisation from below'. Sub-groups and new social movements (NSMs) are seen as diverse, self-mobilising, sceptical, questioning, subversive and challenging agents of change from the bottom-up. (The politics espoused here are in some ways akin to those of Gorz (1993) and anarcho-socialists.) In Beck's account, reflexive modernisation is a consequence of the new type of risks facing society; ones that are potentially catastrophic but defy ready characterisation and management.

Anthropogenic climate change is an example par excellence of the type of risk that we encounter in the risk society – globalised, long term, highly uncertain in many regards and based on multiple feedbacks, in some ways unconscious and beyond knowledge, as well as contested by different experts as a potentially cataclysmic set of events. The threat of climate change hovers over society and urgent action is demanded by politicians, environmentalists, religious leaders and scientists. As Beck puts it:

> Believed risks are the whip used to keep the present day moving along at a gallop. The more threatening the shadows that fall on the present day from a terrible future looming in the distance, the more compelling the shock that can be provoked by dramatizing risk today.
>
> (Beck 2000: 214)

How does CCS fit into Beck's idea of the risk society? In response to the risks of climate change, Beck's reflexive modernisation should result in major protests from NGOs and ad hoc protest NSMs such as the Climate Camp. Yet, as Corry and Riesch point out in Chapter 6 in this book, it is unlikely that such groups would unreservedly support CCS because it is widely perceived as an end-of-pipe technology that legitimises a large, centralised, fossil-fuel power system, controlled by large organisations that are the traditional incumbents of the energy sector (both companies and governmental). The experts and managers of these large energy firms, government ministries and geological surveys are barely trusted by such NGOs and protest groups and their knowledge is, to some extent, discredited and deconstructed. On the face of it, CCS as a major response to ACC does not seem to fit Beck's risk society model particularly well. Indeed, in the summer of 2011, the German Climate Camp explicitly campaigned against CCS. On the other hand, in some ways CCS as a response of the late capitalist society to a perceived threat illuminates some of the paradoxes of the risk society: namely, that the 'technical fix' itself begins to generate new hard-to-assess and control risks.

CCS could be regarded as producing new varieties of manufactured risk of the type that the risk society has reacted in response to. These risks are very difficult to comprehend and to assess, and bear more than a passing resemblance to the risks of nuclear waste storage. Some of the CO_2 storage experts and expert bodies are in fact the same as those who were involved in assessing the risks of nuclear waste storage in geological formations. The threat of proliferating risks result in technologies which, rather than 'solving the problem', end up generating yet further risk, modernity undercutting itself so to speak. CCS is therefore both a response to, and cause of, yet more reflexive modernisation. It could be argued that this is itself a symptom of reflexive modernity – that it will have a tendency to produce the very thing (e.g. risk containment) that it reacts against.

Eco-socialism

Since at least the 1970s, it has been clear that socialist societies have a bad track record on environmental protection and efficient use of resources. Arguably, mixed, coordinated market economies – like Germany and the Netherlands – have done well on environmental policies and in some areas been rather more progressive than liberal market economies such as the US. This difference should not be exaggerated, however, with, for example, the US having pioneered sulphur control in the 1970s, and the UK having taken a leading role on climate change policies in the last decade.

For most of this period, socialist theory has not been at the forefront of environmental thought, although some early contributions, from for example Barry Commoner (1972), can be seen as an emerging 'eco-socialist' tradition of thought, introducing ecological concerns in socialist theorising. (There is certainly more than one eco-socialism, and this text does not do justice to this plurality. In particular, it excludes the more decentralist, cultural post-Marxist theories of, for example, Gorz 1993.)

The basic notion of such theories is that the tireless endeavour of capital to maximise profit (or rather exchange value) and boundless economic growth, will clash with inherent limits set by nature. Capitalist societies treat nature as a resource and a commodity, and will seek to overcome any obstacles in its way including natural limits. Environmental degradation is therefore an inbuilt side effect of capital accumulation and economic growth.

In a society driven by profit maximisation and economic growth, individual firms and other actors can externalise environmental costs as well as pass them on to future generations. In the long run however such environmental degradation threatens to undermine production capacity, for example by contributing to public health problems and producing volatility in commodity markets, and these self-destructive tendencies therefore lead to crises and social upheaval. The capitalist state may seek to correct for these effects using regulation and planning tools and capitalist societies have shown a marked resilience and ability to adapt to crises, by introducing new governance regimes. State regulation under this model is a means of survival for capital, reining in its most destructive tendencies in order to ensure survival.

Later contributions of, for example, Benton (1996) and O'Connor (1996), have further developed eco-socialist theory. A key claim in these contributions is that the tensions of capitalist society tend to undermine production as well as reproduction capacity. Reproduction here includes the natural environment as a necessary and valuable pre-condition of our welfare, as well as human reproduction (thereby forging a link to gender issues and feminism). Later eco-socialist contributions have also articulated the unequal distribution of environmental degradation effects, and the injustices inherent in capital's impact on nature. They see protests against environmental problems (and against gender inequality) as part of a wider landscape of new social movements – agencies with which the labour unions of classical socialism need to collaborate to counter threats to production and reproduction.

To simplify, this perspective predicts two possible scenarios. First, a crisis that is severe enough to destabilise society and lead to revolutionary change. It could be argued that we are living through a crisis right now, with tensions in financial markets threatening policy regimes in individual countries as well as internationally; with the threat of climate change looming ever larger and emerging signs of its impacts; with the stagnant standards of quality of life in many Western countries; and with desperate poverty still a reality in the lives of most people on the planet. Eco-socialism suggests that this crisis could deepen into revolutionary change, but it is impossible to predict what will happen after this point, and difficult to see how that might play out for CCS.

In a second less dramatic scenario, social agents mobilise against capital and the latter is resilient and pragmatic enough to accommodate a degree of reform that is adequate to stave off the threat of a revolution. In this perspective, CCS would be driven by state policy: in the form of regulation and, when necessary, nationalisation. Public support would be given by a coalition of organised labour and climate activists. Policy would be formed in a compromise between the interests of capital (with its short-term thinking, desire to externalise costs and limited ability to defend itself against crises) and the coalition of opponents. Policies would seek to internalise environmental costs. More radical policy suggestions might include no-growth societal regimes, but it seems unlikely that any compromise solution would deliver this. Support for such proposals can still have an impact on negotiations, and serve as a threat to capital (e.g. the spectre of communism after the Second World War and its role in making social democracy viable in Western Europe).

In a governance approach following this theory, the state would impose CCS on a reluctant industry, with all actors in the coalition in agreement that CCS implementation should be supported by clear government policies. The reluctance of industry is already clear from the moderate enthusiasm with which power utilities have embraced CCS to date. We would also suggest that a tendency so far – by capital and the governments it dominates – to talk the talk but not walk the walk on CCS is further evidence of this reluctance, perhaps even with elements of 'green-wash' from industries and a society under threat from climate change.

CCS promotion appears to fit rather well into a potential compromise along eco-socialist lines. It promises to support continued economic growth with recognition that climate change threatens the fossil economy as we know it. Moreover, CCS entails the internalisation of some of the costs of climate change, whether these are ultimately borne by utility owners and employees, consumers or tax payers.

There is however relatively weak evidence to support the viability of this theory of governance when it comes to popular mobilisation. There is some climate activism (e.g. Climate Camp in the UK) and it has attacked the agenda of organised capital in energy generation to a degree (e.g. Kingsnorth), but still exists on a very limited scale. The wider environmental movement has taken up the climate change cause in varying ways (Corry and Riesch, Chapter 6, this book), but the labour movement has been relatively weak since the 1980s onwards, with little sign of any dramatic resurgence. Eco-socialist theory therefore largely provides a gloomy prospect of deepened crisis in the foreseeable future.

Theory and fact

Having reviewed these four theories, we will now very briefly summarise what they each contribute to the CCS governance debate. Technocracy has been the dominant theory of policy styles for CCS to date. The economistic theory of ecological modernisation forms a theoretical backdrop for much of the momentum behind economic instruments in environmental policymaking, and at the same time a concern with the costs and resource inefficiencies of CCS, and a reluctance to introduce stringent policies to force innovation.

CCS has run into resistance and protests (anti-establishment etc.) of a type that is well predicted by risk society theory. However, only limited social agency has been expressed in such activities. And eco-socialist theory would suggest that this could be because the environmental crisis is not yet perceived as severe enough, and that capital tends to favour short-sighted actions.

The evidence to date also poses challenges to each of the theories as explanatory frameworks or governance models for CCS. Ecological modernisation style policy initiatives would be likely to be relatively ineffective in supporting CCS development, due to the significant amount of capital and the challenges of global collective action. And theories of technocracy did not anticipate the instances of effective public protests and wider public scepticism, e.g. at Barendrecht.

The theories of risk society and eco-socialism both suggest that more activism and mobilisation would occur. It is particularly difficult to imagine how CCS could emerge in a risk society-style development, as a bottom-up scenario with small-scale experimentation. But in some sort of accommodation between top-down and bottom-up policy styles, we could perhaps say that risk society supports a degree of incrementalism in CCS development and deployment, suggesting that too large and rapid steps in the scaling-up of technology efforts are likely to be counter-productive.

The point of this analysis is therefore not primarily to use CCS as a test of the four theories above. It is still early days for CCS governance and difficult to conclusively rule out a social theory as a good explanatory model or basis for governance. If CCS is deployed on a broad scale, roughly in the form envisioned today by the IEA and the G8, some of the theories reviewed here (namely risk society and ecological modernisation) would appear to be less suitable as explanations of its popularity in governance. From the point of view of a CCS proponent, this suggests that society would need to change to allow CCS development to progress – probably towards more top-down direction and state command and control policies. But as we argue here, it may well be that CCS has to change to accommodate the rest of society, and perhaps to prioritise negative emissions, a clear revenue stream, or contribute to a hydrogen-based economy.

Nor do we here take on the too-grand challenge of bridging or synthesising the four theories. Their many underlying, conflicting assumptions would make that at least very difficult, probably impossible. The main point rather, is to use these theories to identify challenges for the future evolution of CCS in society. The theories imply different types of futures for CCS development, and we will pragmatically draw on them all to identify actions that could be taken to further, hinder or shape the further development.

Lessons for politics and practice

It was important in planning this book to show how social science can contribute in a practical way to decision-making for governance and beyond. We have therefore sought to include and develop insights of relevance to CCS practitioners, policymakers, project managers and others. This section will first summarise some of the key practical lessons provided by the chapter contributions. We thereafter draw on the analysis of the four theories in the previous section, to further discuss practical actions to promote, hinder, and generally to govern CCS.

Practical lessons from the contributions

A basic insight arising from these studies is that the social matters. The chapter by Bradbury (4) offers practical tips on managing local publics, and in particular that public outreach should not be treated as an isolated add-on activity, but as a core task in project management. The chapter by Stephens and Liu (8) analyses the internal social characteristics of the CCS community and suggests that its size and composition, and how integration is managed, matters for its potential to influence policy. Across the contributions to the book we encounter a wide range of actors, and a key observation is their heterogeneity, their different preferences and allegiances, and types of expertise. Taking CCS forward (or resisting it) requires actors to orientate their views in a complex social landscape, and social science can provide some guideposts along this journey.

A perhaps more challenging set of messages relates to the broad theme of uncertainty and choice. Some of the contributions here have shown how CCS can be understood to be different things at different times, and how actors have interpreted its significance and promise very differently. For example, compare the chapter by Hansson (5) with the contribution by Corry and Riesch (6). Perceptions of promises certainly condition stakeholders' actions and the topic of technology representations is therefore critical to understanding how CCS has developed so far and how it can develop in the future. The sheer variety of CCS representations – and the variety of futures that they imply for the technology – highlights that there are choices beyond the laboratory to be addressed when designing the future of a technology. We can deduce from previous studies of technologies (Mackenzie and Wajcman 1999; Pinch and Bijker 1984) that the future of CCS is not simply determined by an inner logic of the technology, but is open to a degree of choice by a variety of actors. These choices are not merely about whether to promote a technology or not, but also how to promote, develop and shape its 'dominant' form.

The size and variety of the CCS community means that such choices are likely to emerge from a situation characterised by uncertainty, and frequent contestation. Some of the chapters have contributed with insights into how such decision-making can and should take place. A key conclusion from the chapter by Evar and Shackley (10) is that there is always choice with regard to which criteria to base decisions on, and that therefore there is no way to simply identify the right choices from the wrong ones, based on available scientific evidence. Rather, decision-making will be based on a host of criteria, some of which will arguably be less driven by rationales of science and technology than others. The chapter by Shackley and Evar (9) discusses how the context for CCS knowledge production is changing, as the technology transitions from being dominated by R&D and policy formulation, to realisation in the form of expensive pilot and demonstration projects, with increased external scrutiny and scepticism creating the potential for a backlash in social and financial support.

A final theme for practical learning is therefore that technology never develops in splendid isolation, but very much in specific social settings and actors should take this into consideration when deciding on forms and instruments for governance. The chapter by Markusson *et al.* (13) uses a comparative analysis to highlight how project support is – and needs to be – built in different ways, depending on the local and national context. The chapter by Winskel (12) discusses how CCS development challenges dominant innovation policy theories – theories which, he argues, have had some influence on policy; he goes on to outline a revised agenda for researchers' and policymakers' engagement with energy innovation. Bradbury (Chapter 4) reminds us how the characteristics of the local community shape the reception of project initiatives, and how this can matter for project design.

Politics, policies, decision-making

One possible conclusion, in response to the four theories reviewed here, is that CCS is most likely to be spearheaded in China, given its strong state, top-down policy delivery and limited scope for social protest. But reforms to steer in the direction of that type of society elsewhere in the world, will not be seen as a palatable solution by those who are fans of Western-style democracy.

It is also worth remembering that experts in technocratic policy regimes have a poor track record of forecasting future developments, including any negative side-effects. An example is the rhetoric around the introduction of nuclear energy in the 1970s, which was based on power consumption forecasts that turned out to be wrong. We would also point to a lack of predictions of the decline of nuclear and coal and the dash for gas following deregulation of power markets in the 1980s and 1990s (in the UK, the US and elsewhere).

A second possible conclusion is to wait for the environmental crisis to reach heights at which we no longer have choices to make. The gloomiest eco-socialist predictions would suggests that what is happening at the moment is a waiting game, and that only when the crisis is widely enough seen to constitute an urgent threat to our societies will broad public mobilisation be possible. We can only hope that this is wrong and that informed action for governance reform can make a difference. As a somewhat grudging compliment to capitalism, we observe that it has survived crises precipitated by wars, global financial collapses and energy shortages in the past.

More optimistically, some social theories (including variants of eco-socialism) allow for action to increase the awareness of natural limits. The challenge is then a socio-cultural one of establishing an increased awareness of the environmental problems we are faced with, and their causes, rooted in organised resistance against societal threats.

In the rationalist theory of technocracy, this is achieved via information provision and education of the public, or even persuasion. In extreme cases, publics are entirely bypassed or even coerced. Ecological modernisation would instead suggest incentivising people – paying local communities, and countries that might otherwise resist, making it an irresistible and economically rational option to pursue.

If we accept that people's opinions matter, and that they cannot be motivated simply by abstract arguments, but will be shaped by societal contexts and norms, it becomes imperative to have dialogue and seek consensus. If we also accept that there will always exist different and conflicting interests on CCS, the challenge becomes to organise these conflicting views into coherent coalitions, to mobilise for this position for particular interests such as CCS development, and against those who do not agree, to cement a common ground.

Such dialogue, social positioning and alliance formation requires connections to be made between different agendas for action and policy: climate change, sustainable development, economic growth and scientific innovation. And a key question emerges around how flexible coalitions that draw on so many interests can and should be. Can CCS be made to fit with sustainable development work, in spite of its capital intensity and resource inefficiency? Can it come to be seen as a security-enhancing technology, given the threat of climate change-driven conflicts, in spite of its connections with fossil fuel-driven exploitation of some communities? Could it be seen more explicitly as a crisis management technology, in spite of its promise to preserve the reliance on fossil fuels?

It may be that widespread CCS development will only be accomplished by changing the way that the technology is generally framed. There is clearly a risk that part of the diverse coalition will collapse into EOR without ensuring permanent, monitored storage of CO_2. It may also be that the framing will have to abandon permanent geological storage, in favour of CO_2 utilisation, to be economically attractive. The geographic focus of demonstration projects may likewise have to change if CCS is to stave off social and political conflicts arising from equity pressures associated with climate change and contribute towards sustainable development. Crucially, CCS will almost certainly have to be 'domesticated', that is adapted to meet the needs of users, be they large utility companies or offshore hydrocarbon exploration companies, and draw on their past experiences with energy exploration, generation, distribution and risk taking.

There is a balance to be struck here between highly specific goal-orientation and an allowance of multiple and even conflicting perspectives on options for technology innovation and governance. We argue that pressures to conform to a single dominant design, certainly at this relatively early stage of development, may lead the CCS community of interest to fragment and lose momentum. A narrow consensus about what CCS is for and how it should be developed will make the technology unappealing to a large number of actors and thereby divert resources from demonstration activities to other goals. CCS has so far managed to strike that balance, built up a large community of supporters, and exhibited what we call a 'bounded fuzziness' in which the technology has been loosely defined along the lines of several intersecting interests and framings of future potentialities.

Climate change is a problem at a global scale, and inter-twined with most aspects of society. CCS as a climate mitigation solution also has to resolve a tension between being a global technology on one hand, and on the other to adapt to multiple local circumstances and contexts. CCS needs to be flexible enough to fit with varying local industry and policy agendas whilst at the same time contribute to the mitigation of global climate change. Current efforts aimed at including CCS in the CDM mechanism suggest that new solutions would need to be found to address this aim.

Proponents need to be prepared for a bumpy ride. The innovation process looks likely to be messy and political, as CCS has been positioned right in the middle of a conflict between the established fossil economy and the climate change mitigation imperative. There may be considerable learning effects from development and deployment, as sometimes analysed with learning curves (Rubin, Taylor *et al.* 2004). And no doubt it would be agreeable to have a nice and orderly progression of learning and development from pilot to large-scale projects (Gibbins and Chalmers 2008b). Relatively advanced projects have however been faced with obstacles as they near the demonstration phase, for example in the Mongstad case (Evar and Shackley, Chapter 10, this book) and we are likely to see more of this in years to come.

In the same vein, we welcome Winskel's interrogation (Chapter 12, this book) of the currently dominant theory of environmental innovation: transitions theory.

We agree with his critique that transitions theory is a poor explanatory framework for the innovation of CCS technology, and along similar lines we observe that a technology emerging from within the established regime of fossil fuel reliance may turn out to be radical to established interests, although incumbents are likely to seek to co-opt it for conservative purposes. The role of (the promise of) CCS in the energy system is likely to continue to be contested.

Support for CCS comes with a high risk of failure, of turning out to be a dead end or simply an unpalatable option, because of high costs and a risk of negative side effects such as CO_2 leakage and high energy penalties. Drawing on innovation studies as well as on decision-making theory discussed in this book, we would propose that a key strategy to deal with these risks is to take one step at a time into unknown territory and to actively and repeatedly ask what learning about CCS implies for successful governance structures. For a large, 'lumpy' technology like CCS, those steps are by necessity larger and costlier than for wind turbines, but there is a considerable difference between an innovation programme built on incremental up-scaling, and the gung-ho visions of a 'moon landing', expressed in 2006 by the Norwegian government. A strategy of 'lumpy incrementalism' would also have to include planned get-out routes or diversions, to part with early development plans as new barriers are identified. As long as the coalition for CCS remains somewhat unstable and flexible, this remains a possibility.

Research agendas for the future

A few areas of research are so far under-represented both in this book and in the social science of CCS in general. A study of key individuals and environments could add new layers to our understanding of the emergence of the technology. The views on CCS from within industry are also relatively under-theorised and have largely been recorded as survey results and in expert elicitation workshops elsewhere. There is scope for further work on how industry actors are taking up, resisting or contributing to the CCS project, at both strategic and operational levels. We note Bowen (2011) as one of the few examples of work on this topic to date. Similarly, efforts should be devoted to understanding how specific risks and uncertainties are framed across different communities of developers. Given the contested status of CCS, such research is both sensitive to current pressures and potentially very useful for understanding how differentiated views on risks emerge.

We also suggest that our review above of the four social theories of societal greening suggests further avenues of productive research. Examples include: the compatibility of CCS with the ecological modernisation and risk society theses, and the extent of technocracy versus more reflexive governance to date in the CCS endeavour and the roles of labour unions in the formation of CCS related policies.

In terms of methodology, there is scope for more comparative research, as different political economies can be expected to solve public action and coordination problems associated with deployment in different ways. CCS may have to be different things in different localities and countries. Pollak *et al.* (2011) have used

US States as a policy lab to compare such support networks and implementation options. Others who have produced comparative analyses include Román (2010) on India, Brazil and South Africa, van Alphen, Hekkert *et al.* (2010) and Meadowcroft and Langhelle (2009a).

There is therefore clearly scope for further social science research on CCS. We do believe that this volume has demonstrated that such work can be carried out on issues around public understanding as well as a much wider set of topics with a bearing on innovation and governance options. It is therefore our belief that the book shows that qualitative research on CCS can contribute to a field largely dominated by quantitative work. Hopefully you are as excited about the future of the social science of CCS as we are!

REFERENCES

ACCSEPT (Acceptance of CO_2 Capture, Storage, Economics, Policy and Technology) (2007) *Multidisciplinary Analysis and Gap-Filling Strategies Report*, Høvik: Det Norske Veritas.

Adger, W.N., Dessai, S., Goulden, M., Hulme, M., Lorenzoni, I., Nelson, D.R., Naess, L.O., Wolf, J. and Wreford, A. (2009) 'Are there social limits to adaptation to climate change?' *Climatic Change*, 93(3): 335–354.

Adler, E. and Haas, P.M. (1992) 'Conclusion: epistemic communities, world order, and the creation of a reflective research program', *International Organization*, 46(1): 367–390.

Aker (2011) *Positive Research Advances*, Oslo: Aker Clean Carbon AS.

Albaek, E. (1999) 'The never-ending story? The political and legal controversies over HIV and the blood supply in Denmark', in E.A. Feldman and R. Bayer (eds) *Blood Feuds: AIDS, Blood, and the Politics of Medical Disaster,* New York/Oxford: Oxford University Press.

Alberta Minister of Energy (2010) *Carbon Capture and Storage Statutes Amendment Act, 2010,* online, available at: www.assembly.ab.ca/ISYS/LADDAR_files/docs/bills/bill/legislature_27/session_3/20100204_bill-024.pdf (accessed 30 September 2011).

Alberta Minister of Energy (2011) *Carbon Sequestration Tenure Regulation*, online, available at: www.qp.alberta.ca/documents/orders/orders_in_council/2011/411/2011_179.html (accessed 30 September 2011).

Allcott, H. and Mullainathan, S. (2010) 'Behavior and energy policy', *Science*, 327 (5970): 1204–1205.

Allott, K. and Kaszewski, A. (2008) *Evading Capture: Is the UK Power Sector Ready for Carbon Capture and Storage?* Godalming: WWF-UK, online, available at: http://assets.panda.org/downloads/evading_capture.pdf (accessed 31 May 2011).

Anderson, J. and Chiavari, J. (2009) 'Understanding and improving NGO position on CCS', *Energy Procedia*, 1(1): 4811–4817.

Anon (2011) 'Carbon capture plans are "a dangerous distraction"', *Aberdeen Press & Journal,* 22 June, online, available at: www.accessmylibrary.com/article-1G1-259581558/carbon-capture-plans-dangerous.html (accessed 29 September 2011).

Anshelm, J. (2000) *Between Salvation and Judgement Day: On the Nuclear Power's History of Ideas in Sweden 1945-1999* (our translation), Mellan frälsning och domedag - om kärnkraftens politiska idéhistoria i Sverige 1945–1999. Symposium, Eslöv [in Swedish].

Antoniades, A. (2003) 'Epistemic communities, epistemes and the construction of (world) politics', *Global Society*, 17(1): 21–38.

Argyris, C. and Schön, D.A. (1996) *Organisational Learning II. Theory, Method, and Practice*, Reading, MA: Addison-Wesley.

Armeni, C. (2011) 'Legal developments for carbon capture and storage under international and regional marine legislation', in I. Havercroft, R. Macrory and R.B. Stewart (eds) *Carbon Capture and Storage: Emerging Legal and Regulatory Issues*, Oxford: Hart.

Arthur, B.W. (1989) 'Competing technologies, increasing returns, and lock-in by historical events', *The Economic Journal*, 99(394): 116–131.

Arthur, B.W. (1994) *Increasing Returns and Path Dependence in the Economy*, Ann Arbor, MI: University of Michigan Press.

Arts, R., Eiken, O., Chadwick, A., Zweigel, P., van der Meer, L. and Zinszner, B. (2004) 'Monitoring of CO_2 injected at Sleipner using time-lapse seismic data', *Energy*, 29 (9/10): 1383–1392.

Ashworth, P., Boughen, N., Mayhew, M. and Millar, F. (2010a) 'From research to action: now we have to move on CCS communication', *International Journal of Greenhouse Gas Control*, 4(2): 426–433.

Ashworth, P., Bradbury, J., Feenstra, C.F.J., Greenberg, S., Hund, G., Mikunda, T. and Wade, S. (2010b) *Communication, Project Planning and Management for Carbon Storage Projects*, available at http://www.globalccsinstitute.com/publications/communication-project-planning-and-management-for-carbon-capture-and-storage-projects (accessed 30 September 2011).

Ashworth, P., Bradbury, J., Feenstra, C.F.J., Greenberg, S., Hund, G., Mikunda, T., Wade, S. and Shaw, H. (2010c) *Communication/Engagement Toolkit for CCS Projects*, publication which can be found at http://www.globalccsinstitute.com/publications/communication-and-engagament-toolkit-for -CCS-projects (accessed 3 October 2011).

Ashworth, P., Mayhew, M., Millar, F. and Boughen, N. (2007) *An Integrated Roadmap of Communication Activities Around Carbon Capture and Storage (CCS) in Australia and Beyond*, Collingwood, Victoria, Australia: CSIRO.

Ashworth, P., Reiner, D.M., Gardner, J. and Littleboy, A. (2007) *Kyoto or non Kyoto People or Politics: Results of Recent Public Opinion Surveys on Energy and Climate Change*, Greenhouse 2007 Conference, Sydney, 1–5 October.

Australian Department of Resources, Energy and Tourism (2011) *Offshore Petroleum and Greenhouse Gas Storage (Resource Management and Administration) Regulations 2011*, online, available at: www.comlaw.gov.au/Details/C2011C00440 (accessed 30 September 2011).

Bäckstrand, K., Meadowcroft, J. and Oppenheimer, M. (2011) 'The politics and policy of carbon capture and storage: framing an emergent technology', *Global Environmental Change*, 21(2): 275–281.

Bakker, S., de Coninck, H., Ba, L. and Zhou, P. (2007) *Advancing CCS and CDM in Africa, Outcomes of two workshops in Africa*, ECN, online, available at: www.ecn.nl/docs/library/report/2007/e07079.pdf (accessed 29 September 2011).

Balagopal, B., Paranikas, P. and Rose J. (2010) *What's Next for Alternative Energy?* Boston, MA: Boston Consulting Group, online, available at: www.bcg.com/documents/file65187.pdf (accessed 24 September 2011).

Bazerman, M. (2006) 'Climate change as a predictable surprise', *Climatic Change*, 77(1/2): 179–193.

Beck, B., Cunha, P., Ketzer, M., Machado, H., Rocha, P.S., Zancan, F., de Almeida, A.S. and Pinheiro, D.Z. (2011) 'The current status of CCS development in Brazil', *Energy Procedia*, 4: 6148–6151.

Beck, U. (1992) *Risk Society: Towards a New Modernity*, London: Sage.

Beck, U. (1994) 'The reinvention of politics: towards a theory of reflexive modernisation', in U. Beck, A. Giddens and S. Lash (eds) *Reflexive Modernization*, Cambridge: Polity Press.

Beck, U. (2000) 'Risk society re-visited: theory, politics and research programmes', in B. Adam, U. Beck and J. van Loon (eds) *The Risk Society and Beyond: Critical Issues for Social Theory*, London, Sage, pp. 211–229.

Beck, U., Giddens, A. and Lash, S. (1994) *Reflexive Modernity*, Cambridge: Policy Press.

Bellona (2007) *The CO_2 Capture Potential*, online, available at: www.bellona.org/articles/articles_2007/1192429461.67 (accessed 31 May 2011).

Bellona (2008) *How to Combat Global Warming*, Oslo: Bellona Foundation, online, available at: www.bellona.org/filearchive/fil_Bellona_CC8_Report_-_Final_version_-_30_mai.pdf (accessed 31 May 2011).

Bellona (2011a) *Bellona CCS position*, Oslo: Bellona Foundation, online, available at: http://bellona.org/ccs/home/bellona-and-ccs/bellona-ccs-position.html (accessed 31 May 2011).

Bellona (2011b) *Introduction to CCS*, Oslo: Bellona Foundation, online, available at: http://bellona.org/ccs/introduction-to-ccs.html (accessed 31 May 2011).

Bellona (2011c) *Miljøkrig om Mongstad i regjeringen*, online, available at: www.bellona.no/nyheter/nyheter_2011/mongstad_forvirring (accessed 6 September 2011).

Benford, R.D. and Snow, D.A. (2000) 'Framing processes and social movements: an overview and assessment', *Annual Review of Sociology*, 26: 611–639.

Benton, T. (1996) 'Marxism and natural limits: an ecological critique and reconstruction', in T. Benton (ed.) *The Greening of Marxism*, New York, Guilford Press (first published in *New Left Review*, 1989, 178: 51–86).

Bergek, A. and Jacobsson, S. (2002) 'The emergence of a growth industry: a comparative analysis of the German, Dutch and Swedish wind turbine industries', in S. Metcalfe and U. Canter (eds) *Change, Transformation and Development: Schumpeterian Perspectives*, Heidelberg: Physica/Springer, pp. 197–228.

Bergek, A., Jacobsson, S., Carlsson, B., Lindmark, S. and Rickne, A. (2008a) 'Analyzing the functional dynamics of technological innovation systems: a scheme of analysis', *Research Policy*, 37(3): 407–429.

Bergek, A., Jacobsson, S. and Hekkert, M. (2008b) 'Functions in innovation systems: a framework for analysing energy system dynamics and identifying goals for system-building activities by entrepreneurs and policy makers', in T. Foxon, J. Kohler and C. Oughton (eds) *Innovation for a Low Carbon Economy: Economic, Institutional and Management Approaches*, Cheltenham: Edward Elgar.

Berkhout, F., Smith, A. and Stirling, A. (2004) 'Socio-technical regimes and transition contexts', in B. Elzen, F.W. Geels and K. Green (eds) *System Innovation and the Transition to Sustainability: Theory, Evidence and Policy*, Cheltenham: Edward Elgar, pp. 48–75.

BERR (Department for Business, Enterprise and Regulatory Reform) (2007) *Competition for a Carbon Dioxide Capture and Storage Demonstration Project - Project Information Memorandum*, London: BERR.

Bielicki, J.M. and Stephens, J.C. (2008) *Public Perception of Carbon Capture and Storage Technology* Workshop Report, Cambridge, MA: Energy Technology Innovation Policy Group Workshop Series, Harvard Kennedy School.

Bijker, W.E. (1995) *Of Bicycles, Bakelites and Bulbs: Toward a Theory of Sociotechnical Change*, Cambridge MA: MIT Press.

Bijker, W.E. and Law, J. (eds) (1992) *Shaping Technology/Building Society: Studies in Sociotechnical Change*, Cambridge, MA: MIT Press.

Bjørnestad, S. (2011) 'Månelandingen i det Blå', *Aftenposten*, 2 March.

Bliss, A. (2009) 'The greenhouse myths', *The Daily Advocate*, 30 May.

Bloor, D. (1991) *Knowledge and Social Imagery*, 2nd edition, Chicago, IL: Chicago University Press.

Blyth, W., Bradley, R., Bruun, D., Clarke, C., Wilson, T. and Yang, M. (2007) 'Investment risks under uncertain climate change policy', *Energy Policy*, 35(11): 5766–5773.

Boait, F., White, N., Chadwick, A., Noy, D. and Bickle, M. (2011) 'Layer spreading and dimming within the CO_2 plume at the Sleipner Field in the North Sea', *Energy Procedia*, 4: 3254–3261.

Borick, C., Lachapelle, E. and Rabe, B. (2011) 'Climate compared: public opinion on climate change in the United States and Canada, key findings report', *Public Policy Forum*, February 23, online, available at: www.brookings.edu/papers/2011/04_climate_change_opinion.aspx (accessed 3 October 2011).

Borup, M., Brown, N., Konrad, K. and van Lente, H. (2006) 'The sociology of expectations in science and technology', *Technology Analysis and Strategic Management*, 18(3/4): 285–298.

Bottoms, R.R. (1930) *Acidic gas removal using TEA, US Patent No. 1,783,901.*

Bowen, F. (2011) 'Carbon capture and storage as a corporate technology strategy challenge', *Energy Policy*, 39(5): 2256–2264.

Bradbury, J., Greenberg, S. and Wade, S. (2011) *Communicating the Risks of CCS*, report prepared for the Commonwealth Scientific and Industrial Research Organisation, online, available at: https://www.globalccsinstitute.com/sites/default/files/publications/21027/communicating-risks-ccs.pdf (accessed 30 September 2011).

Bradbury, J., Ray, I., Peterson, T., Wade, S., Wong-Parodi, G. and Feldpausch, A. (2009) 'The Role of Social Factors in Shaping Public Perceptions of CCS: Results of Multi-State Focus Group Interviews in the US', *Energy Procedia*, 1(1): 4665–4672.

Brinckerhoff, P. (2011) *Accelerating the Uptake of CCS: Industrial Use of Captured Carbon Dioxide*, Global CCS Institute, online, available at: www.globalccsinstitute.com/publications/accelerating-uptake-ccs-industrial-use-captured-carbon-dioxide (accessed 24 September 2011).

Brown, M., Chandler, J., Lapsa, M. and Sovacool, B. (2007) *Carbon Lock-In: Barriers to Deploying Climate Change Mitigation Technologies*, Oak Ridge, TN: Oak Ridge National Laboratory.

Brown, N. and Michael, M. (2003) 'A sociology of expectations: retrospecting prospects and prospecting retrospects', *Technology Analysis and Strategic Management*, 15(1): 3–18.

Brunner, R.D. (1991) 'Global climate change: defining the policy problem', *Policy Sciences*, 24(3): 291–311.

Brunsting, S., Pol, M., Paukovic, M., Kaiser, M., Zimmer, R., Shackley, S., Mabon, L., Hepplewhite, F., Loveridge, R., Mazurowski, M., Polak-Osiniak, D. and Rybicki, C. (2011) *Characterisation of European CO_2 Storage: Deliverable N° D8.1 Qualitative and*

Quantitative Social Site Characterisations, Rueil Malmaison: Characterisation of European CO_2 Storage, online, available at: www.sitechar-co2.eu/SciPublicationsData.aspx?IdPublication=295&IdType=557 (accessed 18 October 2011).

Bryngelsson, M. and Hansson, A. (2009) 'Energy policy on shaky ground? A study of CCS-scenarios', *Energy Procedia*, 1(1): 4673–4680.

Buehler, R., Messervey, D. and Griffin, D. (2005) 'Collaborative planning and prediction: does group discussion affect optimistic biases in time estimation?' *Organizational Behavior and Human Decision Processes*, 97(1): 47–63.

Buhr, K. and Hansson, A. (2011) 'Carbon dioxide capture and storage in Norwegian and Swedish media', *Global Environmental Change*, 21(2): 336–345.

BUND (Bund für Umwelt and Naturschutz Deutschland) (2006) *BUND-Position: CO_2 Abscheidung in fossilen Kraftwerken*, online, available at: www.bund.net/fileadmin/bundnet/publikationen/energie/20060600_energie_co2_abscheidung_position.pdf (accessed 31 May 2011).

Bulkeley, H. and Newell, P. (2010) *Governing Climate Change*, London: Routledge.

Burtraw, D., Darmstadter, J., Palmer, K. and McVeigh, J. (1999) 'Renewable energy: winner, loser, or innocent victim?' *Resources*, 135: 9–13.

Byerly, R. Jr and Brunner, R.B. (1989) 'Future directions for space policy research', in R. Byerly Jr (ed.) *Space Policy Reconsidered*, Denver, CO: Westview Special Studies in Science, Technology, and Public Policy.

Byrne, J., Hughes, K., Rickerson, W. and Kurdgelashvili, L. (2007) 'American policy conflict in the greenhouse: divergent trends in federal, regional, state, and local green energy and climate change policy', *Energy Policy*, 35(9): 4555–4573.

Callon, M. (1986) 'The sociology of an actor-network: the case of the electric vehicle', in M. Callon, J. Law and A. Rip (eds), *Mapping the Dynamics of Science and Technology: Sociology of Science in the Real World*, Basingstoke: Macmillan, pp. 19–34.

Callon, M. (1998) 'Introduction: the embeddedness of economic markets in economics', in M. Callon (ed.) *The Laws of the Markets*, Oxford: Blackwell.

Cansolv (2011) 'Cansolv boundary dam project begins construction', *Carbon Capture Journal*, July, online, available at: www.carboncapturejournal.com/displaynews.php?News ID=809 (accessed 10 September 2011).

Capros, P., Mantzos, L., Papandreou, V., Tasios, N. and Mantzaras, A. (2007) 'Energy system analysis of CCS technology: PRIMES model scenarios', National Technical University of Athens, Institute of Communication and Computer Systems, online, available at: www.uop.edu.jo/download/PdfCourses/SA/primes.pdf (accessed 30 September 2011).

Carlsson, B. and Stankiewicz, R. (1991) 'On the nature, function and composition of technological systems', *Journal of Evolutionary Economics*, 1(2): 93–118.

Carr, M. (2011) 'Some EU nations are "no go" for carbon capture, RWE says', *Bloomberg*, online, available at: www.bloomberg.com/news/2011-06-02/some-eu-nations-are-no-go-for-carbon-capture-rwe-says-1-.html (accessed 30 October 2011).

CATF 2011 (Clean Air Taskforce) 'CO_2 technology solutions', online, available at: www.catf.us/coal/technology (accessed 6 October 2011).

CCJ (*Climate Capture Journal*) (2011a) 'Bellona – needless postponement at Mongstad', 11 March, online, available at: www.carboncapturejournal.com/displaynews.php?News ID=754 (accessed 30 September 2011).

CCJ (*Climate Capture Journal*) (2011b) 'Statoil calls Bellona's claims "unacceptable"', 8 March, online, available at: www.carboncapturejournal.com/displaynews.php?News ID=749 (accessed 20 September 2011).

CCSA (Carbon Capture and Storage Association) (2011) *A CCSA Strategy for CCS in the UK and Beyond*, London: CCSA.

Chalmers, A. (1999) *What is this Thing Called Science?* 3rd edition, Maidenhead: Open University Press.

Chalmers, H., Jakeman, N., Pearson, P. and Gibbins, J. (2009) 'Carbon capture and storage deployment in the UK: what next after the UK government's competition?' *Proceedings of the IMechE Part A: Journal of Power and Energy*, 223(3): 305–319.

Chapman, J. (2011a) 'CCS would add security and flexibility to energy supply', *Energy World*, April, online, available at: www.energyinst.org/information-centre/ei-publications/energy-world/energy-world-april-2011 (accessed 30 September 2011).

Chapman, J. (2011b) Letter from the Carbon Capture and Storage Association, London, to the Right Honourable Chris Huhne, MP, Secretary of State for Energy and Climate Change, dated 29 June 2011 (available from the CCSA, London).

Chazan, G. (2009) 'Locals try sinking plan to store CO_2 underground', *Wall Street Journal*, 6 October.

Christensen, C.M. (1997) *The Innovator's Dilemma*, Boston MA: Harvard Business School Press.

Client Earth (2008) *Amendments to Introduce a CO_2 Emissions Limit Value in the IED/IPPC Directive*, Brussels, online, available at: www.clientearth.com/climate-energy/climate-energy-publications/ied-ippc-directive-127 (accessed 30 September 2011).

Climate Camp (2008) *Activists Take Site for the Camp for Climate Action and Reveal Location*, 30 July, online, available at: www.indymedia.org.uk/en/2008/07/404961.html (accessed 31 May 2011).

Coal Utilization Technology Evaluation Committee (of METI) (2005) *Nisankatanso-tanso-koteika-gijutsu-kaihatsu: Purojekuto-chukan-hyoka-houkokusho (An)* [Technology development of CO_2 fixation in coal seam: mid-term project evaluation draft report], online, available at: www.meti.go.jp/committee/materials/downloadfiles/g50520a135j.pdf (accessed October 2011) [in Japanese].

Collingridge, D. (1982) *The Social Control of Technology*, London: Pinter/New York: St Martin's Press.

Collingridge, D. (1990) 'Technology organizations and incrementalism: the space shuttle', *Technology Analysis and Strategic Management*, 2(2): 181–200.

Collingridge, D. (1992) *The Management of Scale: Big Organizations, Big Decisions, Big Mistakes*, London/New York: Routledge.

Collingridge, D. and Reeve, C. (1986) *Science Speaks to Power: The Role of Experts in Policy Making*, London: Frances Pinter.

Collins, H.M. (1988) 'Public experiments and displays of virtuosity: the core-set revisited', *Social Studies of Science*, 18(4): 725–748.

Collins, H.M. and Evans, R.J. (2007) *Expertise: A New Analysis*, Chicago, IL: University of Chicago Press.

Commoner, B. (1972) *The Closing Circle: Confronting the Environmental Crisis*, London: Cape.

Correljé, A. and Verbong, G. (2004) 'The transition from coal to gas: radical change in the Dutch gas system', in B. Elzen, F.W. Geels and K. Green (eds) *System Innovation and the*

Transition to Sustainability: Theory, Evidence and Policy, Cheltenham: Edward Elgar, pp. 114–134.

Corry, O. and Reiner, D.M. (2011a) *Carbon Capture and Storage and the Environmental Movement*, Report for The Commonwealth Scientific and Industrial Research Organisation (CSIRO), Cambridge: Cambridge University Judge Business School.

Corry, O. and Reiner, D.M. (2011b) *Evaluating Global Carbon Capture and Storage (CCS) Communication Materials: A Survey of Global CCS Communications*, Report for The Commonwealth Scientific and Industrial Research Organisation (CSIRO), Cambridge: Cambridge University Judge Business School.

CSLF (2010) *Carbon Sequestration Leadership Forum, Technology Roadmap, A Global Response to the Challenge of Climate Change*, online, available at: www.cslforum.org/publications/documents/CSLF_Technology_Roadmap_2010.pdf (accessed 30 October 2011).

Daamen, D. and de Best-Waldhober, M. (2008) 'Uninformed and informed public opinions and choices concerning CO_2 capture and storage', *Proceedings of the GHGT-8 8th International Conference on Greenhouse Gas Control Technologies*, 19–22 June 2008, Trondheim, Norway.

Daamen, D., de Best-Waldhober, M., Damen, K. and Faaij, A. (2006) 'Pseudo opinions on CCS technologies', presented the conference Greenhouse Gas Technologies 8 (GHGT-8), Trondheim, Norway.

Danneels, E. (2004) 'Disruptive technology reconsidered: a critique and research agenda', *Journal of Product Innovation Management*, 21(4): 246–258.

David, P.A. (1985) 'Clio and the economics of QWERTY', *American Economic Review*, 75(2): 332–337.

de Best-Waldhober, M., Daamen, D. and Faaij, A. (2009) 'Informed and uninformed public opinions on CO_2 capture and storage technologies in the Netherlands', *International Journal of Greenhouse Gas Control*, 3(3): 322–332.

de Coninck, H. (2009) *Technology Rules! Can Technology-Oriented Agreements Help Address Climate Change?* Amsterdam: PrimaveraQuint.

de Coninck, H. (2010) 'Advocacy for carbon capture and storage could arouse distrust', *Nature* 463(7279): 293 (21 January) doi:10.1038/463293a; published online 20 January 2010.

de Coninck, H. and Bäckstrand, K. (2011) 'An international relations perspective on the global politics of carbon dioxide capture and storage', *Global Environmental Change*, 21(2): 368–378.

de Coninck, H., Flach, T., Curnow, P., Richardson, P., Anderson, J., Shackley, S., Sigurthorsson, G. and Reiner, D. (2009) 'The acceptability of CO_2 capture and storage (CCS) in Europe: an assessment of the key determining factors: Part 1 – Scientific, technical and economic dimensions', *International Journal of Greenhouse Gas Control*, 3(3): 333–343.

de Coninck, H., Stephens, J.C. and Metz, B. (2009) 'Global learning on carbon capture and storage: a call for strong international cooperation on CCS demonstration', *Energy Policy* 37(6): 2161–2165.

de Koeijer, G., Enge, Y., Sanden, K., Graff, O.F., Falk-Pedersen, O., Amundsen, T. and Overå, S. (2011) 'CO_2 Technology Centre Mongstad – design, functionality and emissions of the amine plant', *Energy Procedia* 4: 1207–1213.

DECC (2010) *Energy Trends: March 2010*, National Statistics and Department of Energy and Climate Change, London: Department of Energy and Climate Change.

DECC (2011) *Planning our Electric Future: A White Paper for Secure, Affordable and Low Carbon Electricity*, London: Department of Energy and Climate Change.

Dery, D. (1984) *Problem Definition in Policy Analysis*, Lawrence, KS: University Press of Kansas.

Desbarats, J., Upham, P., Riesch, H., Reiner, D., Brunsting, S., de Best-Waldhober, M., Duetschke, E., Oltra, C., Sala, R. and McLachlan, C. (2010) *Review of Public Participation Practices for CCS and Non-CCS Projects in Europe*, Brussels/London: Institute for European Environmental Policy.

Devine-Wright, P. (2011) 'Public engagement with large-scale renewable energy technologies: breaking the cycle of NIMBYism', *Wiley Interdisciplinary Reviews: Climate Change*, 2(1): 19–26.

Diamond, J. (2006) *Collapse*, Hemel Hempstead: Penguin Books.

DNV (Det Norske Veritas) (2009) *European CO_2 Capture and Storage (CCS) Demonstration Project Network – Draft Knowledge Sharing Protocol*, online, available at: www.ccsnetwork.eu/eventDocs/Knowledge%20Sharing%20Protocol%20Draft.pdf (accessed 12 February 2010).

DOE (Department of Energy) (2004) *FutureGen, Integrated Hydrogen, Electric Power Production and Carbon Sequestration Research Initiative*, Washington DC: Department of Energy Office of Fossil Energy.

Donaldson, E.C., Chilingarian, G.V. and Yen, T.F. (1989) *Enhanced Oil Recovery: Processes and operations*, Amsterdam: Elsevier.

Donath, J. (2010) 'One German town's fight against CO_2 capture technology', *Der Spiegel*, Spiegel Online, 20 August, online, available at: www.spiegel.de/international/germany/0,1518,710573,00.html (accessed 24 September 2011).

Dooley, J., Dahowski, R. and Davidson, C. (2010) *CO_2-Driven Enhanced Oil Recovery as a Stepping Stone to What?* Springfield, VA: Pacific Northwest National Laboratory, online, available at: www.pnl.gov/main/publications/external/technical_reports/PNNL-19557.pdf (accessed 30 September 2011).

Doppelhammer, M. (2011) 'The CCS Directive, its implementation and the co-financing of CCS and RES demonstration projects under the Emissions Trading System (NER 300 Process)', in I. Havercroft, R. Macrory and R.B. Stewart (eds) *Carbon Capture and Storage: Emerging Legal and Regulatory Issues*, Oxford: Hart.

Douglas, M. and Wildavsky, A. (1982) *Risk and Culture*, Berkeley, CA: University of California Press.

Downs, A. (1972) 'Up and down with ecology: the "issue-attention cycle"', *Public Interest*, 28: 38–50, online, available at: www.nationalaffairs.com/public_interest/detail/up-and-down-with-ecologythe-issue-attention-cycle (accessed 20 October 2011).

Druckman, J.N. and Bolsen, T. (2011) 'Framing, motivated reasoning, and opinions about emergent technologies', *Journal of Communication*, 61(4): 659–688.

DTI (1998) *Conclusions of The Review of Energy Sources for Power Generation and Government Response to Fourth and Fifth Reports of the Trade and Industry Committee*, Cm 4071, London: Department of Trade and Industry.

DTI (2000) *Energy Projections for the UK*, Energy Paper 68, London: Department of Trade and Industry.

DTI (2003) *Our Energy Future: Creating a Low Carbon Economy*, London: Department of Trade and Industry.

DTI (2005) *A Strategy for Developing Carbon Abatement Technologies for Fossil Fuel Use*, London: Department of Trade and Industry.

DTI (2006) *The Energy Challenge: Energy Review*, London: Department of Trade and Industry.

DTI (2007) *Meeting the Energy Challenge: A White Paper on Energy*, London: Department of Trade and Industry.

DTI and Carbon Trust (2004) *Conclusions of the Renewables Innovation Review*, London: Department of Trade and Industry.

EC (EU Commission) (2003) Directive 2003/87/EC of the European Parliament and of the Council of 13 October 2003 establishing a scheme for greenhouse gas emission allowance trading within the Community and amending Council Directive 96/61/EC, *Official Journal of the European Union*, online, available at: http://eur-lex.europa.eu/LexUriServ/LexUriServ.do?uri=OJ:L:2003:275:0032:0032:EN:PDF (accessed 30 September 2011).

EC (EU Commission) (2004) *Directive 2004/35/CE of the European Parliament and of the Council of 21 April 2004 on environmental liability with regard to the prevention and remedying of environmental damage*, *Official Journal of the European Union*, online, available at: http://eur-lex.europa.eu/LexUriServ/LexUriServ.do?uri=OJ:L:2004:143:0056:0075:EN:PDF (accessed 30 September 2011).

EC (European Commission) (2008a) *Proposal for a Directive of the European Parliament and of the council on the geological storage of carbon dioxide and amending Council Directives 85/337/EEC, 96/61/EC, Directives 2000/60/EC, 2001/80/EC, 2004/35/EC, 2006/12/EC and Regulation (EC) No 1013/2006*, Brussels, 23 January, online, available at: http://eur-lex.europa.eu/LexUriServ/LexUriServ.do?uri=COM:2008:0018:FIN:EN:PDF (accessed 30 September 2011).

EC (European Commission) (2008b) 'Commissioner Piebalgs welcomes McKinsey CCS report as a valuable input in the Carbon Capture and Storage debate', online, available at: http://europa.eu/rapid/pressReleasesAction.do?reference=IP/08/1385 (accessed 25 June 2011) European Commission, press release.

EC (European Commission) (2009) 'Directive 2009/31/EC of the European Parliament and of the Council of 23 April 2009 on the geological storage of carbon dioxide and amending Council Directive 85/337/EEC, European Parliament and Council Directives 2000/60/EC, 2001/80/EC, 2004/35/EC, 2006/12/EC, 2008/1/EC and Regulation (EC) No 1013/2006', *Official Journal of the European Union*, online, available at: http://eur-lex.europa.eu/LexUriServ/LexUriServ.do?uri=OJ:L:2009:140:0114:0135:EN:PDF (accessed 30 September 2011).

EC (European Commission) (2011a) *Implementation of Directive 2009/31 on the Geological Storage of Carbon Dioxide - Guidance Documents*, online, available at: http://ec.europa.eu/clima/policies/lowcarbon/ccs_implementation_en.htm (accessed 30 September 2011).

EC (European Commission) (2011b) *PRIMES EU-Wide Energy Model*, online, available at: http://ec.europa.eu/environment/air/pollutants/models/primes.htm (accessed 30 October 2011)

Economist (2009a) 'Climate change: - the illusion of clean coal', *The Economist*, 5 March, online, available at: www.economist.com/node/13235041 (accessed 2 September 2011).

Economist (2009b) 'Carbon capture and storage - trouble in store', *The Economist*, 5 March, online, available at: www.economist.com/node/13226661 (accessed 2 September 2011).

Edquist, C. (2000) 'Systems of innovation approaches: their emergence and characteristics', in C. Edquist (ed.) *Systems of Innovation: Growth, Competitiveness and Employment*, Cheltenham: Edward Elgar, pp. 3–37.

Ehrmann, J.R. and Stinson, B.L. (1999) 'Joint fact-finding and the use of technical experts', in L. Susskind, S. McKearnan and J. Thomas-Larmer (eds) *The Consensus Building Handbook*, Thousand Oaks, CA: Sage Publications, pp. 375–399.

Ellison, B.A. and Newmark, A.J. (2010) 'Building the reservoir to nowhere: the role of agencies in advocacy coalitions', *Policy Studies Journal*, 38(4): 653–678.

Elzen, B., Geels, F.W. and Green, K. (2004) 'Transitions to sustainability: lessons learned and remaining challenges', in B. Elzen, F.W. Geels and K. Green (eds) *System Innovation and the Transition to Sustainability: Theory, Evidence and Policy*, Cheltenham: Edward Elgar, pp. 282–300.

EPA (Environmental Protection Agency) (2010a) 'Federal requirements under the underground injection control (UIC) program for carbon dioxide (CO_2) geologic sequestration (GS) wells (40 CFR Parts 124, 144, 145, 146, and 14)', *US Federal Register*, online, available at: www.federalregister.gov/articles/2010/12/10/2010-29954/federal-requirements-under-the-underground-injection-control-uic-program-for-carbon-dioxide-co2 (accessed 30 September 2011).

EPA (Environmental Protection Agency) (2010b) 'Mandatory reporting of greenhouse gases: injection and geologic sequestration of carbon dioxide (40 CFR Parts 72, 78, and 9)', *US Federal Register*, online, available at: www.federalregister.gov/articles/2010/12/01/2010-29934/mandatory-reporting-of-greenhouse-gases-injection-and-geologic-sequestration-of-carbon-dioxide (accessed 30 September 2011).

EPA/IRIS (2011) *N-Nitrosodimethylamine (CASRN 62-75-9)*, US Environmental Protection Agency, Integrated Risk Information System, online, available at: www.epa.gov/ncea/iris/subst/0045.htm (accessed May 27 2011).

Epstein, S. (1998) *Impure Science: AIDS, Activism, and the Politics of Knowledge*, Berkeley, CA: University of California Press.

ERRG (Energy Research Review Group) (2002) *Chief Scientific Adviser's Energy Research Review Group: Report of the Group – Recommendations to Inform the Performance and Innovation Unit's Energy Policy Review*, London: Office of Science and Technology.

Etzkowitz, H. (2003) 'Innovation in innovation: the Triple Helix of university-industry-government relations', *Social Science Information Sur Les Sciences Sociales*, 42(3): 293–337.

Etzkowitz, H. (2008) *The Triple Helix: University-Industry-Government Innovation in Action*, London: Routledge.

Eurobarometer (2011) *Special Eurobarometer 364: Eurobarometer Survey on Public Awareness and Acceptance of CCS*, Brussels: DG-Research, online, available at: http://ec.europa.eu/public_opinion/archives/ebs/ebs_364_en.pdf (accessed 6 October 2011).

Evar, B. (2011) 'Conditional inevitability: expert perceptions of carbon capture and storage uncertainties in the UK context', *Energy Policy*, 39(6): 3414–3424.

Ezrahi, Y. (1990) *The Descent of Icarus*, Cambridge, MA: Harvard University Press.

Fauset, C. (2008) *Techno-Fixes: A Critical Guide to Climate Change Technologies*, London: Footprint Worker's Cooperative.

Feenstra, C.F.J., Mikunda, T. and Brunsting, S. (2010) 'What happened in Barendrecht? Case study on the planned onshore carbon dioxide storage in Barendrecht, the

Netherlands, ECN and Global CCS Institute', ECN/GCCSI, online, available at: www.csiro.au/files/files/pybx.pdf (accessed 30 October 2011).

Feldpausch-Parker, A.M., Chaudhry, R., Stephens, J.C., Fischlein, M., Hall, D.M., Melnick, L.L., Peterson, T.R., Ragland, C.J. and Wilson, E.J. (2011) 'A comparative state-level analysis of carbon capture and storage (CCS) discourse among US energy stakeholders and the public', *Energy Procedia*, 4: 6368–6375.

Finkenrath, M. (2011) 'Cost and performance of carbon dioxide capture from power generation', *International Energy Agency Working Paper*, Paris: IEA, online, available at: www.oecd-ilibrary.org/docserver/download/fulltext/5kgggn8wk05l.pdf?expires=1325871316&id=id&accname=guest&checksum=4567B73CABDDDA8D804896552978D266 (accessed 30 September 2011).

Flannery, B.P. (2011) 'Comments on "Scaling up carbon dioxide capture and storage: from megatons to gigatons"', *Energy Economics*, 33(4): 605–607.

Flatlandsmo Berglen, T., Tønnesen, D., Dye, C., Karl, M., Knudsen, S. and Tarrasón, L. (2010) *CO_2 Technology Centre Mongstad: Updated Air Dispersion Calculations*, Kjeller: Norwegian Institute for Air Research, online, available at: www.klif.no/nyheter/dokumenter/tcm_soknad_vedlegg8.pdf (accessed 15 October 2011).

Fleck, L. (1979) *The Genesis and Development of a Scientific Fact*, Chicago, IL: University of Chicago Press.

Fleishman, L.A., De Bruin, W.B. and Morgan, M.G. (2010) 'Informed public preferences for electricity portfolios with CCS and other low-carbon technologies', *Risk Analysis*, 30(9): 1399–410.

FoE Cymru (2010) *Coal in Wales*, online, available at: www.foe.co.uk/cymru/english/news/17740.html (accessed 31 May 2011).

Fouche, G. (2010) 'Norway delays Mongstad carbon capture and storage project', *Reuters*, 2 May, online, available at: http://uk.reuters.com/article/2010/05/02/us-norway-carboncapture-idUSTRE6410YM20100502 (accessed 30 September 2011).

Freeman, C. (1987) *Technology Policy and Economic Performance: Lessons from Japan*, London: Pinter Publishers.

Freeman, C. (1991) 'Networks of innovators: a synthesis of research issues', *Research Policy*, 20(5): 499–514.

Freeman, C. (1995) 'The national system of innovation in historical perspective', *Cambridge Journal of Economics*, 19(1): 5–24.

Funtowicz, S.O. and Ravetz, J.R. (1985) 'Three types of risk assessment: a methodological analysis', in C. Wipple and V. Covello (eds) *Risk Analysis in the Private Sector*, New York: Plenum.

FutureGen Alliance (2009) FutureGen Industrial Alliance, Inc. website, online, available at: www.futuregenalliance.org (accessed 25 November 2009).

FutureGen Alliance (2011) *Guidance to Prospective Offerors (10–5–2010)*, online, available at: www.futuregenalliance.org/media/FGA_Guidance_100510_Final.pdf (accessed October 2011).

FutureGen for Illinois (2009) *FutureGen for Illinois*, online, available at: www.futuregenforillinois.com (accessed 25 September 2009).

Gallagher, K., Holdren, J. and Sagar, A. (2006) 'Energy-technology innovation', *Annual Review of Environment and Resources*, 31: 193–237.

Garud, R. and Karnøe, P. (2003) 'Bricolage versus breakthrough: distributed and embedded agency in technology entrepreneurship', *Research Policy*, 32(2): 277–300.

GCCSI (Global CCS Institute) (2009) *Status of Carbon Capture and Storage Projects Globally, Strategic Analysis of the Global Status of Carbon Capture and Storage: Report 1*, online, available at: http://cdn.globalccsinstitute.com/sites/default/files/publications/5751/report-1-status-carbon-capture-and-storage-projects-globally.pdf (accessed 10 October 2011).

GCCSI (Global CCS Institute) (2010) *The Status of CCS Projects - Interim Report 2010*, online, available at: http://cdn.globalccsinstitute.com/sites/default/files/publications/5686/status-ccs-projects-interim-report-2010.pdf (accessed 10 October 2011).

GCCSI (Global CCS Institute) (2011a) *About the Institute*, online, available at www.globalccsinstitute.com/institute (accessed 16 November 2011).

GCCSI (Global CCS Institute) (2011b) *The Global Status of CCS 2010*, online, available at: http://cdn.globalccsinstitute.com/sites/default/files/the_global_status_ccs_2011.pdf (accessed 30 September 2011).

Geels, F.W. (2002) 'Technological transitions as evolutionary reconfiguration processes: a multi-level perspective and a case-study', *Research Policy*, 31(8/9): 1257–1274.

Geels, F.W. (2004a) 'From sectoral systems of innovation to socio-technical systems: insights about dynamics and change from sociology and institutional theory', *Research Policy*, 33(6/7): 897–920.

Geels, F.W. (2004b) 'Understanding system innovations: a critical literature review and a conceptual synthesis', in B. Elzen, F.W. Geels and K. Green (eds) *System Innovation and the Transition to Sustainability: Theory, Evidence and Policy*, Cheltenham: Edward Elgar, pp. 19–47.

Geels, F.W. (2005a) 'The dynamics of transitions in socio-technical systems: A multi-level analysis of the transition pathway from horse-drawn carriages to automobiles (1860–1930)', *Technology Analysis and Strategic Management*, 17(4): 445–476.

Geels, F.W. (2005b) *Technological Transitions and Systems Innovations* Cheltenham: Edward Elgar.

Geels, F.W. (2006a) 'Co-evolutionary and multi-level dynamics in transitions: the transformation of aviation systems and the shift from propeller to turbojet (1930–1970)', *Technovation*, 26(9): 999–1016.

Geels, F.W. (2006b) 'The hygienic transition from cesspools to sewer systems (1840–1930): the dynamics of regime transformation', *Research Policy*, 35(7): 1069–1082.

Geels, F.W. (2007) 'Feelings of discontent and the promise of middle range theory for STS: examples from technology dynamics', *Science, Technology, & Human Values*, 32(6): 627–651.

Geels, F.W. (2010) 'Ontologies, socio-technical transitions (to sustainability), and the multi-level perspective', *Research Policy*, 39(4): 495–510.

Geels, F.W. (2011) 'The multi-level perspective on sustainability transitions: responses to seven criticisms', *Environmental Innovation and Societal Transitions*, 1(1): 24–40.

Geels, F.W. and Schot, J. (2007) 'Typology of sociotechnical transition pathways', *Research Policy*, 36(3): 399–417.

Giampietro, M. and Mayumi, K. (2009) *The Biofuel Delusion: The Fallacy of Large-Scale Agro-Biofuel Production*, London: Earthscan.

Gibbins, J. and Chalmers, H. (2008a) 'Carbon capture and storage', *Energy Policy*, 36(12): 4317–4322.

Gibbins, J. and Chalmers, H. (2008b) 'Preparing for global roll-out: A "developed country first" demonstration programme for rapid CCS deployment', *Energy Policy,* 36(2): 501–507.

Gibbons, M., Limoges, C., Nowotny, H., Schwartzman, S., Scott, P. and Trow, M. (1994) *The New Production of Knowledge: The Dynamics of Science and Research in Contemporary Societies,* London: Sage.

Gibbs, M. (2011) 'The regulation of geological storage of greenhouse gases in Australia', in I. Havercroft, R. Macrory and R.B. Stewart (eds) *Carbon Capture and Storage: Emerging Legal and Regulatory Issues*, Oxford: Hart.

Gitlin, T. (1980) *The Whole World Is Watching: Mass Media in the Making and Unmaking of the New Left*, Berkeley, CA/Los Angeles, CA/London: University of California Press.

Gorz, A. (1993) 'Political ecology: expertocracy versus self-limitation', *New Left Review*, 202: 55–67.

Gough, C. and Shackley, S. (2001) 'The respectable politics of climate change: the epistemic communities and NGOs', *International Affairs*, 77(2): 329–345.

Gough, C., Taylor, I. and Shackley, S. (2002) 'Burying carbon under the sea: an initial exploration of public opinions', *Energy & Environment*, 13(6): 883–900.

Government of Queensland (2009) *Greenhouse Gas Storage Act 2009*, online, available at: www.legislation.qld.gov.au/LEGISLTN/ACTS/2009/09AC003.pdf (accessed 30 September 2011).

Government of South Australia (2011) *Petroleum and Geothermal Energy Act 2000 (amended)*, online, available at: www.legislation.sa.gov.au/LZ/C/A/Petroleum%20and%20Geothermal%20Energy%20Act%202000.aspx (accessed 30 September 2011).

Government of Western Australia (2008) *Barrow Island Act 2003 (amended)*, online, available at: www.ucl.ac.uk/cclp/pdf/Barrow_Island_Act_Western_Australia_2003.pdf (accessed 30 September 2011).

Graham, J. (2011) 'Independent group to review CO_2 leak at Weyburn', *The Canadian Press*, online, available at: www.albertasurfacerights.com/articles/?id=835 (accessed 20 October 2011).

Green Alliance (2008) *A Last Chance for Coal: Making Carbon Capture and Storage a Reality*, online, available at: www.green-alliance.org.uk/grea_p.aspx?id=3278 (accessed 31 May 2011).

Greenberg, S. and Wade, S. (2011) *Social Site Characterisation: From Concept to Application,* report prepared for the Commonwealth Scientific and Industrial Research Organisation, online, available at: www.globalccsinstitute.com/publications/social-site-characterisation-concept-application (accessed 30 September 2011).

Greenpeace (2011) *Top 10 Things we Need to See in the Budget to Solve our Energy Problems*, online, available at: www.greenpeace.org.uk/blog/climate/top-10-things-we-need-see-budget-solve-our-energy-problems-20110321 (accessed 31 May 2011).

Greenpeace International (2008) *False Hope: Why Carbon Capture and Storage Won't Save the Climate*, Amsterdam: Greenpeace International, online, available at: www.greenpeace.org/usa/en/media-center/reports/false-hope-why-carbon-capture (accessed 24 September 2011).

Gunderson, L. and Holling, C. (2002) *Panarchy: Understanding Transformations in Human and Natural Systems*, Washington DC: Island Press.

Gusfield, J.R. (1981) *The Culture of Public Problems: Drinking-Driving and the Symbolic Order*, Chicago, IL: University of Chicago Press.

Haas, P.M. (1990) 'Obtaining environmental protection through epistemic consensus', *Millennium: Journal of International Studies*, 19(3): 347–363.

Haas, P.M. (1992) 'Introduction: epistemic communities and international policy coordination', *International Organization*, 46(1): 1–35.

Hajer, M. (1995) *The Politics of Environmental Discourse: Ecological Modernization and the Policy Process*, Oxford: Oxford University Press.

Hajer, M. (2002) 'Discourse analysis and the study of policy-making', *European Political Science* 2(1): 61–65.

Ham, C. and Hill, M. (1984) *The Policy Process in the Modern Capitalist State*, Hemel Hempstead: Harvester Wheatsheaf.

Hamilton, R., Herzog, H.J. and Parsons, J.E. (2009) 'Costs and US public policy for new coal power plants with carbon capture and sequestration', *Energy Procedia*, 1(1): 4487–4494.

Hammond, J. and Shackley, S. (2010) *Towards a Public Communication and Engagement Strategy for Carbon Dioxide Capture and Storage Projects in Scotland: A Review of Research Findings, CCS Project Experiences, Tools, Resources and Best Practices*, Report commissioned by the Scottish Carbon Dioxide Capture, Transport and Storage Development Study Working Paper SCCS 2010-08, online, available at: www.sccs.org.uk/publications.html (accessed 30 September 2011).

Hansson, A. (2008) *Kolets återkomst: Koldioxidavskiljning i vetenskap och politik*, Linköping: Linköping Studies in Arts and Science.

Hansson, A. and Bryngelsson, M. (2009) 'Expert opinions on carbon dioxide capture and storage: a framing of uncertainties and possibilities', *Energy Policy*, 37(6): 2273–2282.

Hård, M. (1993) 'Beyond harmony and consensus: a social conflict approach to technology', *Science, Technology, & Human Values*, 18(4): 408–432.

Hargadon, A. (2010) 'Technology policy and global warming: why new innovation models are needed', *Research Policy*, 39(8): 1024–1026.

Harvey, D. (2005) *A Brief History of Neoliberalism*, Oxford: Oxford University Press.

Havercroft, I. and Purdy, R. (2007) *Carbon Capture and Storage: A Legal Perspective*, paper presented at Expert Group Meeting: Carbon Capture and Storage and Sustainable Development, United Nations, New York, 10–11 September, online, available at: www.un.org/esa/sustdev/sdissues/energy/op/ccs_egm/presentations_papers/havercroft_paper_legal.pdf (accessed 30 September 2011).

Hawkins, D, Peridas, G. and Steelman, J. (2009) 'Twelve years after Sleipner: moving CCS from hype to pipe', *Energy Procedia*, 1(1): 4403–4410.

HCSTC (2003) *Towards a Non-Carbon Fuel Economy: Research, Development and Demonstration*, London: House of Commons Science and Technology Committee, online, available at: www.publications.parliament.uk/pa/cm200203/cmselect/cmsctech/55/55.pdf (accessed 24 September 2011).

HCSTC (2006) *Meeting UK Energy and Climate Needs: The Role of Carbon Capture and Storage*, London: House of Commons Science and Technology Committee, online, available at: www.publications.parliament.uk/pa/cm200506/cmselect/cmsctech/1036/1036.pdf (accessed 30 September 2011).

Heiskanen, E., Kivisaari, S., Lovio, R. and Mickwitz, P. (2009) 'Designed to travel? Transition management encounters environmental and innovation policy histories in Finland', *Policy Sciences*, 42(4): 409–427.

Hekkert, M.P., Suurs, R.A.A., Negro, S.O., Kuhlmann, S. and Smits, R. (2007) 'Functions of innovation systems: a new approach for analysing technological change', *Technological Forecasting and Social Change*, 74(4): 413–432.

Hellin, M. (2008) 'Carbon capture and storage: technology, discourse and politics', unpublished research master's essay, Carleton University, Ottawa, Canada.

Helm, D. (2003) *Energy, the State and the Market: British Energy Policy since 1979*, Oxford: Oxford University Press.

Hendricks, C., Mace, M.J. and Coenraads, R. (2005) *Impacts of EU and International Law on the implementation of geological storage of carbon capture and geological storage in the European Union*, London: FIELD and Utrecht: Ecofys, online, available at: http://pdf.wri.org/ccs_impact_of_eu_law_on.pdf (accessed 30 September 2011).

Herzog, H.J. (2011) 'Scaling up carbon dioxide capture and storage: from megatons to gigatons', *Energy Economics*, 33(4): 597–604.

Hillman, K.M., Suurs, R.A.A., Hekkert, M.P. and Sanden, B.A. (2008) 'Cumulative causation in biofuels development: a critical comparison of the Netherlands and Sweden', *Technololgy Analysis and Strategic Management*, 20(5): 593–612.

HMG (Her Majesty's Government) (2008) *Climate Change Act*, Norwich: The Stationery Office, online, available at: www.legislation.gov.uk/ukpga/2008/27/contents (accessed 10 October 2011).

HMG (Her Majesty's Government) (2009) *The UK Low Carbon Transition Plan: National Strategy for Climate and Energy*, Norwich: The Stationery Office, online, available at: www.decc.gov.uk/publications/basket.aspx?FilePath=White+Papers%2fUK+Low+Carbon+Transition+Plan+WP09%2f1_20090724153238_e_%40%40_lowcarbontransitionplan.pdf&filetype=4#basket (accessed 30 September 2011).

Hoffert, M.I., Wey, Y.C., Callegari, A.J. and Broecker, W.C. (1979) 'Atmospheric response to deep-sea injections of fossil-fuel carbon dioxide', *Climatic Change*, 2(1): 53–68.

Holloway, S. (ed.) (1996) *The Underground Disposal of Carbon Dioxide*, final report of the Joule II Project, No. CT92_0031, Nottingham: British Geological Survey.

Howlett, M. (1997) 'Issue-attention and punctuated equilibria models reconsidered: an empirical examination of the dynamics of agenda-setting in Canada', *Canadian Journal of Political Science/Revue canadienne de science politique*, 30(1): 3–29.

Howlett, M. and Migone, A. (2011) 'Charles Lindblom is alive and well and living in punctuated equilibrium land', *Policy and Society*, 30(1): 53–62.

Howlett, M., Ramesh, M. and Perl, A. (2009) *Studying Public Policy: Policy Cycles and Policy Subsystems*, Oxford: Oxford University Press.

Huber, J. (2008) 'Pioneer countries and the global diffusion of environmental innovations: theses from the viewpoint of ecological modernisation theory', *Global Environmental Change*, 18(3): 360–367.

Hughes, C. (2009) *CCS: Legislating to Quantify Risk and Increase the Financial Viability of CCS Projects*, Think Piece, London: Faculty of Laws, University College London, online, available at: www.ucl.ac.uk/cclp/pdf/Calum_thinkpiece_final.pdf.

Hughes, T.P. (1983) *Networks of Power: Electrification in Western Society, 1880–1930*, Baltimore, MD/London: Johns Hopkins University Press.

Hughes, T.P. (1987) 'The evolution of large technological systems', in W.E. Bijker, T.P. Hughes and T. Pinch (eds) *The Social Construction of Technological Systems: New Directions in the Sociology and History of Technology*, Cambridge, MA: MIT Press, pp. 51–82.

Hughes, T.P. (1989) *American Genesis: A Century of Invention and Technological Enthusiasm, 1870–1970*, New York: Viking Penguin.

Huijts, N.M.A. (2003) 'Public perceptions of CO_2 storage, the role of affect and trust in attitude formation', master's thesis, Department of Technology Management, Eindhoven, the Netherlands: Eindhoven University of Technology.

Huijts, N.M.A, Midden, C.J.H. and Meijnders, A.L. (2007) 'Social acceptance of carbon dioxide storage', *Energy Policy*, 35(5): 2780–2789.

Hulme, M. (2009) *Why We Disagree About Climate Change*, Cambridge: Cambridge University Press.

Hultman, N.E. and Koomey, J.G. (2007) 'The risk of surprise in energy technology costs', *Environmental Research Letters*, 2(3), online, available at: http://iopscience.iop.org/1748-9326/2/3/034002 (accessed 30 September 2011).

Hund, G. and Judd, K. (2009) 'Stakeholder acceptance issues concerning advanced coal power plants with carbon capture and storage: lessons learned from FutureGen', *Energy Procedia 1(1): Proceedings of the 9th International Conference on Greenhouse Gas Control Technologies (GHGT-9)*, 16–20 November 2008, Washington, DC.

ICEPT (2003) *Innovation in long term renewables options in the UK: Overcoming barriers and 'systems failures'*, London: Imperial College Centre for Energy Policy and Technology, online, available at: http://webarchive.nationalarchives.gov.uk/+/http://www.berr.gov.uk/files/file22072.pdf (accessed 30 October 2011).

ICEPT and E4tech (2003) *The UK Innovation Systems for New and Renewable Energy Technologies: A Report to the DTI Renewable Energy Development and Deployment Team*, London: ICEPT (Imperial College Centre for Energy Policy and Technology) and E4tech Consulting, online, available at: http://webarchive.nationalarchives.gov.uk/+/http://www.berr.gov.uk/files/file22069.pdf (accessed 20 September 2011).

IEA (1998) *Enhancing the Market Deployment of Energy Technology A Survey of Eight Technologies*, Paris: International Energy Agency.

IEA (2003) *Creating Markets for Energy Technologies*, Paris: International Energy Agency.

IEA (2004) *Prospects for CO_2 Capture and Storage*, Paris: Organisation for Economic Co-operation and Development and International Energy Agency.

IEA (2008) *Energy Technology Perspective - Blue Map Scenario*, Paris: International Energy Agency, online, available at: www.iea.org/techno/etp/ETP_2008_Exec_Sum_English.pdf (accessed 30 September 2011).

IEA (2009) *Technology Roadmap: Carbon Capture and Storage*, Paris: International Energy Agency, online, available at: www.iea.org/roadmaps/ccs_roadmap.asp (accessed 30 September 2011).

IEA (2010a) *Carbon Capture and Storage –Model Regulatory Framework*, Paris: International Energy Agency, online, available at: www.iea.org/ccs/legal/model_framework.pdf (accessed 30 September 2011).

IEA (2010b) *Energy Technology Perspectives: Scenarios and Strategies to 2050*, Paris: International Energy Agency.

IEA (2011a) *Addressing Climate Change: Policies and Measures*, Paris: International Energy Agency.

IEA (2011b) *World Energy Model*, Paris: International Energy Agency, online, available at: www.iea.org/weo/model.asp (accessed 30 October 2011).

In Salah CO_2 (2010) *In Salah: Industrial-scale CCS in action*, online, available at: www.insalahco2.com (accessed 30 September 2011).

Industrial Structure Council (2009) *Nisankatanso-tanso-koteika-gijutsu-kaihatsu Purojekuto-hyoka (Jigo) Houkokusho (An)* [Technology development of CO_2 fixation in coal seam: post-project evaluation draft report], online, available at: www.meti.go.jp/committee/materials2/downloadfiles/g90128a09j.pdf (accessed October 2011) [in Japanese].

Infoshop News (2011) *Germany: climate and energy action camp 2011*, 1 July, online, available at: http://news.infoshop.org/article.php?story=2011070115482134 (accessed 30 September 2011).

Inglehart, R.F. (2008) 'Changing values among Western publics from 1970 to 2006', *West European Politics*, 31(1/2): 130–146.

IPCC (Intergovernmental Panel on Climate Change) (1990) *Working Group I: The Scientific Assessment of Climate Change*, Cambridge: Cambridge University Press.

IPCC (Intergovernmental Panel on Climate Change) (2005) *Carbon Dioxide Capture and Storage. IPCC Special Report on Carbon Dioxide Capture and Storage*, prepared by the Working Group III of the IPCC, Cambridge: Cambridge University Press.

Ishii, A. and Langhelle, O. (2011) 'Toward policy integration: assessing carbon capture and storage policies in Japan and Norway', *Global Environmental Change*, 21(2): 358–367.

Jablonska, H.B. (2010) *Svar på oppdragsbrev om kunnskapsstatus for utslipp av aminer til luft*, Oslo: Climate and Pollution Agency.

Jablonska, H.B. (2011) *Status om ny kunnskap om aminutslipp fra CO_2-fangst*, Oslo: Climate and Pollution Agency.

Jaccard, M. and Tu, J. (2011) 'Show some enthusiasm, but not too much: carbon capture and storage development prospects in China', *Global Environmental Change*, 21(2): 402–412.

Jacobsson, S. and Bergek, A. (2004) 'Transforming the energy sector: the evolution of technological systems in renewable energy technology', *Industrial and Corporate Change*, 13(5): 815–849.

Jacobsson, S. and Johnson, A. (2000) 'The diffusion of renewable energy technology: an analytical framework and key issues for research', *Energy Policy*, 28(9): 625–640.

Jacobsson, S., Sandén, B. and Bångens, L. (2004) 'Transforming the energy system: the evolution of the German technological system for solar cells', *Technology Analysis and Strategic Management*, 16(1): 3–30.

Jamasb, T. and Köhler, J. (2007) 'Learning curves for energy technology: a critical assessment', in M. Grubb, T. Jamasb and M.G. Pollitt (eds) *Delivering a Low Carbon Electricity System: Technologies, Economics and Policy*, Cambridge: Cambridge: University Press.

Jänicke, M. (2008) 'Ecological modernisation: new perspectives', *Journal of Cleaner Production*, 16(5): 557–565.

Jasanoff, S., Markle, G., Pinch, T. and Petersen, J. (eds) (2002) *Handbook of Science, Technology and Society*, London: Sage.

Jenkins-Smith, H. and Sabatier, P. (1994) 'Evaluating the advocacy coalition framework', *Journal of Public Policy*, 14(2): 175–203.

Johnsson, F., Reiner, D., Itaoka, K. and Herzog, H. (2010) 'Stakeholder attitudes on carbon capture and storage: an international comparison', *International Journal of Greenhouse Gas Control*, 4(2): 410–418.

Jørgensen, U. and Karnøe, P. (1995) 'The Danish wind-turbine story: technical solutions to political visions', in A. Rip, T.J. Misa and J. Schot (eds) *Managing Technology in Society: The Approach of Constructive Technology Assessment*, London: Pinter, pp. 57–82.

Juniper, T. (2011) interviewed on 7 January.

Kapila, R., Chalmers, H. and Leach, M. (2009) *Investigating the Prospects for Carbon Capture and Storage Technology in India*, Edinburgh: Scottish Carbon Capture and Storage, University of Edinburgh, online, available at: www.geos.ed.ac.uk/sccs/India__CCS_Report-Oct2009.pdf (accessed 30 September 2011).

Karlström M. and Sandén, B.A. (2004) 'Selecting and assessing demonstration projects for technology assessment: the case of fuel cells and hydrogen systems in Sweden', *Innovation: Management, Policy & Practice*, 6(2): 286–293.

Karnøe, P. and Garud, R. (2003) 'Co-creation processes in the emerging Danish wind turbine field', unpublished draft book chapter.

Kasperson, J.X., Kasperson, R.E., Perkins, B.J., Renn, O. and White, A.L. (2005) 'Media risk signals and the proposed Yucca Mountain Nuclear Waste Repository, 1985–1989', in J.X. Kasperson and R.E. Kasperson (eds) *The Social Contours of Risk*, London: Earthscan.

Kemp, A. and Kasim, A.S. (2010) 'A futuristic least-cost optimisation model of CO_2 transport and storage in the UK/UK continental shelf', *Energy Policy*, 38(7): 3652–3667.

Kemp, R., Schot, J. and Hoogma, R. (1998) 'Regime shifts to sustainability through processes of niche formation: the approach of strategic niche management', *Technology Analysis and Strategic Management*, 10(2): 175–195.

Kern, F. and Smith, A. (2008) 'Restructuring energy systems for sustainability? Energy transition policy in the Netherlands', *Energy Policy*, 36(11): 4093–4103.

Knorr Cetina, K. (1999) *Epistemic Cultures*, Harvard, MA: Harvard University Press.

Knudsen, S., Karl, M. and Randall, S. (2009) *Summary Report: Amine Emissions to Air during Carbon Capture. Phase I: CO_2 and Amines Screening Study for Effects to the Environment*, Kjeller: Norwegian Institute for Air Research, online, available at: http://co2.nilu.no/LinkClick.aspx?fileticket=m5iQC0a%2FPqg%3D&tabid=2549&mid=5547&language=pl-PL (accessed 30 September 2011).

Konrad, K., Truffer, B. and Voss, J.-P. (2008) 'Multi-regime dynamics in the analysis of sectoral transformation potentials: evidence from German utility sectors', *Journal of Cleaner Production*, 16(11): 1190–1202.

Krasner, S.D. (1983) *International Regimes*, Ithaca, NY: Cornell University Press.

Kristensen, K. and Kokkvoll Tveit, A. (2010) 'Unlikely that Mongstad CCS demonstration plant should be delayed, despite Statoil reservations', online, available at: www.bellona.org/articles/articles_2010/mongstad_delay (accessed 27 May 2010).

Låg, M., Lindeman, B., Instanes, C., Brunborg, G. and Schwarze, P. (2011) *Health Effects of Amines and Derivatives Associated with CO_2 Capture*, Oslo: The Norwegian Institute of Public Health, online, available at: www.klif.no/publikasjoner/2806/ta2806.pdf (accessed 20 September 2011).

Lakoff, G. (1995) 'Metaphor, morality, and politics, or, why conservatives have left liberals in the dust', *Social Research*, 62(2): 177–213.

Langhelle, O. and Meadowcroft, J. (2009) 'Carbon capture and storage in comparative perspective', in J. Meadowcroft and O. Langhelle, *Caching the Carbon: The Politics and Policy of Carbon Capture and Storage*, Cheltenham: Edward Elgar, pp. 236–266.

Latour, B. (1993) *We Have Never Been Modern*, Cambridge MA: Harvard University Press.

Law, J. (1987) 'Technology and heterogeneous engineering: the case of Portuguese expansion', in W.E. Bijker, T.P. Hughes, and T. Pinch (eds) *The Social Construction of*

Technological Systems: New Directions in the Sociology and History of Technology, Cambridge MA: MIT Press, pp. 111–134

Law, J. and Callon, M. (1992) 'The life and death of an aircraft: a network analysis of technical change', in W.E. Bijker and J. Law (eds) *Shaping Technology/Building Society: Studies in Sociotechnical Change*, Cambridge MA: MIT Press, pp. 21–52.

Lebow, R.N. (2009) 'Constitutive causality: imagined spaces and political practices', *Millennium – Journal of International Studies*, 38(2): 211–239.

Lempert, R.J. and Collins, M.T. (2007) 'Managing the risk of uncertain threshold response: comparison of robust, optimum, and precautionary approaches', *Risk Analysis*, 27(4): 1009–1026.

Lenfle, S. (2011) 'The strategy of parallel approaches in projects with unforeseeable uncertainty: the Manhattan case in retrospect', *International Journal of Project Management*, 29(4): 359–373.

Levi-Faur, D. and Jordana, J. (2005a) 'Globalizing regulatory capitalism', *The Annals of the American Academy of Political and Social Science*, 598: 184–190.

Levi-Faur, D. and Jordana, J. (2005b) 'The rise of regulatory capitalism: the global diffusion of a new order', *The Annals of the American Academy of Political and Social Science*, 598: 200–217.

Lindblom, C. (1959) 'The science of "muddling through"', *Public Administration Review*, 19(2): 79–88.

Lindblom, C. (1979) 'Still muddling, not yet through', *Public Administration Review*, 39(6): 517–526.

Loiter, J.M. and Norberg-Bohm, V. (1999) 'Technology policy and renewable energy: public roles in the development of new energy technologies', *Energy Policy*, 27(2): 85–97.

London Protocol (2006) 'Resolution LP.1(1) on the Amendment to Include CO_2 Sequestration in Sub-Seabed Geological Formations in Annex 1 to the London Protocol [LC 28/15, Annex 6 (Adopted on 2 November 2006)]', online, available at: www.imo.org/blast/blastDataHelper.asp?data_id=17614&filename=01.pdf (accessed 17 November 2011).

London Protocol (2009) 'Resolution LP.3(4) on the Amendment to Article 6 of the London Protocol (adopted on 30 October 2009)', online, available at: www.ucl.ac.uk/cclp/pdf/ResolutionLP3-4.pdf (accessed 17 November 2011).

Lounsbury, H. (2007) 'Burying global warming: Gaylord set to test storage of greenhouse gas deep underground', Bay City, MI: *The Bay City Times*, 25 March, online, available at: http://216.109.210.162/userdata/Press/BURYING%20GLOBAL%20WARMING Gaylord%20set%20to%20test%20storing%20greenhouse%20gas%20deep%20 underground.pdf (accessed 10 October 2011).

Lovins, A.B. and Sheikh, I. (2008) 'The nuclear illusion', *Ambio*, Nov 08 preprint draft, online, available at: www.nirs.org/alternatives/e0801ambionucillusion.pdf (accessed 30 September 2011).

Lundvall, B.-A. (ed.) (1992) *National Systems of Innovation: Towards a Theory of Innovation and Interactive Learning*, London: Pinter.

Lundvall, B.-A., Johnson, B., Andersen, E.S. and Dalum, B. (2002) 'National systems of production, innovation and competence building', *Research Policy*, 31(2): 213–231.

Macher, J.T. and Richman, B.D. (2004) 'Organisational responses to discontinuous innovation: a case study approach', *International Journal of Innovation Management*, 8(1): 87–114.

MacKenzie, D. and Wajcman, J. (eds) (1999) *The Social Shaping of Technology*, 2nd edition, Buckingham/Philadelphia, PA: Open University Press.

MacKerron, G. (2009) 'Lessons from the UK on urgency and legitimacy in energy policymaking', in I. Scrase and G. MacKerron (eds) *Energy for the Future: A New Agenda*, Basingstoke: Palgrave Macmillan, pp. 76–88.

Macrory, R. (2011) *12 Months in CCS Legislation: A Period of Consolidation and Reflection*, GCCSI, online, available at: www.globalccsinstitute.com/community/blogs/authors/rmacrory/2011/05/10/12-months-ccs-legislation-period-consolidation-and-refle (accessed 30 September 2011).

Majone, G. (1989) *Evidence, Argument and Persuasion in the Policy Process*, New Haven, CT: Yale University Press.

Malone, E.L., Dooley, J.J. and Bradbury, J.A. (2010) 'Moving from misinformation of public attitude surveys on carbon dioxide capture and storage toward realistic stakeholder involvement', *International Journal of Greenhouse Gas Control*, 4(2): 419–425.

Mander, S. and Gough C. (2006) 'Media framing of new technology: the case of carbon capture and storage', presented at the conference Greenhouse Gas Technologies 8, Trondheim, Norway.

Marchetti, C. (1977) 'On geo-engineering and the CO_2 problem', *Climatic Change*, 1(1): 59–68.

Markard, J. and Truffer, B. (2008) 'Technology innovation systems and the multi-level perspective: towards an integrated framework', *Research Policy*, 37(4): 596–615.

Markusson, N. and Haszeldine, S. (2010) '"Capture ready" regulation of fossil fuel power plants: betting the UK's carbon emissions on promises of future technology', *Energy Policy*, 38(11): 6695–6702.

Markusson, N., Ishii, A. and Stephens, J.C. (2011) 'The social and political complexities of learning in carbon capture and storage demonstration projects', *Global Environmental Change*, 21(2): 293–302.

Martin, J.-M. (1996) 'Energy technologies: systemic aspects, technological trajectories, and institutional frameworks', *Technological Forecasting and Social Change*, 53(1): 81–95.

Maskell, P. (2004) Learning in the village economy of Denmark', in P. Cooke, M. Heidenreich and H.-J. Braczyk (eds) *Regional Innovation Systems: The Role of Governance in a Globalised World*, London: Routledge.

McCright, A.M. and Dunlap, R.E. (2011) 'The politicization of climate change and polarization in the American public's view on global warming', *The Sociological Quarterly*, 52: 155–194.

McDowall, W. and Eames, M. (2006) 'Forecasts, scenarios, visions, backcasts and roadmaps to the hydrogen economy', *Energy Policy*, 34(11): 1236–1250.

McFarland, J.R., Reilly, J.M. and Herzog, H.J. (2004) 'Representing energy technologies in top-down economic models using bottom-up information', *Energy Economics*, 26(4): 685–707.

McKibbin, W.J. and Wilcoxen, P.J. (2009) 'Uncertainty and climate change policy design', *Journal of Policy Modeling*, 31(3): 463–477.

McKinsey and Company (2008) *Carbon Capture and Storage: Assessing the Economics*, n.p.: McKinsey and Company.

McVeigh, J., Burtraw, D., Darmstadter, J. and Palmer, K. (1999) *Winner, Loser, or Innocent Victim? Has Renewable Energy Performed As Expected?*, Washington DC: Resources for the Future.

Meadowcroft J. and Langhelle, O. (eds) (2009a) *Caching The Carbon: The Politics and Policy of Carbon Capture and Storage*, Cheltenham: Edward Elgar.

Meadowcroft, J. and Langhelle, O. (2009b) 'The politics and policy of carbon capture and storage', in J. Meadowcroft and O. Langhelle (eds) *Caching The Carbon: The Politics and Policy of Carbon Capture and Storage*, Cheltenham: Edward Elgar.

Meadowcroft, J. and Langhelle, O. (2009c) 'The politics and policy of CCS: the uncertain road ahead', in J. Meadowcroft and O. Langhelle (eds) *Caching The Carbon: The Politics and Policy of Carbon Capture and Storage*, Cheltenham: Edward Elgar.

Metcalfe, S. (2010) 'Examining the trilateral networks of the Triple Helix: intermediating organizations and academy-industry-government relations', *Critical Sociology*, 36(4): 503–519.

METI (2006) *CCS 2020*, Tokyo: Ministry of Economy, Trade and Industry, online, available at: www.meti.go.jp/committee/materials/downloadfiles/g60525a09j.pdf (accessed March 2010) [in Japanese].

METI and The General Environmental Technos (2008) *Nisankatanso-tanso-koteika-gijutsu-kaihatsu" Hyoka-shiryo* [Technology development of CO_2 fixation in coal seam: evaluation material], online, available at: www.meti.go.jp/policy/tech_evaluation/c00/C0000000H20/080729_CO2kaishuu/CO2kaishuu_6-2.pdf (accessed October 2011) [in Japanese].

Michael, M. (2000) 'Futures of the present: from performativity to prehension', in N. Brown, B. Rapert and A. Webster (eds) *Contested Futures: A Sociology of Prospective Techno-Science*, Aldershot: Ashgate.

Miller, E., Bell, L. and Buys, L. (2007) 'Public understanding of carbon sequestration in Australia: socio-demographic predictors of knowledge, engagement and trust', *Australian Journal of Emerging Technologies and Society*, 5(1): 15–33.

Misa, T.J. (1994) 'Retrieving sociotechnical change from technological determinism', in M.R. Smith and L. Marx (eds) *Does Technology Drive History? The Dilemma of Technological Determinism*, Cambridge MA: MIT Press.

MIT (2007) *The Future of Coal: Options for a Carbon Constrained World,* Cambridge, MA: Massachussetts Institute of Technology.

MIT (2011) *Power Plant Carbon Dioxide Capture and Storage Projects*, online, available at: http://sequestration.mit.edu/tools/projects/index.html (accessed 31 October 2011).

Mol, A. (2000) 'The environmental movement in an era of ecological modernisation', *Geoforum*, 31(1): 45–56.

Mol, A., Sonnenfeld, D. and Spaargaren, G. (eds) (2009) *The Ecological Modernization Reader: Environmental Reform in Theory and Practice*, London: Routledge.

Morone, J.G. and Woodhouse, E.J. (1986) *Averting Catastrophe: Strategies for Regulating Risky Technologies*, Berkeley, CA: University of California Press.

Mottier, V. (1995) 'The politics of sex: truth games and the Hite Reports', *Economy and Society*, 24(4): 520–539.

Mowery, D.C., Nelson, R.R. and Martin, B.R. (2010) 'Technology policy and global warming: why new policy models are needed (or why putting new wine in old bottles won't work)', *Research Policy*, 39(8): 1011–1023.

MPE (Ministry of Petroleum and Energy) (2007) *St.prp. nr. 49 - Samarbeid om håndtering av CO_2 på Mongstad*, Oslo: Det Kongelige Olje- og Energidepartementet.

MPE (Ministry of Petroleum and Energy) (2009a) *Prop. 44 S - Proposisjon til Stortinget om endringar i statsbudsjettet for 2009 under Olje- og energidepartementet*, Oslo: Det Kongelige Olje- og Energidepartementet.

MPE (Ministry of Petroleum and Energy) (2009b) *St.prp. nr. 38 - Investering i teknologisenter for CO_2-håndtering på Mongstad*, Oslo: Det Kongelige Olje- og Energidepartementet.

MPE (Ministry of Petroleum and Energy) (2010) *Prop. 35 S - Proposisjon til Stortinget - Endringar i statsbudsjettet for 2010 under Olje- og energidepartementet*, Oslo: Det Kongelige Olje- og Energidepartementet.

MPE (Ministry of Petroleum and Energy) (2011) *Meld. St. 9 - Melding til Stortinget: Fullskala CO_2-håndtering*, Oslo: Det Kongelige Olje- og Energidepartementet.

Munshi, K. (2004) 'Social learning in a heterogeneous population: technology diffusion in the Indian green revolution', *Journal of Development Economics*, 73(1): 185–213.

Myhrvold, T., Helle, K., Johnsen, K. and Hussain, A. (2009) 'Development of a guideline for the qualification of CO_2 capture technology', *Energy Procedia*, 1(1): 1527–1534.

Mythen, G. (2004) *Ulrich Beck: A Critical Introduction to the Risk Society*, London: Pluto Press.

Nader, L. (1981) 'The politics of energy: toward a bottom-up approach', *Radcliffe Quarterly*, 67: 5–6.

Neij, L. (1997) 'Use of experience curves to analyse the prospects for diffusion and adoption of renewable energy technology', *Energy Policy*, 25(13): 1099–1107.

Nelkin, D. (1975) 'The political impact of technical expertise', *Social Studies of Science*, 5(1): 35–54.

Nelson, R.R. (ed.) (1993) *National Systems of Innovation: A Comparative Study*, Oxford: Oxford University Press.

Nelson, R.R. and Winter, S.G. (1982) *An Evolutionary Theory of Economic Change*, Cambridge, MA: Harvard University Press.

NETL (National Energy Technology Laboratory) (2010) *Best Practices for Public Outreach and Education for Carbon Storage Projects*, Washington DC: US Department of Energy.

Newby-Clark, I., Ross, M., Buehler, R., Koehler, D. and Griffin, D. (2000) 'People focus on optimistic scenarios and disregard pessimistic scenarios while predicting task completion times', *Journal of Experimental Psychology: Applied*, 6(3): 171–182.

Nill, J. and Kemp, R. (2009) 'Evolutionary approaches for sustainable innovation policies: from niche to paradigm?' *Research Policy*, 38(4): 668–680.

NMF (Norwegian Ministry of Finance) (2009) *St.prp. nr. 67 – Tilleggsbevilgninger og omprioriteringer i statsbudsjettet 2009*, Oslo: Finansdepartementet.

NMF (Norwegian Ministry of Finance) (2010) *Prop. 125 S – Tilleggsbevilgninger og omprioriteringer i statsbudsjettet 2010*, Oslo: Finansdepartementet.

NMTI (Norwegian Ministry of Trade and Industry) (2006) *St.meld. nr. 13 - Et aktivt og langsiktig eierskap*, Oslo: Nærings- og Handelsdepartmentet.

NOAH (Friends of the Earth Denmark) (2009) *NOAH's Position on CCS as a Climate Change Tool (long version)*, online, available at: http://ccs-info.org/pos_long.pdf (accessed 31 May 2011).

NOAH (Friends of the Earth Denmark) (2011) 'Welcome to CCS-info NOAH, Friends of the Earth Denmark's website on CCS', online, available at: http://ccs-info.org/index.html (accessed 6 October 2011).

Norton, A. and Elson, D. (2002) *What's Behind the Budget? Politics, Tights and Accountability in the Budget Process*, London: Overseas Development Institute.

NPD (Norwegian Petroleum Directorate) (2010) *The Petroleum Sector – Norway's Largest Industry*, online, available at: www.npd.no/en/Publications/Facts/Facts-2010/Chapter-1 (accessed 30 October 2011).

NyTeknik, (2009) 'Protester stoppar danskt CO_2-lager', online, available at: www.nyteknik.se/nyheter/energi_miljo/energi/article583207.ece (accessed 30 October 2011).

NZEC (UK-China Near Zero Emissions Coal project) (2007) *The UK-China Near Zero Emissions Coal project*, online, available at: www.nzec.info/en (accessed 30 September 2010).

O'Connor, J. (1996) 'The second contradiction of capitalism', in T. Benton (ed.) *The Greening of Marxism*, New York: Guilford Press.

Odenberger, M. and Johnsson, F. (2010) 'Pathways for the European electricity supply system to 2050: the role of CCS to meet stringent reduction targets', *International Journal of Greenhouse Gas Control*, 4(2): 327–340.

Odenberger, M., Kjärstad, J. and Johnsson, F. (2008) 'Ramp-up of CO_2 capture and storage within Europe', *International Journal of Greenhouse Gas Control*, 2(4): 417–438.

Ofgem (Office of Gas and Electricity Markets) (2010) *Project Discovery: Options for delivering secure and sustainable energy supplies*. London: Ofgem.

Oltra, C., Sala, R., Sola, T., Di Massa, M. and Rowe, G. (2010) 'Lay perceptions of carbon capture and storage technology', *International Journal of Greenhouse Gas Control*, 4(4): 698–706.

O'Riordan, T. (2009) 'Reflections on the pathways to sustainability', in N. Adger and A. Jordan (eds) *Governing Sustainability*, Cambridge: Cambridge University Press.

OSPAR Commission (2007a) *Amendments of Annex II and Annex III to the Convention in relation to the storage of carbon dioxide streams in geological formations*, Ostend: Annex 4 of the OSPAR Convention for the Protection of the Marine Environment of the North-East Atlantic Meeting of the OSPAR Commission Ostend: 25–29 June 2007), online, available at: www.ucl.ac.uk/cclp/pdf/OSPAR2007-Annex-4.pdf (accessed 17 November 2011).

OSPAR Commission (2007b) *OSPAR Decision 2007/1 to Prohibit the Storage of Carbon Dioxide Streams in the Water Column or on the Sea-bed* (Annex 5, OSPAR Convention for the Protection of the Marine Environment of the North-East Atlantic Meeting of the OSPAR Commission, Ostend: 25–29 June 2007), online, available at: www.ucl.ac.uk/cclp/pdf/OSPAR2007-Annex-5.pdf (accessed 17 November 2011).

OSPAR Commission (2007c) *OSPAR Decision 2007/2 on the Storage of Carbon Dioxide Streams in Geological Formations* (Annex 6, OSPAR Convention for the Protection of the Marine Environment of the North-East Atlantic Meeting of the OSPAR Commission Ostend: 25–29 June 2007), online, available at: www.ucl.ac.uk/cclp/pdf/OSPAR2007-Annex-6.pdf (accessed 17 November 2011).

Othman, M. (2009) 'Strategic planning on carbon capture from coal fired plants in Malaysia and Indonesia: a review', *Energy Policy*, 37(5): 1718–1735.

Otway, H. (1992) 'Public wisdom, expert fallibility: toward a contextual theory of risk', in S. Krimsky and D. Golding (eds) *Social Theories of Risk*, Westport CT: Praeger.

Otway, H. and von Winterfeld, D. (1982) 'Beyond acceptable risk: on the social acceptability of technologies', *Policy Sciences*, 14(3): 247–256.

Pacala, S. and Socolow, R. (2004) 'Stabilization wedges: solving the climate problem for the next 50 years with current technologies', *Science*, 305(5686): 968–972.

Page, S.C., Williamson, A.G. and Mason, I.G. (2009) 'Carbon capture and storage: fundamental thermodynamics and current technology', *Energy Policy*, 37(9): 3314–3324.

Palmgren, C.M., Morgan, M.G., Bruine de Bruin, W. and Keith, D.W. (2004) 'Initial public perceptions of deep geological and oceanic disposal of CO_2', *Environmental Science and Technology*, 38(24): 6441–6450.

Pan, Z. and Kosicki G. (2003) 'Framing analysis: an approach to news discourse', *Political Communication*, 10(1): 55–75.

Paredis, E. (2011) 'Sustainability transitions and the nature of technology', *Foundations of Science*, 16(2): 195–225.

Perrow, C. (2010) 'Comment on Mowery, Nelson and Martin', *Research Policy*, 39(8): 1030–1031.

Peters, B.G. and Hogwood, B.W. (1985) 'In search of the issue-attention cycle', *The Journal of Politics*, 47(1): 238–253.

Pietzner, K., Schumann, D., Tvedt, S.D., Torvatn, H.Y., Næss, R., Reiner, D.M., Anghel, S., Cismaru, D., Constantin, C., Daamen, D.D.L., Dudu, A., Esken, A., Gemeni, V., Ivan, L., Koukouzas, N., Kristiansen, G., Markos, A., Ter Mors, E., Nihfidov, O.C., Papadimitriou, J., Samoila, I.R., Sava, C.S., Stephenson, M.H., Terwel, B.W., Tomescu, C.E. and Ziogou, F. (2011) 'Public awareness and perceptions of carbon dioxide capture and storage (CCS): insights from surveys administered to representative samples in six European countries', *Energy Procedia*, 4: 6300–6306.

Pinch, T.J. and Bijker, W.E. (1984) 'The social construction of facts and artefacts: or how the sociology of science and the sociology of technology might benefit each other', *Social Studies of Science*, 14(3): 399–441.

Pinch, T.J. and Bijker, W.E. (1992) 'The social construction of facts and artifacts: or how the sociology of science and the sociology of technology might benefit each other', in W.E. Bijker and J. Law (eds) *Shaping Technology/Building Society*, Cambridge, MA: MIT Press, pp. 17–50.

PIU (Performance and Innovation Unit) (2002) *The Energy Review*, London: Cabinet Office.

Pollak, M., Johnson, J. and Wilson, E. (2009) 'The geography of CCS regulatory development in the U.S.', *Energy Procedia*, 1(1): 4543–4550.

Pollak, M., Phillips, S. and Vajjhala, S. (2011) 'Carbon capture and storage policy in the United States: a new coalition endeavors to change existing policy', *Global Environmental Change*, 21(2): 313–323.

Polson, D. and Curtis, A. (2010) 'Dynamics of uncertainty in geological interpretation', *Journal of the Geological Society*, 167(1): 5–10.

Porter, T. (1995) *Trust in Numbers*, Princeton, NJ: Princeton University Press.

Porter, M. and van der Linde, C. (1995) 'Towards a new conception of the environment-competitiveness relationship', *Journal of Economic Perspectives*, 9(4): 97–118.

Praetorius, B. and Schumacher, K. (2009) 'Greenhouse gas mitigation in a carbon constrained world: the role of carbon capture and storage', *Energy Policy*, 37(12): 5081–5093.

PTRC (Petroleum Technology Research Centre) (2011) *Weyburn-Midale CO_2 Project*, online, available at: www.ptrc.ca/weyburn_overview.php (accessed 30 September 2011).

Rabe, B. (2007) 'Beyond Kyoto: climate change policy in multilevel governance systems', *Governance*, 20(3): 423–444.

Rai, V., Victor, D. and Thurber, M. (2009) 'Carbon capture and storage at scale: lessons from the growth of analogous energy technologies', *Energy Policy*, 38(8): 4089–4098.

Raven, R. (2007) 'Niche accumulation and hybridisation strategies in transition processes towards a sustainable energy system: an assessment of differences and pitfalls', *Energy Policy*, 35(4): 2390–2400

Raven, R. and Verbong, G. (2009) 'Boundary crossing innovations: case studies from the energy domain', *Technology in Society*, 31(1): 85–93.

Rajamani, L. (2011) 'India and climate change: contextualising India's approach to CCS technologies', in I. Havercroft, R. Macrory and R.B. Stewart (eds) *Carbon Capture and Storage: Emerging Legal and Regulatory Issues*, Oxford: Hart.

Rayner, S. and Cantor, R. (1987) 'How *fair* is safe enough?' *Risk Analysis*, 7(1): 3–9.

RCEP (2000) *Energy: The Changing Climate*, London: Royal Commission on Environmental Pollution.

Reese, S. (2001) 'Prologue – Framing Public Life: A Bridging Model for Media Research', in S. Reese, O. Gandy and A. Grant (eds) *Framing Public Life: Perspectives on Media and our Understanding of the Social World*, Mahwah, NJ: A. Lawrence Erlbaum Associates.

Reiner, D.M., Curry, T.E., deFigueiredo, M.A., Herzog, H.J., Ansolabehere, S.D., Itaoka, K., Johnsson, F. and Odenberger, M. (2006) 'American exceptionalism? Similarities and differences in national attitudes toward energy policy and global warming', *Environmental Science and Technology*, 40(7): 2093–2098.

Riahi, K., Barreto, L., Rao, S. and Rubin, E.S. (2004a) 'Towards fossil-based electricity systems with integrated CO_2 capture: implications of an illustrative long-term technology policy', *Proceedings of 7th International Conference on Greenhouse Gas Control Technologies (GHGT-7) Vol. 1: Peer-Reviewed Paper and Overviews*, n.p.: Elsevier, pp. 921–929.

Riahi, K., Rubin, E.S., Taylor, M.R., Schrattenholzer, L. and Hounshell, D. (2004b) 'Technological learning for carbon capture and sequestration technologies', *Energy Economics*, 26(4): 539–564.

Rip, A. and Kemp, R. (1998) 'Technological change', in S. Rayner and E.L. Malone (eds) *Human Choice and Climate Change*, Columbus OH: Battelle Press.

Rip, A., Misa, T.J. and Schot, J. (eds) (1995) *Managing Technology in Society: The Approach of Constructive Technology Assessment*, London: Pinter.

Roberts, T. and Mander, S. (2011) 'Assessing public perceptions: benefits, challenges and methods', *Energy Procedia*, 4: 6307–6314.

Robertson, K., Findsen, J. and Messner, S. (2006) *International Carbon Capture and Storage Projects Overcoming Legal Barriers National Energy Technology Laboratory*, Washington, DC: Department of Energy/National Energy Technology Laboratory.

Rochefort, D.A. and Cobb, R.W. (eds) (1994) *The Politics of Problem Definition: Shaping the Policy Agenda*, Lawrence, KS: UniversityPress of Kansas.

Rohracher, H. and Ornetzeder, M. (2002) 'Green buildings in context: improving social learning processes between users and producers', *Built Environment*, 28(1): 73–84.

Røkke, N.A. (2011) 'Carbon capture and storage of gas-fired power station flue-gases in Norway', Trondheim: SINTEF, online, available at: www.sintef.no/home/Press-Room/Research-News/
Carbon-capture-and-storage-of-gas-fired-power-station-flue-gases-in-Norway.

Román, M. (2011) 'Carbon capture and storage in developing countries: a comparison of Brazil, South Africa and India', *Global Environmental Change*, 21(2): 391–401.

Rosental, C. (2004) 'Fuzzyfying the world: social practices of showing the properties of fuzzy logic', in M.N. Wise (ed.) *Growing Explanations: Historical Perspectives on Recent Science*, Durham, NC: Duke University Press.

Rosental, C. (2005) 'Making science and technology results public – a sociology of demos', in B. Latour and P. Weibel (eds) *Making Things Public: Atmospheres of Democracy*, Cambridge, MA: MIT Press.

Rubin, E.S., Chen, C. and Rao, A.B. (2007) 'Cost and performance of fossil fuel power plants with CO_2 capture and storage', *Energy Policy*, 35(9): 4444–4454.

Rubin, E.S., Hounshell, D., Yeh, S., Taylor, M., Schrattenholzer, L., Riahi, K., Barreto, L. and Rao, S. (2004) *The Effect of Government Actions on Environmental Technology Innovation: Applications to the Integrated Assessment of Carbon Sequestration Technologies (Paper 96)*, Pittsburgh, PA: Department of Engineering and Public Policy, Carnegie Mellon University, online, available at: http://repository.cmu.edu/epp/96 (accessed 30 September 2011).

Rubin, E.S., Taylor, M., Yeh, S. and Hounshell, D. (2004) 'Learning curves for environmental technology and their importance for climate policy analysis', *Energy*, 29(9/10): 1551–1559.

Rubin, E.S., Yeh, S., Antes, M., Berkenpas, M. and Davison, J. (2007) 'Use of experience curves to estimate the future cost of power plants with CO_2 capture', *International Journal of Greenhouse Gas Control*, 1(2): 188–197.

Russell, S. (1986) 'The social construction of artefacts: a response to Pinch and Bijker', *Social Studies of Science*, 16(2): 331–346.

Russell, S. (1993) 'Writing energy history: explaining the neglect of CHP/DH in Britain', *British Journal of the History of Science*, 26(1): 33–54.

Russell, S. and Williams, R. (2002) 'Social shaping of technology: frameworks, findings and implications for policy', in K.H. Sørensen and R. Williams (eds) *Shaping Technology, Guiding Policy*, Cheltenham: Edward Elgar.

Russell, S., Markusson, N. and Scott, V. (2011) 'What will CCS demonstrations demonstrate? A research agenda', *Mitigation and Adaptation of Strategies for Global Change* (in press).

Sabatier, P. (1988) 'An advocacy coalition framework of policy change and the role of policy-oriented learning therein', *Policy Sciences*, 21(2/3): 129–168.

Sabatier, P. (1998) 'The advocacy coalition framework: revisions and relevance for Europe', *Journal of European Public Policy*, 5(1): 98–130.

Sabatier, P. and Jenkins-Smith, H. (eds) (1993) *Policy and Learning: An Advocacy Coalition Approach*, Boulder, CO: Westview Press.

Sabatier, P. and Weible, C.M. (2007) 'The advocacy coalition framework: innovations and clarifications', in P. Sabatier (ed.) *Theories of the Policy Process*, 2nd edition, Boulder, CO: Westview Press.

Saether, A.K. (2011) 'Statoil is stabbing CCS in the back', online, available at: www.bellona.org/articles/articles_2011/mongstad_back_stab (accessed 27 May 2011).

Sagar, A. and Gallagher, K.S. (2004) 'Energy technology demonstration and deployment', in J.P. Holdren, W.K. Reilly, J.W. Rowe, P. Sharp and J. Grumet (eds) *Ending the Energy Stalemate: A Bipartisan Strategy to Meet America's Energy Challenges*, Washington DC: National Commission on Energy Policy.

Sandelson, M. (2010a) 'Mongstad minister could be reshuffled', *The* Foreigner, online, available at: http://theforeigner.no/pages/news/mongstad-minister-could-be-reshuffled (accessed 12 September 2011).

Sandelson, M. (2010b) 'Mongstad minister to be questioned', *The* Foreigner, online, available at: http://theforeigner.no/pages/news/mongstad-minister-to-be-questioned (accessed 12 September 2011).

SBSTA (Subsidiary Body for Scientific and Technological Advice) (2010) *Views Related to Carbon Dioxide Capture and Storage in Geological Formations as a Possible Mitigation Technology – Submission from Parties (Addendum)*, 8, UNFCCC, online, available at: http://unfccc.int/resource/docs/2010/sbsta/eng/misc02a01.pdf (accessed 30 September 2011).

SBSTA (Subsidiary Body for Scientific and Technological Advice) (2011) *Synthesis of Views on Modalities and Procedures for Carbon Dioxide Capture and Geological Storage in Formations as Clean Development Mechanisms* (Advanced Version 4 August), UNFCCC, online, available at: http://unfccc.int/resource/docs/2011/sbsta/eng/inf07.pdf (accessed 30 September 2011).

SCCS (Scottish Carbon Capture and Storage) (2011) *Progressing Scotland's CO_2 Storage Opportunities*, Edinburgh: University of Edinburgh, online, available at: www.sccs.org.uk/progress-to-co2-storage-scotland (accessed 30 September 2011).

Scharpf, F.W. (1997) *Games Real Actors Play: Actor-Centered Institutionalism in Policy Research*, Boulder, CO: Westview Press.

Schrag, D. (2009) 'Coal as a low-carbon fuel?' *Nature Geoscience*, 2(12): 818–820.

Schulman, P.R. (1980) *Large-Scale Policy-Making,* New York: Elsevier.

Schwarz, M. and Thompson, M. (1990) *Divided We Stand: Redefining Politics, Technology and Social Choice,* Philadelphia, PA: University of Pennsylvania Press.

Scrase, I. and Watson, J. (2009) 'CCS in the UK: squaring coal use with climate change?' in J. Meadowcroft and O. Langhelle (eds) *Caching the Carbon: The Politics and Policy of Carbon Capture and Storage*, Cheltenham: Edward Elgar, pp. 158–185.

Shackley, S. (1996) 'Global climate change and modes of international science and policy', in A. Elzinga and C. Landstrom (eds) *Internationalism and Science*, London: Taylor Graham.

Shackley, S. (1998) 'Introduction to special section on the use of models in appraisal and policy-making', *Impact Assessment and Project Appraisal*, 16(2): 81–89.

Shackley, S. and Gough, C. (eds) (2006) *Carbon Capture and its Storage: An Integrated Assessment*, London: Ashgate Publishing.

Shackley, S. and Green, K. (2007) 'Exploring transitions to sustainable energy: decarbonisation of the UK', *Energy*, 32: 221–236.

Shackley, S. and Thompson, M. (2011) 'Lost in the mix: will the technologies of carbon dioxide capture and storage provide us with a breathing space as we strive to make the transition from fossil fuels to renewables?' *Climatic Change*, online version 5 May This is now published as Shackley & Thompson (2012), Climatic Change 110: 101-121.

Shackley, S., McLachlan, C. and Gough C. (2005) 'The public perception of carbon capture and storage in the UK: results from focus groups and a survey', *Climate Policy*, 4(4): 377–398.

Shackley, S., Reiner, D., Upham, P., de Coninck, H.C., Sigurthorsson, G. and Anderson, J. (2009) 'The acceptability of CO_2 capture and storage (CCS) in Europe: an assessment of the key determining factors, Part 2: The social acceptability of CCS and the wider impacts and repercussions of its implementation', *International Journal of Greenhouse Gas Control*, 3(3): 344–356.

Shackley, S., Waterman, H., Godfroij, P., Reiner, D., Anderson, J., Draxlbauer, K. and Flach, T. (2007) 'Stakeholder perceptions of CO_2 capture and storage in Europe: results from a survey', *Energy Policy*, 35(10): 5091–5108.

Shaner, K. (2008) 'Residents voice skepticism, concerns about carbon sequestration', Greenville, OH: *The Daily Advocate*, 14 August.

Shapin, S. (1984) 'Pump and circumstance: Robert Boyle's literary technology', *Social Studies of Science*, 14(4): 481–520.

Sharp, J. (2008a) *Carbon Capture and Storage: The Views of the Canadian Public*, Drayton Valley, Alberta: Pembina Institute and ISEEE.

Sharp, J. (2008b) *CCS Thought Leader Opinion Surveying Conclusions*, Drayton Valley, Alberta: Pembina Institute and ISEEE.

Sharp, J.D., Jaccard, M.K. and Keith, D.W. (2009) 'Anticipating public attitudes toward underground CO_2 storage', *International Journal of Greenhouse Gas Control*, 3(5): 641–651.

Shell (2008) 'Going underground with greenhouse gas', *Shell World*, 15 August, online, available at: www-static.shell.com/static/aboutshell/downloads/swol/july_oct_2008/ccs_15082008.pdf (accessed 20 October 2011).

Smil, V. (2000) 'Perils of long-range energy forecasting: reflections on looking far ahead', *Technological Forecasting and Social Change*, 65(3): 251–264.

Smith, A., Voß, J.-P. and Grin, J. (2010) 'Innovation studies and sustainability transitions: the allure of the multi-level perspective and its challenges', *Research Policy*, 39(4): 435–448.

Smith, J.W. (1991) *The High Tech Fix: Sustainable Ecology or Technocratic Megaprojects for the 21st Century?* Aldershot: Avebury.

Sørensen, K.H. (1996) *Learning Technology, Constructing Culture: Socio-Technical Change as Social Learning*, STS Working Paper No. 18/96, Trondheim: Centre for Technology and Society, University of Trondheim, online, available at: www.rcss.ed.ac.uk/SLIM/public/phase1/knut.html (accessed 30 September 2011).

Sørensen, K.H. and Levold, N. (1992) 'Tacit networks, heterogeneous engineers, and embodied technology', *Science, Technology, & Human Values*, 17(1): 13–35.

Spaargaren, G. and Mol, A. (1992) 'Sociology, environment and modernity: ecological modernisation as a theory of social change', *Society and Natural Resources*, 5(4): 323–344.

Spence, A., Venables, D., Pidgeon, N., Poortinga, W. and Demski, C. (2010) 'Public perceptions of climate change and energy futures in Britain: summary findings of a survey conducted in January–March 2010', *Understanding Risk Working Paper* 10-01, Cardiff School of Psychology, University of Cardiff, online, available at: www.ipsos-mori.com/Assets/Docs/Polls/climate-change-public-perceptions-of-climate-change-report.pdf (accessed 23 October 2011).

Spinardi, G. (2008) 'Ballistic missile defence and the politics of testing: the case of the US ground-based midcourse defence', *Science and Public Policy*, 35(10): 703–715.

Stangeland, A. (2007) 'A model for the CO_2 capture potential', *International Journal of Greenhouse Gas Control*, 1(4): 418–429.

Stangeland, A. (2008) *How to Combat Global Warming: An Ambitious but Necessary Approach to Reduce Greenhouse Gas Emissions*, Oslo: The Bellona Foundation, online, available at: www.bellona.org/reports/How_to_Combat_Global_Warming (accessed 20 September 2011).

Statoil (2009a) *Sleipner Vest*, online, available at: www.statoil.com/en/TechnologyInnovation/ProtectingTheEnvironment/CarboncaptureAndStorage/Pages/CarbonDioxideInjectionSleipnerVest.aspx (accessed 30 September 2011).

Statoil (2009b) *Annual Report 2008*, online, available at: www.statoil.com/AnnualReport2008/en/Sustainability/Climate/Pages/5-3-2-4_Sn%C3%B8hvitCCS.aspx (accessed 30 September 2011).

StatoilHydro (2009) *CO_2 Masterplan Mongstad*, online, available at: www.statoil.com/en/NewsAndMedia/News/2009/Pages/11FebMongstad.aspx (accessed 30 September 2011).

Stephens, J.C. (2006) 'Growing interest in carbon capture and storage (CCS) for climate change mitigation', *Sustainability: Science, Practice, and Policy*, 2(2): 4–13.

Stephens, J.C. (2009) 'Technology leader, policy laggard: carbon capture and storage (CCS) development for climate mitigation in the U.S. political context', in J. Meadowcroft and O. Langhelle (eds) *Caching the Carbon: The Politics and Policy of Carbon Capture and Storage*, Cheltenham: Edward Elgar Publishing.

Stephens, J.C. and Jiusto, S. (2010) 'Assessing innovation in emerging energy technologies: socio-technical dynamics of carbon capture and storage (CCS) and enhanced geothermal systems (EGS) in the USA', *Energy Policy*, 38(4): 2020–2031.

Stephens, J.C. and Verma, P. (2006) 'The role of environmental advocacy groups in the advancement of carbon capture and storage (CCS)', paper presented at the Fifth Annual Conference on Carbon Capture and Sequestration, 8–11 May. Alexandria, Virginia.

Stephens, J.C., Hansson, A., Liu, Y., de Coninck, H.C. and Vajjhala, S. (2011) 'Characterizing the international carbon capture and storage community', *Global Environmental Change*, 21(2): 379–390.

Stern, N., Peters, S., Bakhshi, V., Bowen, A., Cameron, C., Catovsky, S., Crane, D., Cruickshank, S., Dietz, S., Edmonson, N., Garbett, S.L., Hamid, L., Hoffman, G., Ingram, D., Jones, B., Patmore, N., Radcliffe, H., Sathiyarajah, R., Stock, M., Taylor, C., Vernon, T., Wanjie, H. and Zenghelis, D. (2006) *Stern Review: The Economics of Climate Change*, London: HM Treasury, online, available at: http://webarchive.nationalarchives.gov.uk/+/http:/www.hm-treasury.gov.uk/sternreview_index.htm (accessed 30 September 2011).

Stoltenberg, J. (2007) *Statsministerens nyttårstale 2007*, The Office of the Prime Minister, Oslo.

Stone, D. (1996) *Capturing the Political Imagination: Think Tanks and the Policy Process*, New York: Frank Cass.

Stone, D.A. (1988) *Policy Paradox and Political Reason*, New York: HarperCollins.

Stone, D.A. (1989) 'Causal stories and the formation of policy agendas', *Political Science Quarterly*, 104(2): 281–300.

Strachan, N. (2011) 'UK energy policy and UK energy modelling: fit for purpose?', *Energy Policy*, 39(3): 1037–1040.

Sumner, J., Bird, L. and Smith, H. (2009) *Carbon Taxes: A Review of Experience and Policy Design Considerations*, technical report, Golden, CO: National Renewable Energy Laboratories, online, available at: www.nrel.gov/docs/fy10osti/47312.pdf (accessed 30 September 2011).

Surber, J. (2009) 'Waiting to exhale', Greenville, OH: *The Daily Advocate*, 3 June.

Suurs, R.A.A. and Hekkert, M.P. (2008) *Cumulative Causation in the Formation of a Technological Innovation System: The Case of Biofuels in the Netherlands*, Utrecht: Utrecht

University, Department of Innovation Studies in its series Innovation Studies Utrecht (ISU) Working Paper No. 08-04.

Tancau, R. (2010) 'WikiLeaks: US knew about Mongstad before Norwegian Parliament', *The Foreigner*, online, available at: http://theforeigner.no/pages/news/wikileaks-us-knew-about-mongstad-before-norwegian-parliament (accessed 12 September 2011).

TCM (Test Centre Mongstad) (2008) *Konsekvensutredning for European CO_2 for Test Centre Mongstad*, online, available at: www.klif.no/nyheter/dokumenter/mongstad_tcm_konsekvensutredning300508.pdf (accessed 16 September 2011).

TCM (Test Centre Mongstad) (2010) *Søknad om tillatelse til utslipp fra CO_2*, online, available at: www.klif.no/nyheter/dokumenter/tcm_soknad030910.pdf (accessed 16 September 2011).

Teaford, J. (2009) 'Hear and Respond', Greenville, OH: *The Daily Advocate*, 9 August.

Terwel, B.W., Harinck, F., Ellemers, N. and Daamen, D.D. (2009) 'Competence-based and integrity-based trust as predictors of acceptance of carbon dioxide capture and storage (CCS)', *Risk Analysis*, 29(8): 1129–1140.

The Climate Group (2010) *CCS: Mobilising Private Sector Finance*, London: The Climate Group, online, available at: www.theclimategroup.org/publications/2010/9/20/ccs-mobilising-private-sector-finance (accessed 30 September 2011).

The General Environmental Technos (former Kansai-sogo-kankyo-senta) (2003) *Nisankatanso-tanso-koteika-gijutsu-kaihatsu: Gemba-yobi-jikken ni kansuru Jimoto-setsumeikai-gijiroku* [Technology development of CO_2 fixation in coal seam: minutes of the town meeting for the neighbourhood communities on explanation of the on-site preparatory experiment], 22 May (unpublished document).

The Norwegian Parliament (2010) *Innstilling fra finanskomiteen om revidert nasjonalbudsjett for 2010, tilleggsbevilgninger og omprioriteringer i statsbudsjettet for 2010 og endringer i skatte- og avgiftsreglene mv.*, Oslo: Stortinget, Finanskomiteen.

Thompson, M., Ellis, R. and Wildavsky, A. (1990) *Cultural Theory*, Boulder, CO: Westview Press.

Tichy, G. (2004) 'The over-optimism among experts in assessment and foresight', *Technological Forecasting and Social Change*, 71(4): 341–363.

Tjernshaugen, A. (2008) 'Political commitment to CO_2 capture and storage: evidence from government RD&D budgets', *Mitigation, Adaptation, Strategy Global Change*, 13(1): 1–21.

Tjernshaugen, A. and Langhelle, O. (2009) 'Technology as political glue: CCS in Norway', in J. Meadowcroft and O. Langhelle (eds) *Caching the Carbon: The Politics and Policy of Carbon Capture and Storage*. Cheltenham: Edward Elgar Publishing.

TNS Opinion & Social (2011) *Special Eurobarometer 364: Public Awareness and Acceptance of CO_2 Capture and Storage*, Brussels: European Commission Directorate-General for Energy.

Torgerson, D. (1997) 'Policy professionalism and the voices of dissent: the case of environmentalism', *Polity*, 29(3): 345–374.

Torvanger, A. and Meadowcroft, J. (2011) 'The political economy of technology support: making decisions about CCS and low carbon energy technologies', *Global Environmental Change*, 21(2): 303–312.

Tsouris, C., Aaron, D. and Williams, K. (2010) 'Is carbon capture and storage really needed?' *Environmental Science and Technology*, 44(11): 4042–4045.

UNFCCC (United Nations Framework Convention on Climate Change) (2010) 'Decision 7/CMP. 6: Carbon dioxide capture and storage in geological formations as clean development mechanism project activity', online, available at: http://unfccc.int/resource/docs/2010/cmp6/eng/12a02.pdf#page=27 (accessed 30 September 2011).

Unruh, G.C. (2000) 'Understanding carbon lock-in', *Energy Policy*, 28(12): 817–830.

Unruh, G.C. (2002) 'Escaping carbon lock-in: the climate policy implications of carbon lock in', *Energy Policy*, 30(4): 317–325.

Unruh, G.C. and Carrillo-Hermosilla, J. (2006) 'Globalizing carbon lock-in', *Energy Policy*, 34(10): 1185–1197.

Upham. P. and Roberts, T. (2010) *Public Perceptions of CCS: The Results of NEAR CO_2 European Focus Groups*, Manchester: University of Manchester, online, available at: www.communicationnearco2.eu/fileadmin/communicationnearco2/user/docs/Near_CO2_WP4_report_final.pdf (accessed 30 September 2011).

Utgikar, V.P. and Scott, J.P. (2006) 'Energy forecasting: predictions, reality and analysis of causes and error', *Energy Policy*, 34(17): 3087–3092.

Utterback, J.M. (1994) *Mastering the Dynamics of Innovation: How Companies can Seize Opportunities in the Face of Technological Change*, Cambridge, MA: Harvard Business School Press.

van Alphen, K., Hekkert, M.P. and Turkenburg, W.C. (2010) 'Accelerating the deployment of carbon capture and storage technologies by strengthening the innovation system', *International Journal of Greenhouse Gas Control*, 4(2): 396–409.

van Alphen, K., Noothout, P.M., Hekkert, M.P. and Turkenburg, W.C. (2010) 'Evaluating the development of carbon capture and storage technologies in the United States', *Renewable & Sustainable Energy Reviews*, 14(3): 971–986.

van Alphen, K., van Ruijven, J., Kasa, S., Hekkert, M. and Turkenburg, W. (2009) 'The performance of the Norwegian carbon dioxide, capture and storage innovation system', *Energy Policy*, 37(1): 43–55.

van Alphen, K., van Voorst tot Voorst, Q., Hekkert, M.P. and Smits, R.E.H.M. (2007) 'Societal acceptance of carbon capture and storage technologies', *Energy Policy*, 35(8): 4368–4380.

van den Broek, M., Yeh, S., Hoefnagals, R., Junginger, M., Rubin, E. and Faaij, A. (2010) 'Carbon capture and storage (CCS) technologies', in M. Junginger, W. van Sark and A. Faaij (eds) *Technological Learning in the Energy Sector: Lessons for Policy, Industry and Science*, Cheltenham: Edward Elgar, pp. 160–175.

Van der Heijden, H.-A. (2010) 'Environmental movements and political opportunities: the case of China', *Social Movement Studies: Journal of Social, Cultural and Political Protest*, 9(1): 51–68.

Van Lente, H. and Rip, A. (1998) 'Expectations in technological developments: an example of prospective structures to be filled in by agency', in C. Disco and B.J.R. van der Meulen (eds) *Getting New Technologies Together*, Berlin: Walter de Gruyter.

van Noorden, R. (2010) 'Buried trouble', *Nature*, 463(7283): 871–873.

van Notten, P.W.F., Sleegers, A.M. and van Asselt, M.B.A. (2005) 'The future shocks: on discontinuity and scenario development', *Technological Forecasting and Social Change*, 72(2): 175–194.

Van Zwanenberg, P. and Millstone, E. (2005) *BSE: Risk, Science and Governance*, Oxford: Oxford University Press.

Vehre, A. (2009a) 'Seismic testing causes concern', Greenville, OH: *The Daily Advocate*, 29 May.

Vehre, A. (2009b) 'Locals against proposed sequestration project', Greenville, OH: *The Daily Advocate*, 3 June.

Verbong, G., Christiaens, W., Raven, R. and Balkema, A. (2010) 'Strategic Niche Management in an unstable regime: Biomass gasification in India. Environmental Science & Policy 13(4): 272–281.

Vergragt, P.J. (2009) 'CCS in the Netherlands: glass half empty or half full?', in J. Meadowcroft and O. Langhelle (eds) *Caching the Carbon: The Politics and Policy of Carbon Capture and Storage*, Cheltenham: Edward Elgar.

Vergragt, P.J., Markusson, N. and Karlsson, H. (2011) 'Carbon capture and storage, bio-energy with carbon capture and storage, and the escape from the fossil-fuel lock-in', *Global Environmental Change*, 21(2): 282–292.

Verweij, M. and Thompson, M. (2006) *Clumsy Solutions for a Complex World: Governance, Politics and Plural Perceptions*, Basingstoke: Palgrave Macmillan.

Victorian Government (2008) *Greenhouse Gas Geological Sequestration Act 2008*, online, available at: www.legislation.vic.gov.au/Domino/Web_Notes/LDMS/PubStatbook.nsf/51dea49770555ea6ca256da4001b90cd/7E4801FE0E8E3A55CA2574F80019A141/$FILE/08-61a.pdf (accessed 30 September 2011).

Wallquist, L., Visschers, V.H.M. and Siegrist, M. (2009) 'Lay concepts on CCS deployment in Switzerland based on qualitative interviews', *International Journal of Greenhouse Gas Control*, 3(5): 652–657.

Wallquist, L., Visschers, V.H.M., Dohle, S. and Siegrist, M. (2011) 'Adapting communication to the public's intuitive understanding of CCS', *Greenhouse Gases: Science and Technology*, 1(1): 83–91.

Wartmann, S., Jaworski, P., Klaus, S. and Beyer, C. (2009) *Scenarios on the Introduction of CO_2 Emission Performance Standards for the EU Power Sector*, Nuremberg: Ecofys, online, available at: www.ecofys.com/en/publication/65 (accessed 30 September 2011).

Watson, J. (2004) 'Selection environments, flexibility and the success of the gas turbine', *Research Policy*, 33(8): 1065–1080.

Webler, T. and Tuler, S.P. (2010) 'Getting the engineering right is not always enough: researching the human dimensions of the new energy technologies', *Energy Policy*, 38(6): 2690–2691.

Weiss, A. and Woodhouse, E. (1992) 'Reframing incrementalism: a constructive response to the critics', *Policy Sciences*, 25(3): 255–273.

Weiss, J.A. (1989) 'The powers of problem definition: the case of government paperwork', *Policy Sciences*, 22(2): 97–121.

Wells, K. and Elgin, B. (2011) 'What's killing carbon capture?' *Bloomberg Business Week*, 21 July, online, available at: www.businessweek.com/magazine/whats-killing-carbon-capture-07212011.html (accessed 29 September 2011).

Wenger, E. (2000) 'Communities of practice and social learning systems', *Organization*, 7(2): 225–246.

Whitton, P.N. (2009) 'Carbon capture: the technologies', in *Carbon Capture and Storage: Time to Deliver*, London: Institution of Civil Engineers.

Wildavsky, A.B. (1975) *Budgeting: A Comparative Theory of Budgetary Processes*, Boston, MA: Little, Brown.

Williams, R. (2008) 'Compressed foresight and narrative bias: pitfalls in assessing high technology futures', in E. Fisher, S. Selin, and J. Wetmore (eds) *Yearbook of Nanotechnology in Society, Volume 1: Presenting Futures*, pp. 265–289, DOI: 10.1007/978-1-4020-8416-4_22.

Williams, R. and Russell, S. (1988) *Opening the Black-Box and Closing it Behind You: On Microsociology in the Social Analysis of Technology*: PICT Working Paper, No. 3, Edinburgh: Research Centre for Social Sciences, Edinburgh University.

Williams, R., Stewart, J. and Slack, R. (2005) *Social Learning in Technological Innovation: Experimenting with Information and Communication Technologies*, Cheltenham: Edward Elgar Publishing.

Wilson, E.J. and Gerard, D. (2007) *Carbon Capture and Sequestration: Integrating Technology, Monitoring and Regulation*, Oxford: Blackwell.

Wilson, E.J., Friedmann, S.J. and Pollak, M.F. (2007) 'Research for deployment: incorporating risk, regulation, and liability for carbon capture and sequestration', *Environmental Science & Technology*, 41(17): 5945–5952.

Wilson, E.J., Zhang, D. and Zheng, L. (2011) 'The socio-political context for deploying carbon capture and storage in China and the U.S.', *Global Environmental Change*, 21(2): 324–335.

Winner, L. (1993) 'Social constructivism: opening the black box and finding it empty', *Science as Culture*, 18(3): 427–452.

Winskel, M. (2002a) 'Autonomy's end: nuclear power and the privatization of the British electricity supply industry', *Social Studies of Science*, 32(3): 439–467.

Winskel, M. (2002b) 'When systems are overthrown: the "dash for gas" in the British electricity supply industry', *Social Studies of Science*, 32(4): 563–598.

Winskel, M., Markusson, N., Jeffrey, H., Jablonski, S., Candelise, C., Ward, D. and Howarth, P. (2008) *Technology Change and Energy Systems: Learning Pathways for Future Sources of Energy*, Working Paper, London: UKERC, online, available at: www.ukerc.ac.uk/support/tiki-download_file.php?fileId=1074 (accessed 24 September 2011).

Winskel, M., McLeod, A., Wallace, R. and Williams, R. (2006) 'Energy policy and institutional context: marine energy innovation systems', *Science and Public Policy*, 33(5): 365–376.

Wong-Parodi, G. and Ray, I. (2009) 'Community perceptions of carbon sequestration: insights from California,' *Environmental Research Letters*, 4(3), 034002, online, available at: http://iopscience.iop.org/1748-9326/4/3/034002 (accessed 30 September 2011).

Woodhouse, E.J. and Collingridge, D. (1993) 'Incrementalism, intelligent trial-and-error, and the future of political decision theory', in H. Redner (ed.) *An Heretical Heir of the Enlightenment: Science, Politics and Policy in the Work of Charles E. Lindblom*, Boulder, CO: Westview.

WRI (World Resources Institute) (2010) *CCS and Community Engagement, Guidelines for Community Engagement in Carbon Dioxide Capture, Transport and Storage Projects*, Washington DC: WRI.

WWF Germany (2008) *Versuchsballon zum Klimaschutz*, online, available at: www.wwf.de/presse/details/news/ccs_versuchsballon_zum_klimaschutz (accessed 31 May 2011).

Wynne, B. (1991) 'Knowledge in context', *Science, Technology, & Human Values*, 16(1): 1–17.

Wynne, B. (1992) 'Uncertainty and environmental learning: global environmental change – reconceiving science and policy in the preventive paradigm', *Global Environmental Change*, 2(2): 111–127.

Wynne, B. (1996) 'May the sheep safely graze? A reflexive view of the expert-lay knowledge divide', in S. Lash, B. Szerszynski and B. Wynne (eds) *Risk, Environment and Modernity: Towards a New Ecology*, London: Sage.

Yang, C.-J. and Oppenheimer, M. (2007) 'A "Manhattan Project" for climate change?' *Climatic Change*, 80(3): 199–204.

Yorkshire Post (2011) 'Yorkshire hopes snubbed over flagship carbon storage project', online, available at: www.yorkshirepost.co.uk/news/around-yorkshire/local-stories/yorkshire_hopes_snubbed_over_flagship_carbon_storage_project_1_3429904 (accessed 30 October 2010).

ZEP (The Zero Emissions Platform) (2008a) *EU Demonstration Programme for CO_2 Capture and Storage: ZEP's Proposal*, online, available at: www.zero-emissionplatform.eu/website/docs/ETP%20ZEP/EU%20Demonstration%20Programme%20for%20CCS%20-%20ZEP's%20Proposal.pdf (accessed 25 June 2011).

ZEP (The Zero Emissions Platform) (2008b) *The EU Flagship Programme for CO_2 Capture and Storage (CCS) ZEP Recommendations: Implementation and Funding*, online, available at: www.zero-emissionplatform.eu/website/docs/ETP%20ZEP/ETP%20ZEP%20Recommendations.pdf (accessed 25 June 2011).

ZEP (The Zero Emissions Platform) (2011a) EU CCS demonstration projects website, available at: www.zeroemissionsplatform.eu/projects/eu-projects.html (accessed 30 October 2011).

ZEP (The Zero Emissions Platform) (2011b) *The Costs of CO_2 Capture, Transport and Storage*, online, available at: www.zeroemissionsplatform.eu/library/publication/165-zep-cost-report-summary.html (accessed 30 September 2011).

ZERO (2011) *NGO Positions on CCS*, online, available at: www.zero.no/ccs/environmental-organisations (accessed 6 October 2011).

Zimmerman, M.B. (1982) 'Learning effects and the commercialization of new energy technologies: the case of nuclear power', *The Bell Journal of Economics*, 13(2): 297–310.

Ziogou, F., Gemeni, V., Koukouzas, N., Schumann, D., Anghel, S., Cismaru, D., Constantin, C., Dudu, A., Ivan, L., Nihfidov, O.C., Samoila, I.R., Sava, C.S., Tomescu, C.E., Daamen, D.D.L., Esken, A., Kristiansen, G., Markos, A., ter Mors, E., Næss, R., Papadimitriou, J., Pietzner, K., Reiner, D.M., Terwel, B.W., Torvatn, H.Y. and Tvedt, S.D. (2010) *Scrutinizing the Impact of CCS Communication on Opinion Quality: Recommendations for the Communication of CCS*, online, available at: www.ccs-communications.gr/reports/WP4%20report.pdf (accessed 6 October 2011).

INDEX